Fundamentals of Economics for Applied Engineering
Second Edition

Fundamentals of Economics for Applied Engineering
Second Edition

S Kant Vajpayee and MD Sarder

CRC Press
Taylor & Francis Group
Boca Raton London New York

CRC Press is an imprint of the
Taylor & Francis Group, an **informa** business

CRC Press
Taylor & Francis Group
6000 Broken Sound Parkway NW, Suite 300
Boca Raton, FL 33487-2742

© 2020 by Taylor & Francis Group, LLC
CRC Press is an imprint of Taylor & Francis Group, an Informa business

No claim to original U.S. Government works

Printed on acid-free paper

International Standard Book Number-13: 978-0-367-18946-4 (Hardback)

Visit the Taylor & Francis Web site at
http://www.taylorandfrancis.com

and the CRC Press Web site at
http://www.crcpress.com

Printed and bound by CPI Group (UK) Ltd, Croydon, CR0 4YY

Contents

7 Rate of Return   179

8 Benefit-Cost Ratio   209

Preface

Like many others, we have taught engineering economics numerous times. In spite of our best efforts to make it enjoyable and interesting, we found many of our students struggling in this course. Some of them might have blamed us as instructors. However, all along we felt that the real problem has been the textbook. We adopted other texts too, and found them equally unsatisfactory, primarily due to the overemphasis on engineering science. Therefore, we decided to write this textbook!

The *Fundamentals of Economics for Applied Engineering* textbook provides a one-semester introduction to the fundamentals of engineering economics. This text provides an overview of the basic theory and mathematics determining operational business decisions that engineering technology, engineering, and applied science students will encounter every day when they enter the workforce. This text will also prove useful to students focused on business management, physics, chemistry, computer sciences, and mathematics. A basic knowledge of economics allows students to balance costs with production in their future endeavors. This knowledge base will provide a context for students in the larger world around them, regardless of their specialty.

The *Fundamentals of Economics for Applied Engineering* textbook presents the material in plain language that is easy to comprehend. By the material being presented plainly to students, it allows them to quickly grasp the concepts of economics even if they are not familiar with commonly used economics verbiage. In addition to the plain language of the text, treatment of basic concepts has been simplified and kept straightforward by thoroughly explaining difficult fundamentals and principles. The plain language and presentation reflect the intended audience of the textbook, which is students interested in "how to apply" economic principles. Core concepts are highlighted by practical examples throughout the text with additional instruction in Microsoft Excel. The use of Microsoft Excel examples is a method of teaching students practical application considering that many users of this text will input and calculate business data using similar technology. The underlying principle of this method is to provide a base of knowledge backed by practical examples and instruction of how to navigate through such a software system dedicated to business operations.

The key features of the text are:

1. The writing style is novel. We have tried to talk to the reader, as if delivering a lecture.
2. Most of the problems relate to engineering projects and are as close to the real world as the treatment of the material would allow. To enliven the discussions, a few personal financing problems have also been included.
3. The treatment has been kept simple and straightforward, focusing on the learning of concepts. Material likely to be difficult to the average student has been thoroughly explained.

4. The time constraint of a one-semester three-credit course has been kept in mind in determining the extent of coverage.

5. The textbook is targeted at readers interested in "how to apply" the economic principles. Theoretical derivations and equations have been kept to the bare minimum—no calculus is required to understand the concepts. Concepts and first-principles have been emphasized instead.

6. The textbook covered contemporary issues of decision-making such as risk analysis and provided enough example problems in each chapter to illustrate the fundamentals. Chapter-end exercises comprise discussion and multiple-choice questions along with numerical problems.

The textbook comes with a solution manual, instructor resources, and student resources including MS Excel templates for easy-to-run senility analysis using charts. The manual offers solutions of the chapter-end exercises and a suggested course outline and hints on how to render engineering economics interesting. It also contains our contact information for instructors adopting the book. Feedback from them and other readers, especially students, is welcome for improving future editions.

S Kant Vajpayee
MD Sarder

Authors

Dr. S Kant Vajpayee is a professor emeritus at the University of Southern Mississippi. He is a distinguished educator, who has served in academia for more than 30 years. He has authored multiple textbooks in the area of engineering economics and computer integrated manufacturing, and he has taught engineering economics since the 1980s.

Dr. MD Sarder is a professor and chair of Engineering Technologies at Bowling Green State University. He has worked at the U.S. Air Force Academy as a distinguished research fellow, and as an associate professor and graduate program director at the University of Southern Mississippi. He has authored three books, and has developed a BS program and an MS program in logistics, a research center on logistics trade and transportation, and numerous undergraduate and graduate level courses. He founded a division of Logistics and Supply Chain within the Institute of Industrial & Systems Engineers and has served as the editor in chief for the International Journal of Logistics and Transportation Research.

Symbols[1]/Abbreviations

△BCR	Incremental benefit-cost ratio
△ROR	Incremental rate of return
+ve	Positive
−ve	Negative
/ ÷	(division)
μ	Mean
σ	Standard deviation
π	22/7 = 3.1416
A	End-of-period cashflows in a uniform series
A_1	The first cashflow in a geometric series
AB	Annual benefit
ACRS	Accelerated cost recovery system
AGV	Automated guided vehicle
AM	Annual maintenance cost
APR	Annual percentage rate
APY	Annual percentage yield
AW	Annual worth
AWs	Annual worths, plural of AW
B/C	Benefit-cost ratio
BCD	Benefit-cost difference
BCR	Benefit-cost ratio
BS	Bachelor of Science
CAM	Computer-aided machining or manufacturing
CD	Certificate of deposit
CNC	Computer numerical control
CPA	Chartered public accountant
CPI	Consumer price index
DDB	Double declining balance
EPA	Environment Protection Agency
EUAB	Equivalent uniform annual benefit
EUAC	Equivalent uniform annual cost
e	Base of natural logarithm = 2.171828
F	Future sum of money, future payment, final payment
f	inflation rate, function of
FW	Future worth

---

[1] Most are as per the *Manual of Standard Notation for Engineering Economy Parameters and Interest Factors* by the American Society for Engineering Education, Engineering Economy Division.

FWs	Future worths, plural of FW
G	Arithmetic gradient (fixed increase in arithmetic series)
g	Geometric gradient (rate of increase in geometric series)
GIGO	Garbage in garbage out
I	Total interest for the period
i	Interest rate for the compounding period ($= r/m$)
i_{eff}	Effective annual interest rate
i_f	"real" interest rate that accounts for inflation
IRA	Individual retirement account
IROR	Internal rate of return
IRS	Internal Revenue Service
ISO	International Organization for Standardization
k	Prefix for kilo, meaning 1,000
LCM	Least common multiple
m	Number of compounding periods per year
MACRS	Modified accelerated cost recovery system
MAPI	Machinery and Allied Products Institute
MARR	Minimum attractive rate of return
MBA	Master of Business Administration
N	Useful life
n	Number of compounding periods, useful life
NASA	National Aeronautics and Space Administration
NPW	Net present worth
OSHA	Occupational Safety and Health Administration
P	Present sum of money, initial investment, principal
PC	Personal computer
PW	Present worth
PWs	Present worths, plural of PW
r	Nominal interest rate (per year)
ROR	Rate of return
SI	System International
SL	Straight line (depreciation method or rate)
SOYD	Sum-of-years digits
U.S.	The United States
x	Independent variable
y(x)	Dependent variable y as a function of x

Part I

Fundamentals

The first part of the text is its *foundation*. In its four chapters, we discuss the basic principles essential for comprehending and solving engineering economics problems. This part prepares the reader to follow the rest of the coverage. As mentioned in the preface, mitigating the "fear" and "anxiety" of engineering economics as a course is a major objective of this text. Part I contributes immensely to the attainment of this objective.

Chapter 1 introduces engineering economics. It emphasizes that engineering economics is basically decision-making. The concepts of cashflow tables and diagrams are presented in Chapter 2. Chapters 3 and 4 discuss the time-valued equivalency of cashflows in terms of single payment and multiple payments, which is the core of engineering economics.

Chapter 1

Introduction

At the dawn of the 21st century as the world shrinks to what is being called the global village, more and more goods and services cross geographic boundaries, with the result that their production, distribution, and consumption are becoming truly international. The globalization and fierce competition of the marketplace demand that economic decisions be both precise and accurate. Recent management trends of re-engineering, downsizing, restructuring, total quality management (TQM), and continuous improvement testify to the fact that the competition for the market of goods and services has really swung into the top gear. Economic factors play a much more crucial role in industry today than at any time in the past.

As a distinct group of professionals, engineers and technologists continually strive to enhance the productivity and quality of the products and services they are involved with. They are often asked to do more with less (resources). It is becoming increasingly more important that these professionals be thoroughly skilled in engineering economics so that they can contribute rationally to capital investment and other cost-related decisions. To this end, an undergraduate course in engineering economics is almost mandatory for all majors of engineering and engineering technology.

This text discusses the basic principles of engineering economics and illustrates their applications to industrial projects involving costs and benefits. Though addressed primarily to engineers and engineering technologists, it will also be useful to engineering managers, industrial technologists, business managers, and applied scientists.

1.1 ENGINEERING ECONOMICS

Economics is concerned with decision-making relating to design, production, distribution, and consumption of goods and services. *Engineering economics* is a specialty of economics which focuses on engineering projects. It deals with the economic aspects of product, equipment, service, or technical support. Its knowledge is "a must" for engineers and engineering technologists—in fact for all technical and management professionals.

If you have $20,000 to invest and are deciding which company shares to buy, then the decision involves *business economics*, or simply economics. If, on the other hand, you decide to use this sum as a capital to make and market a better mouse trap, then the associated sets of decisions fall in the arena of *engineering economics*. Engineering economics presumes some technical knowledge on the part of the decision maker, while (business) economics does not. Since both economics and engineering economics share the same fundamental principles, a person skilled in one *may be able* to venture into the other.

1.2 ECONOMIC DECISION-MAKING

We frequently make economic decisions for a variety of problems. But, only the well-thought, rational decisions lead to successful conclusions. If the problem is simple, we may be able to think it through in our heads. Consider that you need to buy a pencil. The local bookstore sells pencils for 20 cents each, or a pack of five for 60 cents. You immediately ask yourself: Should I buy a pack? You mentally determine that the pack works out cheaper (12 cents each, instead of 20 cents if bought separately) and, let us say, decide to buy a pack. In so deciding you probably also considered:

1. Should I spend 60 cents in buying the pack, rather than 20 cents for one?
2. Can I spare the 60 cents now? Do I have another immediate need for this sum, for example, for a can of coke?
3. What shall I do with the other four pencils if I buy the pack? Shall I be able to use them later? Oh! Yes, my daughter may need them since her school begins next week.

Decision-making problems as simple as this are obviously too trivial for a textbook. At the other extreme, we sometimes face enormously complex problems. Besides the economic considerations, such problems also involve non-economic factors that may be political, social, or ethical in nature. Most of these factors are non-quantifiable in monetary terms. Consider, for example, the task of preparing the national budget. Besides its size, some relevant questions are: How to finance the expenditures? How much to borrow? Whom to tax? How will the budget affect people living below the poverty line? Such questions raise issues that are much more than economic. In such problems political and social considerations may overtake the economic ones, rendering decision-making really difficult. Moreover, of the several alternatives or solutions, none may be satisfactory to the decision maker and/or the people affected by the decision. Such enormously complex problems too are beyond the scope of this text.

In between the *trivial* and the *enormously complex*, there exists an array of moderately difficult problems that are primarily economic in nature. Such engineering-related problems fall within the realm of engineering economics and are the subject matter of this text. Their solutions require that we

1. State and comprehend the problem.
2. Collect and analyze the associated data.
3. Carry out the calculations.
4. Decide which alternative offers the best solution.

Engineering economics is much more than carrying out the calculations for a problem. It involves all the four tasks listed. That is why it is said that *engineering economics* is basically *decision-making*. Consider, for example, Linda, an engineering supervisor, who has been facing frequent breakdown of the plant's only stamping machine. At the annual budget time, she has to decide whether to replace the machine this year or to keep it for another year. If she decides to replace it, which of the five makes of machine available in the market to buy, and thus budget for? Such problems fall within the scope of this text, and are discussed later in Chapter 14.

Economic decision-making is a rational process. In general, it involves the following nine steps.

1. Recognition of the problem
2. Goal(s) of the solution
3. Data collection/gathering
4. Research for feasible alternatives
5. Criterion for selecting the best alternative
6. Mathematical modeling and associated calculations
7. *What-if* analyses with the model
8. Selection of the best alternative
9. Post-implementation follow-up

Some of the these steps may be absent in engineering projects, not because they are not required but because they have already been completed by someone. For example, the recognition of the problem might have been done by the plant manager. From a business viewpoint, step 1 may sometimes represent an opportunity, rather than being a problem, to invest capital for realizing profits. The goal in step 2 may be cost savings, increased throughput, higher profit, better quality, and so on.

In this text, we focus primarily on steps 5, 6, and 8 through discussions, examples, and chapter-end exercises. The other six steps are usually learned at the job. The last step helps the engineer to learn from the decision so that better decisions can be made in future. Step 7 is performed mostly with commercial software. The textual problem statements have been structured to contain sufficient relevant data that in industry are normally generated at steps 1 through 4. In the real-world many a data may not be readily available; their collection (step 3) consumes significant time and effort.

1.3 INTEREST RATE

Interest is the cost of using someone else's money. In that sense, it may be thought of as rent. The money you borrow from the bank belongs to savers whom the bank pays interest. In turn, the bank charges you interest. What you pay to the bank as interest is more than what the bank pays to the saver. The bank is thus the "middle man," pocketing the difference in interests.

Interest rate is a measure of the interest for a given loan during the loan period. Consider that you borrow $200 for one year from the local bank at an interest rate of 5% per year. For the year the interest will be 5% of $200, i.e., $10; your payback will thus be $200 principal plus $10 interest—a total of $210. The term *payback* is used here in a literal sense to mean the total of what is paid back to the bank. In Chapter 5, it is used in its criterion sense as practiced by engineering economists.

The prevailing interest rate is a function of both time and place. The rate may be different today from what it was last month or year. It may be different in the U.S. from that in India or Kenya. It may even differ from bank to bank in the same town. The interest rate also depends on the credit-worthiness of the borrower (individual or company), i.e., on the risk involved in lending as judged by the lender. The overall economic climate and government policies also play critical roles in the prevailing range of interest rates.

Interest rate is usually denoted by i and expressed in percentage (%). An annual interest rate of 8% is stated as $i = 8\%$ per year. Let us say that you borrow a sum of $350 for one year. This sum is called the *principal*. The relevant calculations for the year-end payback (principal plus interest) are:

$$\text{Interest rate per year} = 8\%$$

$$= 8 \text{ percent}$$

$$= 8 \text{ per hundred}$$

$$= \$8 \text{ per hundred dollars}$$

$$\text{Number of hundreds in the } \$350 - \text{principal} = 3.5$$

$$\text{Therefore, interest for one year} = \$8 \times 3.5 = \$28$$

$$\text{Year} - \text{end total payback} = \text{principal} + \text{interest}$$

$$\$378 = \$350 + \$28$$

In these calculations, we used the number of hundreds in the loan ($350) because the interest rate was expressed in percent. If the 8% rate was expressed in fraction as 8/100 = 0.08, i.e., $0.08 for every dollar of the principal, we would have multiplied 0.08 with $350, getting the same result of $28 for the interest, as shown as follows.

$$\text{Interest rate per year} = 8 \text{ percent}$$

$$= 8 \text{ per hundred}$$

$$= 8/100 \text{ per unit}$$

$$= 0.08 \text{ per unit}$$

$$= \$0.08 \text{ per unit dollar}$$

$$\text{Number of units in the } \$350 - \text{principal} = 350 \text{ dollars}$$

$$\text{Interest for one year} = (\$0.08/\text{dollar}) \times 350 \text{ dollars} = \$28$$

To summarize,

The interest rate is usually denoted by i and expressed in percent (%) per year. To calculate the amount of interest earned for a given principal, first convert i into its fractional value by dividing by 100. Then multiply the fractional value of i with the principal to determine the interest for the year.

1.4 SIMPLE INTEREST

Interest is calculated for each agreed period of the loan duration. The period may be a year, month, quarter of a year, or any other time window. The loan duration comprises one or more periods. Interest charged for a loan is called *simple* if the earned interest does not become a part of the principal for the next period. In other words, the interest earned during a period *does not* earn interest in the subsequent period. Instead, it is paid to the lender. If it is not paid out every period, it simply accumulates without earning any interest. At the end of the loan, the principal and the accumulated interests are paid together. Thus, the interest earned during each period leaves the principal unaffected; as a result, the principal does not increase during the loan duration.

If you borrow $300 for a period of 3 years at 7% annual interest rate, then under *simple interest* your total payback at the end of the loan period will be $363. Of this, $300 is the principal and $63 the accumulated interest[1] for 3 years.

In equation form, total simple interest I is expressed as

$$I = P\,i\,n \qquad (1.1)$$

where,

P = principal
i = interest rate (in fraction) per period
n = number of periods

In this illustrative example, where period was expressed in years,

$P = \$300$

$i = 0.07\,(\text{fractional value of } 7\%)$ per year, and

$n = 3\,\text{years}$

Therefore, from Equation (1.1)

	A	B	C
1	P	$300.00	
2	i	0.07	
3	n	3	
4	I	$63.00	=B1*B2*B3
5			

Equation (1.1) involves four variables: I, P, *i*, and n. Given the values of any three, the fourth can be evaluated.

Equation (1.1) can be extended to determine the final payback F by adding the total interest I to the principal P. Thus,

$F = P + I$

$\quad = P + P\,i\,n \qquad (1.2)$

$\quad = P(1 + i\,n)$

Equation (1.2) too involves four variables: F, P, *i*, and n. Given the values of any three, the fourth can be evaluated.

---

[1] Interest for one year = Interest rate × Units of dollar= 0.07 × $300 = $21 Interest for 3 years = $21 × 3

Equations (1.1) and (1.2) are handy and should normally be used to work out problems involving simple interest. But many times it is relatively easier to follow the first-principles[2] rather than use the equation; simply trace the procedure that was used to derive the equation.

Interest rate is usually quoted for the year (*annual*). Sometimes, it may be quoted for the month or other units of time duration such as quarter (one-fourth of the year, i.e., three months). Whatever the time period, *i and n must be expressed in compatible units.* For example, if i = 1% per month, then the value of n for a three-year investment is 36 months (not 3 years). For this rate and duration, the total interest for an investment of $300 will, from Equation (1.1), be

	A	B	C	D
1	I=Pin			
2		P	$300.00	
3		i	0.01	
4		n	36	
5		I =	$108.00	=C2*C3*C4
6				

Simple interest rates are rare in the financial marketplace. Hardly anyone transacts money at simple interest. Relatives and close friends may at times display their generosity by loaning money at simple interest. In the business world, however, *simple interest* is non-existent.

Example 1.1: Payment Schedule

After earning a BS (bachelor of science) degree in construction engineering, John decided to start his own business. His rich, generous uncle has agreed to loan him $80,000 at an annual 4% simple interest. John is to pay his uncle the yearly interest on the loan anniversary and return the principal on the fifth anniversary. What is John's loan payment schedule?

SOLUTION:

By payment schedule we mean the scheme by which interest and the principal are paid back. In other words, how much will be paid when. In this case, the interest for the year is paid on the loan anniversary, while the principal is paid back on the fifth anniversary along with the interest for the fifth year.

We need to determine the interest for the year, since it is due on each anniversary. Since no portion of the principal is paid until the fifth anniversary, the principal remains $80,000 during the entire loan duration.

We can use Equation (1.1) to determine the annual (n = 1) interest. The given data are:

$$P = \$80{,}000 \quad i = 4\% \quad n = 1$$

---

[2] What has been discussed as an illustration in the second paragraph of Section 1.4 and its footnote is a good example of first-principles. By first-principles we mean a solution approach based on fundamentals, not on equation(s). For example, to determine the year-end interest we multiply the principal with the fractional value of annual interest rate. The alternative approach is to use Equation (1.1) with n = 1. You may wonder: What actually is the difference between first-principles and the equation? Not much. In fact, an equation is derived following the first-principles. Thus, by first-principles we mean the procedure that yields the equation. Of the two, use the one that renders the solution easier.

Remember to use the fractional value of the interest rate, i.e., i = 4/100 = 0.04.
  Therefore,

	A	B	C	D
1	I (interest for year)=	P	i	n
2	$3,200.00	$80,000.00	0.04	1
3	=B2*C2*D2			

Thus, John will pay his uncle $3,200 as interest on each of the first four loan anniversaries. The payment on the fifth anniversary will be

$$= Principal + Interest\ for\ the\ fifth\ year$$

$$= \$80,000 + \$3,200$$

$$= \$83,200$$

Thus, John's payment schedule is: $3,200 on each of the first four loan anniversaries, and $83,200 on the fifth anniversary.

In this example, a table might have been appropriate to summarize the loan payment schedule, as follows. Note that the table columns have been given proper headings.

	A	B
1	Loan Anniversary	Payment Due
2	1	$3,200.00
3	2	$3,200.00
4	3	$3,200.00
5	4	$3,200.00
6	5	$83,200.00

Engineers and technologists usually express themselves through charts, graphs, tables, or modern tools such as animation. These modes of expression summarize the data and highlight the important pieces of information. Even when the problem (project) does not specifically ask for a table, graph, or chart, it is desirable to use them to enhance the presentation of results.

1.5 COMPOUNDING

In most cases, if the interest is not withdrawn, or paid out when due, it is allowed to augment the principal at the end of the (interest accounting) period. The total of the principal and the interest for the period becomes the new principal, which earns interest during the following period. Consequently, *the interest earns interest*. This augmenting effect of interest on the principal is called *compounding*.
  The dictionary meaning of the word compounding is putting together or combining. Here, it means combining together of the interest earned for a period with the *beginning* principal for the period. The total of the two yields the *ending* principal for the period, which becomes the beginning principal for the subsequent period. The following discussion illustrates the concept of compounding.
  Consider that you borrow $200 for 2 years at 20% yearly interest with annual compounding. Thus, the beginning principal P is $200, and the value of *i* is 0.20. Since compounding is annual, the interest period is year. The interest for the first period or year, payable at its

end = $200 × 0.20 = $40. This interest is combined with the beginning (original) principal $200. Thus, the principal at the end of the first year = $200 + $40 = $240. This $240 now becomes the principal at the beginning of the second period. Therefore, the interest for the second year = $240 × 0.20 = $48. Thus, the final (total) amount F payable at the end of the second year is

= Principal at the beginning of second year

+Interest for the second year

= $240 + $48

= $288

Had the $200 loan been at simple interest (not compounded), the value[3] of F, payable at the end of the second year, would have been

= $200 + Total simple interest earned over two years

= $200 + ($200 × 0.20) × 2

= $280

The final payment of $280 under simple interest is less than when compounding was done ($288). The interest earned by the lender under compounding is $8 more, which you paid as a borrower. Thus, compounding is beneficial to the lender, and expensive to the borrower— exactly by the same amount.

All borrowing and lending transactions in today's marketplace use compound interest. That is why all the discussions from now onwards are based on compounding. *Assume compounding of interest everywhere unless mentioned otherwise.*

Example 1.2

Maya owns and operates a business of used robots. From the profit realized in the current fiscal year, she decides to save a sum of $30,000 for use later to expand her business. The sum is invested in a three-year certificate-of-deposit (CD) with the local credit union at an annually compounded interest rate of 6% per year. How much will this CD mature to?

SOLUTION:

The CD investment is for a term of 3 years at a yearly interest rate of 6%, compounded annually. We solve this problem following the first-principles.
   Thus, Maya's $30,000 CD will mature to $35,730.48.

Example (1.2) has illustrated how problems involving compounding of interest can be worked out following the first-principles approach. However, this approach becomes lengthy and cumbersome for compounding involving several periods. In such cases, equations are easier to use. We derive such equations in Chapter 3 and illustrate their use there and beyond.

---

[3] One can use Equation (1.2), wherefrom F = P (1 + in) = $200 (1 + 0.20 × 2) = $280.

Example 1.3

In Example 1.2, how much more did Maya's CD earn in comparison to that under simple interest?

SOLUTION:

Total interest earned in Example 1.2

	A	B	C
1	F-P=Total Interest Earned		
2	F	$35,730.48	
3	P	$30,000	
4	Total Interest	$5,730	=B2-B3
5			

Had Maya's CD earned simple interest, total interest I at the end of 3 years, from Equation (1.1), would have been

	A	B	C	D
1	I (interest for year)=	**P**	*i*	**n**
2	$5,400.00	$30,000.00	0.06	3
3	=B2*C2*D2			

Thus, compounding yielded[4] an additional earning = $5,730.48 – $5,400 = $330.48.

1.6 TIME VALUE OF MONEY

As seen in the previous three sections, a sum of money has the potential to "grow" due to the interest it can earn. This fact is stated by saying that *money has time value*. The current sum is the principal (or *present* sum) P, which increases due to interest to a final (or future) sum F. The increase in P is F – P, which equals the total interest I. The value of I depends on the principal P, period n, and interest rate *i*, and whether or not the periodic interest is compounded. In Chapter 3, we learn more about the time value of money and its effect on cashflows.

1.7 THE SIX-STEP PROCEDURE

Engineering economics problems are primarily numerical in
    nature—the values of certain parameters are given and the value of the unknown is determined. Over the years, I have noticed three common weaknesses in students enrolled in engineering economics course:

1. Difficulty in comprehending the problem statement
2. Confusion with the mathematics[5] involved and
3. How to solve, i.e., the sequence of solution steps

Most numerical problems in an engineering economics course are stated at a language level expected of college students. The language level of the problem statement in Example 1.2

---

[4] Alternatively, you can work out the difference between the final sum F under compounding and that under simple interest. Since F with compounding is already known to be $35,730.48, use Equation (1.2) to determine F under simple interest. Thus, Additional earning = $35,730.48 – P(1 + in)= $35,730.48 – 30,000(1 + 0.06 × 3)= $35,730.48 – $35,400= $330.48

[5] The mathematics involved in engineering economics is mostly algebra.

is typical. Occasionally, students find a problem statement incomprehensible. Other likely difficulties are: Not knowing how to proceed, which data to use, which equation(s) to use; worrying[6] about not using all the given data; unable to manipulate the equation; and so on.

Given next is a six-step procedure[7] that should alleviate some of the difficulties you may face in solving engineering economics problems.

Step 1 *Comprehension*
>   Read and reread the problem statement until you understand what is being asked. If it is long, mentally divide it into logical, comprehensible modules.

Step 2 *Summarize the Given Data*
>   Write down the given data against their customary symbols in the format: *Symbol = Data*. For example, P = $300, n = 5 years. Problem statements may at times contain more data than you actually need to solve the problem. In project statements of the real-world, the desired data may be hidden in or entangled with the non-essential data; so be careful while gathering and summarizing them.

Step 3 *What is the Unknown?*
>   Identify the unknown, and write down its symbol if any. This is the parameter needing evaluation for the given set of data.

Step 4 *Search for the Process*
>   Ponder on how to proceed to solve the problem. Am I better off following the first-principles? Is there an equation that relates the unknown to the knowns (the given data)? Look for a form

$$\text{Unknown} = f(\text{knowns})$$

>   where f stands for "function of."
>   The unknown-knowns relationship may not be obvious. Search for indirect relationships, if necessary. Logically trace the relationship(s) among the given parameters to answer: *How to reach the "unknown" from the "knowns."* If needed, rearrange the parameters of the equation to get all the knowns on the right side of the equation.

Step 5 *Solve for the Unknown*
>   Manipulate the knowns with the tools available, such as log tables, scientific calculators, compound interest tables, or software.

Step 6 *Confirm the Answer, if Possible*
>   This final step is essential to achieve confidence in the result. Use your "gut feeling," subjective judgment, mental arithmetic, and/or scribbles to check whether the answer seems right. For example, interest rate *i* cannot be negative, period n must be +ve, and F cannot be less than P. If the answer looks suspicious, research the given data and go through the solution once more.

You can remember the six steps through the acronym CD-UP-SC, whose letters stand for:

C Comprehend – the problem statement
D Data   – gather the given data
U Unknown – what needs to be evaluated
P Process   – equation or first-principles
S Solve   – solve to get the answer
C Confirm   – check the answer

---

[6] Problem statements may contain redundant data.
[7] The procedure can in fact prove useful in solving any numerical problem.

Example 1.4, a modified version of Example 1.1, illustrates how to apply the six-step procedure.

Example 1.4

After finishing a BS in computer engineering technology, Joshua has set up a computer repair business. He needs to buy a diagnostic system that will expedite repair. The system costs $9,500 and is likely to generate $3,750 per year. Since his current loans are excessive, no financial institution is interested in lending him any more money. He discussed the difficulty of raising the capital with his rich uncle who happens to be generous. His uncle agrees to lend him $9,500 at 4% simple interest provided Joshua pays him the interest annually on time. How much will Joshua earn annually by investing in the system after paying the interest to his uncle?

SOLUTION:

Let us go through each of the six steps.

Step 1 *Comprehend*
Note that the loan earns simple interest. It is not clear when Joshua is to pay back his uncle the $9,500 loan; the problem statement is vague about it. But do we really need this information to solve the problem? Not really. Note that the interest for the year is paid annually on the loan anniversary. This payment must be subtracted from the annual income of $3,750.

Step 2 *Data*
The given data are:

$$P = \$9,500 \quad i = 4\% \quad n = 1$$

Remember to use the fractional value of the interest rate, i.e., $i = 4/100 = 0.04$.

Step 3 *Unknown*
The unknown here is the annual earning after paying for the interest.

Step 4 *Process*
Since the annual income from the use of the system is given, we need to determine the annual interest to evaluate the unknown. Shall we follow the first-principles? Or is there an equation that relates the earned interest I with P, i, and n? Yes indeed, there is one—Equation (1.1) discussed earlier. Which of the two—first-principles or Equation (1.1)—yields an easier solution? Here, it does not matter; both are equally effective. So let us use Equation (1.1).

Step 5 *Solve*
The interest for the year can be found from Equation (1.1),
with n = 1, as

	A	B	C	D
1	I (interest for year)=	P	i	n
2	$380.00	$9,500.00	0.04	1
3	=B2*C2*D2			

Thus, Joshua will pay his uncle each year $380 on the loan anniversary. Since his income from the diagnostic system is $3,750 per year, the annual saving will be

$$= \$3,750 - \$380$$

$$= \$3,370$$

Step 6 *Confirm*

Let us check if the answer seems correct. As it is a simple problem, we can do it in our "heads." The principal is $9,500. If the interest rate were 1%, the interest for the year would be $95. This $95 can be rounded to $100 to aid the mental arithmetic. So for an interest rate of 4% (four times as high), the approximate interest for the year should be four times of $100, i.e., $400. Thus, the value of I at step 5 as $380 seems alright[8] since it is closer to the mentally-worked-out approximate value of $400.

The usefulness of the six-step procedure might not have been quite obvious in the previous illustration. Try applying it to other problems of the text. Frequent use of the procedure will hone your skills and confirm its usefulness.

Do we have to follow the six-step procedure in a very formal way, as illustrated? Not really. But keeping the procedure in mind while solving problems is likely to be helpful.

1.8 SUMMARY

Engineering economics plays an increasingly more critical role in industry as competition for the production and distribution of goods and services intensifies under the impact of globalization. It is basically a decision-making process, comprising nine steps. In engineering economics, one learns to solve moderately difficult engineering problems involving costs and benefits. Interest can be thought of as the rent for using someone's money; interest rate is a measure of the cost of this use. Interest rates are of two types: simple or compounded. Under simple interest only the principal earns interest. Simple interest is non-existent in today's financial marketplace. Under compounding, the interest earned during a period augments the principal; thus, interest earns interest. Compounding of interest is beneficial to the lender. Due to its capacity to earn interest, money has time value. The time value of money is important in making decisions pertaining to engineering projects. A six-step procedure that is helpful in solving engineering economics problems has been presented in Chapter 1.

DISCUSSION QUESTIONS

1.1 Why is the study of economics by engineers and technologists more important today than in the past?

1.2 State in not more than one hundred words an engineering economics problem of your choice.

1.3 Your local government is considering building a new road. Discuss whether or not this is an engineering economics problem.

1.4 Is the six-step procedure discussed in Section 1.7 for solving engineering economics problems helpful? Justify your answer.

1.5 Why do financial institutions never offer loans at simple interest?

1.6 The prevailing interest rate is a function of what factors?

1.7 List and discuss the four requirements of engineering economics decision-making.

1.8 Compare and contrast the differences in the investments made by governments versus investment by industries.

---

[8] We could have approximated the principal itself to $10,000 to aid mental arithmetic. At 4% rate, this would have given $400 as annual interest.

MULTIPLE-CHOICE QUESTIONS

1.9 The decision to go on a two-week vacation represents
   a. an engineering problem
   b. an economics problem
   c. an engineering economics problem
   d. none of the above

1.10 Nasim is overjoyed on winning $500,000 in a lottery. Of this windfall, he decides to spend $20,000 on his marriage next month, buy a $300,000 motel as an investment, and use the remainder to start a business of making plastic toothpicks. This decision falls under
   a. business economics
   b. engineering economics
   c. home economics
   d. none of the above

1.11 Which of the following is more like an engineering economics problem?
   a. Earning a BS degree
   b. Buying a car for business as well as personal use
   c. Getting married
   d. Preparation of the US federal budget

1.12 Problems most suitable for engineering economics analyses
   1. are sufficiently important
   2. can't be worked out in our heads
   3. focus primarily on economic factors
   4. cost a lot to solve
   a. 1 and 3
   b. 2 and 4
   c. 1, 2, and 3
   d. 1, 2, 3, and 4

1.13 Simple interest means that the interest
   a. rate is quoted in round figures, not in decimals
   b. must be paid at the end of each interest period
   c. does not augment the principal
   d. is paid whenever convenient to the borrower

1.14 Time value of money means that
   a. time is money
   b. time not used sensibly translates into lost money
   c. the principal is capable of earning interest
   d. it is hard to enjoy "the good life" without money

1.15 Engineering economics decision-making involves nine steps. Which of the following is NOT one of them?
   a. recognition of the problem
   b. determination of the feasible alternatives
   c. financing of the project
   d. selection of the best alternative

1.16 Interest is classified as:
   a. revenue
   b. expense
   c. asset
   d. liability

1.17 The principle of time value of money states a sum of money has the potential for what?
   a. growth
   b. shrinkage
   c. loss
   d. remaining constant

1.18 The augmenting effect of interest on the principal is a result of what type of interest?
   a. simple
   b. time value of money
   c. compounding
   d. none of the above

1.19 What is the simple interest on the principal of $5,000 over four periods at an interest rate of 5%?
   a. $1,500
   b. $2,000
   c. $1,200
   d. $1,000

1.20 What is the principal balance of investment with compounding interest in year three with a beginning principal of $30,000 and an interest rate of 7%?
   a. $45,000
   b. $36,752
   c. $34,347
   d. $28,675

1.21 Supposed you took out a loan for $20,000 with a semiannually compounding interest rate of 6%. Assuming no payments were made, what would be the balance at the beginning of the third year?
   a. $28,370
   b. $23,543
   c. $25,250
   d. $21,345

1.22 A loan of $6,000 was repaid at the end of 8 months with a check for $6,350. What was the annual rate of interest charged?
   a. 7.05%
   b. 8.83%
   c. 9%
   d. 6.75%

1.23 An investment company pays 5% annual interest; how much should you deposit now in order to have $10,000 at the end of 6 years?
   a. $7,692
   b. $6,000
   c. $7,557
   d. $8,976

NUMERICAL PROBLEMS

1.24 A young engineer has just started her small construction company. Due to paucity of funds, she is not able to buy a truck the company desperately needs. Her grandpa has an old truck and is willing to help her. The truck's market value is $3,500, but he sells it to her on credit for $3,000. She agrees to pay him each year 3% simple annual interest for the next 5 years. On the fifth anniversary of the loan she will also pay him $3,000. Prepare the payment schedule.

1.25 To help his young engineer son, Kashi lends him $50,000 for 7 years at simple interest of 3.5% per year. How much will Kashi receive from his son when the loan will be due? Assume that the interest is paid together with the principal as a lump sum.

1.26 Your father deposits $10,000 in a savings account and earns simple interest. If he is paid $205 quarterly as interest, what is the annual interest rate?

1.27 How long will it take $3,000 invested at simple annual interest of 5% to become $5,000?

1.28 How many years will it take a sum to double if it earns simple interest at 9% per year?

1.29 Determine the annual rate of simple interest if $360 is earned as interest at the end of 15 months for an investment of $3,500.

1.30 Jose borrows $40,000 to buy a new machine at an annually compounded interest rate of 9% per year. The loan is to be paid in full on its fourth anniversary. How much will the payment be?

1.31 Mumtaz invested $2,000 in a five-year CD at an annual interest rate of 5%, compounded yearly. Due to an unforeseen situation, she had to withdraw all the money on the third anniversary of the investment. How much did she get if there was a $25 penalty for early withdrawal?

1.32 A small tool and die company takes a loan at 12% annual interest rate to buy an injection molding machine. The company pays back the loan (capital and interest) on the first loan anniversary in the sum of $62,350. If compounding is annual, how much was the loan?

1.33 Engineering Unlimited takes a loan at 16% yearly interest rate, compounded quarterly, for purchasing a robotic system. The company pays back the loan (capital and interest) through two payments: $50,000 at the end of six months, and $65,000 on the first loan anniversary. How much was the loan?

1.34 You need to borrow $1,000 to complete your education, intending to pay it back along with the interest next year. Credit union A charges 14.5% interest compounded annually, while B charges 14% compounded quarterly. Which credit union should you borrow from?

1.35 If you deposited $7,000 dollars into a savings account with a compounding interest rate of 6%, how long until you have doubled your money?

1.36 You applied for and received a credit card with a compounding interest rate of 13.49%; this rate is compounded at the end of every month. The day you received the card in the mail you maxed out your $2,500 limit. How much money would you owe the company after 8 months?
   a. Now compute the simple interest for the same time period.
   b. What are the savings for simple versus compounding interest?

1.37 What was the beginning principal of a balance totaling $13,400 at the end of 10 years bearing an annual interest rate of 8%?

1.38 Suppose that interest on an investment is compounded continuously. You made an investment with a 7.5% interest rate. How much would you have to invest today in order to have a balance of $50,000 at the end of 9 years?

1.39 How long will it take for an initial investment of $25,000 to grow to $45,000 if the simple interest rate is 6.49%?

1.40 How long would it take for you to double your investment of $27,000 if:
   a. 10%, compounded quarterly?
   b. 10%, compounded annually?
   c. 10%, compounded bi-annually?
   d. which of these is the best investment?

1.41 The company you work for took a loan out for $125,000 in January of last year on a new drill press to decrease processing time in the warehouse. Assume that the interest rate in 5.5% compounded quarterly. Due to a boom in production your company was able to pay the loan plus interest in full in September of this year. What amount did they repay the bank?

FE EXAM PREP

1.42 A $100,000 sorting machine is financed at a rate of 4% per year. How much interest is owed at the end of year 1?
    a. $ 8,000
    b. $ 4,000
    c. $10,000
    d. $12,000

1.43 A forklift dealer offers a 6.5% financing rate compounded annually. If the initial loan amount is $60,000, the balance at the end of 3 years would closely resemble which of the following?
    a. $73,456
    b. $72,475
    c. $65,000
    d. $68,639

1.44 The study of engineering economics is important because engineers
    a. must create designs that are economically efficient and feasible
    b. work closely with finance and accounting departments
    c. provide a framework of comparison between alternative solutions
    d. all of the above

1.45 A company may invest in engineering projects in order to
    a. increase revenues
    b. decrease costs
    c. reduce production capabilities
    d. fully expend all available capital

1.46 Which of the following would involve economic analysis?
    a. determining the best start date for an engineering project
    b. determining the useful life of an asset
    c. selecting between two software packages for implementation
    d. all of the above

1.47 A dealership offers a 4.5% interest rate compounded annually. If you took a loan out for $12,500, what would the balance be at the end of year 1?
    a. $14,765
    b. $10,000
    c. $12,500
    d. $13,062

1.48 A $75,000 conveyor system is financed at a rate of 8% per year. How much interest is owed at the end of year 1?
    a. $ 8,000
    b. $ 6,000
    c. $10,000
    d. $12,000

1.49 A dealer offers a 10% financing rate compounded semiannually. If the initial loan amount is $45,000, the balance at the end of 2 years would closely resemble which of the following?
   a. $73,456
   b. $72,475
   c. $65,000
   d. $68,639

1.50 A $23,000 air filtration system is financed at a rate of 5% per year. How much interest is owed at the end of year 1?
   a. $ 800
   b. $ 600
   c. $1,150
   d. $1,200

1.51 A dealer offers a 6% financing rate compounded annually. If the initial loan amount is $4,000, the balance at the end of 2 years would closely resemble which of the following?
   a. $ 736
   b. $4,480
   c. $6,500
   d. $8,639

Chapter 2

Cashflows

LEARNING OUTCOMES

- The concept of cashflows
- Cashflow table
- Sign convention in a cashflow table
- Auxiliary table
- How to prepare a cashflow table
- End-of-period assumption
- Cashflow diagrams
- How to sketch a cashflow diagram

Engineering projects involve transactions of money. Money spent on the project is called *cost, disbursement, expenditure,* or *cash outflow,* and the money earned (or saved[1]) from the project is antonymously called *benefit, receipt, revenue,* or *cash inflow.* These transactions or cashflows occur at different times during the project life. The terms *cash outflow* and *cash inflow* denote the directions in which cash "flows" with reference to the project account. These are simply accountant's terms for *cost* and *benefit* respectively. Engineers and technologists are more conversant with the terms cost and benefit rather than with cash outflow and cash inflow.

An engineering project with its *estimated* costs and *expected* benefits is basically an engineering economics problem whose solution requires a clear understanding of the associated cashflows. The problem is easier to comprehend and solve if the cashflows and their timings are properly tracked. This is done by

(a) Tabulating the cashflows and their timings
(b) Diagramming the cashflows

The result of (a) is a *cashflow table,* whereas that of (b) is a *cashflow diagram.* In this chapter, we learn how to express the cashflows in such a table or diagram.

Cashflow tables and diagrams are tools. They help us visualize and understand the problem better. They facilitate comprehension of the problem; they may not be essential for solving it. Experienced engineering economists can mentally visualize the cashflows and their timings, and proceed directly to solve the problem without needing the table or the diagram. But even for them mental comprehension becomes unmanageable for complex problems, necessitating the use of tables and/or diagrams. For beginners, the use of tables and diagrams is highly recommended. It is essential that sufficient skills be mastered in developing cashflow table(s) and diagram(s) as a prelude to the solution.

---

[1] If a project saves money through higher productivity, better product quality, or other ways, the saving is in fact earning.

2.1 CASHFLOW TABLE

A cashflow table is basically a two-column table. Its first column shows the timing of the cashflows, while the second column shows the amounts of cashflow. The column headings of a typical cashflow table are *year* and *amount* (or *cashflow*).

Consider that you borrow $2,000 today and pay back this loan over the next 3 years as annual payments of $800 each on the loan anniversaries. The corresponding cashflow table will be:

YEAR	0	1	2	3
CASHFLOW	$2,000	–$800	–$800	–$800

Let us discuss the entries in the table, especially the sign convention. In the first column, year 0 represents "now" or "today," when the project begins, which in this case is your borrowing (receiving) $2,000 from the lending institution. Year 1 represents the time period from now to this time next year, i.e., the end of the first year (loan anniversary). Similarly, the entries of 2 and 3 in the first column mark the ends of year 2 and year 3. The entry of 800 for year 1 is the cashflow (payment) on the first loan anniversary. Note that no cashflows *during* the year, except *on* the loan anniversary when you pay back $800. The same thing happens on the second and third anniversaries.

Let us now look at the second column. Note that the currency symbol $ has been prefixed only to the first cashflow. The other cashflows are assumed to be in the same currency. This avoids currency symbol swamping the tabulated data.

The other point to note in the second column is the convention of algebraic sign for the cashflows. Each data in a cashflow table must be preceded by either a negative (–) sign or no sign (understood to be positive). Amounts received (inflows) are given no signs and those paid out (outflows) are given – signs. The best way to avoid any confusion with the sign convention is to remember that money received (inflows) increases the balance in the account, i.e., has a positive effect on the account. The reverse is true when transactions are paybacks or costs (outflows).

The cashflow of $2,000 in the table actually has a positive sign; but, since positive numbers are not signed, we did not attach a + sign in front of $2,000. The signs in front of the other three cashflows (800) are negative. This follows the convention that a negative number must be preceded by a – sign. Each payment of $800 reduced your account balance by that much, i.e., had a negative effect on your account. That is why these paybacks have been preceded by a – sign in the table.

2.1.1 Whose Cashflow Table?

An important consideration while preparing a cashflow table is to ask the question: Whose cashflow table is it? The previous table was yours (the borrower's). How will the cashflow table of the lender look like? From the lender's viewpoint, say the bank's, the cashflow table will be:

YEAR	0	1	2	3
CASHFLOW	–$2,000	$800	$800	$800

Note the change in cashflow signs as compared to the previous table. The logic of the sign convention, however, remains the same and can explain the signs in this table too. With reference to the bank's account, the $2,000 loaned to you is an outflow that reduced the bank's account balance. That is why it is prefixed with a – sign. Your loan payments of $800 increase (+ve effect) the bank's account balance and therefore are labeled positive (no sign).

In general, cashflow signs in the borrower's (one party) cashflow table are just the opposite of those in the lender's (the other party).

To summarize:

Cash Inflows/Receipts/Benefits/Revenues NO SIGN
Cash Outflows/Disbursements/Costs/Expenditures –ve SIGN

2.1.2 Development of Cashflow Table

When the cashflows of an engineering project (economics problem) are complex, it may be desirable to develop and use auxiliary tables as an aid to comprehension. Such tables are later summarized as the (final two-column) cashflow table. The important question while developing auxiliary tables are: How many columns to have and for what purpose? The obvious answer is: As many as necessary to develop the cashflow table with minimum confusion. The decision on number and size of the auxiliary tables is one of the creative steps in the solution process. The following example illustrates the point.

Example 2.1: Cashflow Table

Jack, an engineering freshman at a senior college, borrows $5,000 from the local bank at 8% annual interest to buy a used car. He will use his financial aid to pay back the loan during the 4 years at the college. He plans to pay $1,000 of the principal on each of the first three anniversaries of the loan along with the interest for the year. On the fourth loan anniversary (his graduation year), he plans to pay back the remainder of the loan along with the interest. Develop his cashflow table.

SOLUTION:

To help us prepare Jack's cashflow table, we will use an auxiliary table containing four columns. Why four columns? Because the transactions have two components: one principal and the other interest. We need two columns for these, and another (third) column to enter their total. There has to be a column for the time period. Thus, a four-columned auxiliary table will suffice.

To keep the auxiliary table development easier, we will exclude year 0 for now and add it later. The skeleton of the auxiliary table will look like:

YEAR	PRINCIPAL	INTEREST	TOTAL
1			
2			
3			
4			

The first column is always for the period; year in this case. In the second and third columns we will enter the principal paid and the interest for the year, while in the last column the total paid. Note that each column has an appropriate heading.

Let us now do the calculations and post the data in the table.

Year 1: We enter $1,000 in column 2 (principal) as part payment of the principal for the first year. Interest for the first year = $5,000 × 0.08 = $400. Thus, $400 is posted in the interest column. The total of these two, i.e., $1,400 is entered in the last column. With these entries the table looks like:

YEAR	PRINCIPAL	INTEREST	TOTAL
1	$1,000	$400	$1,400
2			
3			
4			

*Year 2:*Again, $1000 in column 2. The principal for which interest is earned during the second year = $5,000 – $1,000 = $4,000 (since $1,000 of the principal was paid on the first loan anniversary). Therefore, the interest for the second year = $4,000 × 0.08 = $320. Thus, $320 is posted in the interest column. The total of these two, i.e., $1,320 is entered in the last column.

In a similar way, the data for the other two rows can be calculated and posted.

With all the data entered in, the auxiliary (or payback) table becomes:

YEAR	PRINCIPAL	INTEREST	TOTAL
1	$1,000	$400	$1,400
2	$1,000	$320	$1,320
3	$1,000	$240	$1,240
4	$2,000	$160	$2,160

Note the second-column entry of $2,000 for year 4. This is the final payment of the principal ($5,000 – $3,000), since only $3,000 had been paid by year 3.

This auxiliary table showing the payback schedule can now be summarized to yield Jack's cashflow table. In general, only that column of the auxiliary table which shows the total is retained in the (final) cashflow table.

To complete the table, a new row showing the loan against year 0 is inserted now. With appropriate cashflow signs, the resulting cashflow table is:

YEAR	0	1	2	3	4
CASHFLOW	$5,000	–$1,400	–$1,320	–$1,240	–$2,160

Some engineering economists prefer to keep the calculations intact by retaining all the columns of the auxiliary table in the cashflow table. Many prefer to see only two columns in a cashflow table, and therefore exclude the other columns of the auxiliary table(s), and the calculations therein, as done above. If the auxiliary table(s) and the associated data are likely to impede the visualization of net cashflows, it is preferred to have only two columns in the *(final)* cashflow table. A two-column cashflow table offers definite advantage when several auxiliary tables are involved, as in complex problems.

There is no standard way of preparing an auxiliary table. Engineering economists use different columns in the auxiliary table(s) for the same problem. But, *the resulting two-column cashflow table for a problem must come out to be the same.* Consider, for example, the following auxiliary table developed by another economist, based on his own creative thinking, for solving the same problem (Example 2.1). Note the different headings, one more column, and the inclusion of year 0 in the auxiliary table in the beginning, as compared to the previous solution of Example 2.1.

YEAR	PAYBACKS	INTEREST	P+I	PRINCIPAL REMAINING
0				$5,000
1	$1,000	$400	$1,400	$4,000
2	$1,000	$320	$1,320	$3,000
3	$1,000	$240	$1,240	$2,000
4	$2,000	$160	$2,160	$0

The purpose of the data in the last column of this table is to enable the tracking of remaining principal, so that interest calculation for the following year can be done with less confusion. When summarized carefully as a two-column cashflow table, this auxiliary table also yields the same result as earlier.

From the discussions here and in Example 2.1:

1. One or more auxiliary tables may be used to facilitate calculations and to minimize confusion.
2. Once the cashflow data have been posted in the auxiliary table(s), they should be summarized as a cashflow table showing the timing of each transaction and the corresponding amount.

Sometimes the total of the transactions is also shown in the table as *net cashflow*, as done as follows for Example 2.1.

YEAR	0	1	2	3	4	NET CASHFLOW
CASHFLOW	$5,000	−$1,400	−$1,320	−$1,240	−$2,160	−$1,120

This net flow of $1,120 in Jack's cashflow table represents the (total) interest Jack *paid* (− sign) to the bank for the loan. In the lender's cashflow table, the net cashflow will be $1,120 (+ sign, opposite of that in Jack's table), representing the interest *earned* by the bank.

The development of the cashflow table in Example 2.1 has been simple. The next example considers an engineering project whose cashflow table is relatively complex to develop.

Example 2.2: Cashflow Table

The international environmental management standard ISO 14000 requires that Jenny, the plant supervisor, must install suitable pollution control equipment. She is considering two of them, one based on the principle of neutralization and the other of precipitation. They cost $750,000 and $450,000 respectively; both are expected to last 5 years and have salvage values of $150,000 and $105,000 respectively. The chemical needed for the neutralization-based equipment costs $50,000 per year, while for the other $95,000 per year. Their annual maintenance costs are $500 and $750 respectively. Develop a cashflow table to help Jenny visually compare the data pertaining to the two alternatives.

SOLUTION:

Since several costs are involved, auxiliary tables are desirable in this case. The question is: How many and what columns should they have? Let us begin with the first auxiliary table. Others will be prepared as and when necessary. As discussed earlier, the auxiliary tables will be summarized into a final two-column cashflow table.

The first column of the auxiliary table will of course be for the period; year in this case. Since two types of equipment are under consideration, there should at least be two additional columns, one for each. Let us code neutralization equipment as A and precipitation equipment as B. This will keep the table headings concise. The skeleton of the table then becomes:

YEAR	1	2	3	4	5
A					
B					

Now we need to decide how many columns should there be under A and B. The initial costs of A and B may be kept excluded from the auxiliary table; they will be added to the cashflow table later against year 0, as done in Example 2.1. Some economists prefer to include these costs in the beginning itself by inserting another row for year 0. In that case, the previous table gets modified to:

YEAR	0	1	2	3	4	5
A	−$750,000					
B	−$450,000					

The signs for the cost entries and the currency sign $ should be obvious based on the discussions in Section 2.1 and Example 2.1.

If we decide not to include the row for year 0 at this point of the development, our auxiliary table remains unchanged as:

YEAR	1	2	3	4	5
A					
B					

Now, for both A and B, the cashflow for each year comprises the following two elements:

1. cost of chemical, and
2. maintenance cost

Further, there is a third element at the end of the useful life (fifth year), namely salvage value as cash inflow (benefit).

Thus, we should have at least three columns, one for each of the three cashflow elements. There should be another column for the total of these three elements. The four columns may be given proper headings, for example, *Chem.* for element 1 (chemical), *Main.* for element 2 (maintenance), *Sal.* for element 3 (salvage), and *Total* for the last column. The auxiliary table skeleton thus looks like:

YEAR	A				B			
	Chem.	*Main.*	*Sal.*	*Total*	*Chem.*	*Main.*	*Sal.*	*Total*
1								
2								
3								
4								
5								

Note that the table had to be widened to accommodate the new columns. The width need of the auxiliary table was not known in the beginning. Thus, as you proceed, you may have to keep adjusting the width of the table, sometimes needing wider paper.

We are now ready to post the data in the appropriate columns. Be careful while doing it; any mistake at this stage of development may lead to an erroneous cashflow table, and eventually to a wrong solution. With the data posted in, the auxiliary table looks like:

YEAR	A				B			
	Chem.	Main.	Sal.	Total	Chem.	Main.	Sal.	Total
1	−$50,000	−$500			−$95,000	−$750		
2	−$50,000	−$500			−$95,000	−$750		
3	−$50,000	−$500			−$95,000	−$750		
4	−$50,000	−$500			−$95,000	−$750		
5	−$50,000	−$500	$150,000		−$95,000	−$750	$105,000	

Note that the costs have been prefixed with a − sign, and benefits with no sign (i.e., + sign), as discussed earlier. We shall follow these sign conventions throughout. Also note that the salvage values are entered at the fifth year, when they are realized as lump sum revenues.

Looking at this table, it is obvious that the space for data under the *Total* column is tight. This can be circumvented by using a wider sheet of paper, or by making two auxiliary tables; one for A and the other for B. Let us adopt the latter. In that case, the two auxiliary tables will be:

YEAR	A			
	Chem.	Main.	Sal.	Total
1	−$50,000	−$500		
2	−$50,000	−$500		
3	−$50,000	−$500		
4	−$50,000	−$500		
5	−$50,000	−$500	$150,000	

YEAR	B			
	Chem.	Main.	Sal.	Total
1	−$95,000	−$750		
2	−$95,000	−$750		
3	−$95,000	−$750		
4	−$95,000	−$750		
5	−$95,000	−$750	$105,000	

Each of these auxiliary tables needs to be completed for the net cashflows under the *Total* column. The usefulness of signed cashflows becomes evident while totaling the various items. Following the algebraic addition rule, the two auxiliary tables complete in every respect become:

YEAR	A			
	Chem.	Main.	Sal.	Total
1	−$50,000	−$500		−$50,500
2	−$50,000	−$500		−$50,500
3	−$50,000	−$500		−$50,500
4	−$50,000	−$500		−$50,500
5	−$50,000	−$500	$150,000	$99,500

YEAR	B			
	Chem.	Main.	Sal.	Total
1	−$95,000	−$750		−$95,750
2	−$95,000	−$750		−$95,750
3	−$95,000	−$750		−$95,750
4	−$95,000	−$750		−$95,750
5	−$95,000	−$750	$105,000	$ 9,250

These auxiliary tables can now be summarized as a two-column cashflow table. Inserting the row for year 0 to show the initial equipment costs, and decoding the equipment type, the final cashflow table with proper headings becomes:

YEAR	Neutralization Equip	Precipitation Equip
0	−$750,000	−$450,000
1	−$ 50,500	−$ 95,750
2	−$ 50,500	−$ 95,750
3	−$ 50,500	−$ 95,750
4	−$ 50,500	−$ 95,750
5	$ 99,500	$ 9,250

Example 2.2 illustrated the use of auxiliary tables, the need for additional care in posting the relevant data, and the creative tasks involved in developing a cashflow table. If you can think of other ways (perhaps better!) to develop the cashflow table for Example 2.2, try them.

Cashflow tables, such as those in Examples 2.1 and 2.2, are useful by themselves since they help the analyst visualize the problem. At times they are used to sketch cashflow diagrams, discussed later in Section 2.2, though the diagrams may be developed directly from the problem statement. Both the table and the diagram contain and display data that are used for further analysis of the problem, as illustrated in other chapters.

2.1.3 End-of-Period Assumption

One basic practice in engineering economic analyses is that the cashflows be posted at the end of the period (year, quarter, month, or whatever it is for the given problem). Even when a transaction takes place at other times within the period, we assume the cashflow to have occurred at the end of the period.

Consider, for example, the annual maintenance costs of the two pieces of equipment in Example 2.2 earlier. These have been posted in the table as a cost at the end of the year, say on 31 December. We know that a piece of equipment can break down any time during the year (say in June), and that it may be repaired immediately (next week) so that it can be put back to use without delay. Thus, the money earmarked for maintenance is used as and when the equipment breaks down. But, by convention, we post all the maintenance costs for the period together as one lump sum at period-end.

Isn't this practice wrong? In a sense, it is. But, if we were to consider the cashflows exactly when they occur, the posting of the data may become cumbersome due to their numerous timings. And then, how far do we go? Do we consider the cost every quarter, every month, every week, or every day? That is why we make the assumption, and in doing so sacrifice some accuracy to achieve practicality, that cashflows be treated as if they occurred *at the end of the period.*

Another reason for the period-end assumption is that the funds earmarked for use during the period are kept "liquid" for easy access, usually in checking accounts that bear no interest. So, any consideration of the actual timings of the cashflows within the period does not matter. Moreover, the period considered is usually pegged to the accounting cycle.

2.2 CASHFLOW DIAGRAM

A cashflow diagram is a graphical sketch of the cashflows and their timings. Since the cashflows may be contained in their table, a cashflow diagram can be sketched using the corresponding cashflow table. The cashflow table and the corresponding diagram display two views of the same data.

A cashflow diagram depicts the magnitude and direction of the cashflows as vertical lines at specific points (markers) along a timeline. The timeline is horizontal and shows the time periods at which cashflows occur. It is appropriately divided and marked. The left end of the timeline usually represents 0, meaning *now* or *today*. At each of the relevant time markers a vertical line, also called vector, proportional to the cashflow amount is drawn. The vector is upward, above the timeline, for cashflows that are benefits (inflows), and downward for costs (outflows). The vectors are arrow-headed. Fig. 2.1 shows a typical cashflow diagram.

2.2.1 Development of Cashflow Diagram

The drawing or sketching of a cashflow diagram involves: (i) laying out the timeline and marking it, and (ii) drawing the vertical cashflow vectors. The use of the term vector, rather than line, is more appropriate since it indicates the direction too (of the cashflow). Altogether four steps are involved in the development of a cashflow diagram, as explained next. The steps will be clearer in the illustrative example that follows.

Step 1 *Set out (sketch) a horizontal line.* The length of the line should be sufficient to accommodate the time periods pertinent to the problem. The line may be 4 to 6 inches long. Since the lengths along the line represent time, it may be called the timeline.

Step 2 *Divide the timeline equally beginning at the left end*, so that all the periods can be accommodated. Erase any excess length at the right end, and mark the divisions to represent the periods. The left-end point is usually marked 0 to represent *now* (or some reference time period).

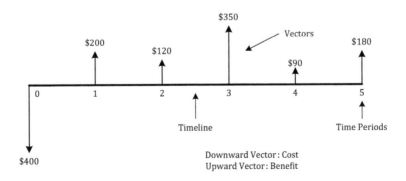

Figure 2.1   A typical cashflow diagram.

Step 3 *Work out a suitable scale to size the cashflows.* Use the largest cashflow to determine an appropriate scale so that the other vectors will fit within the space available.

Step 4 *Sketch the vectors to represent the cashflows.* Keep their lengths in proportion to the cashflow amounts. Take care of the direction. Benefits (inflows) are drawn upward above the timeline, while costs (outflows) downward. You can remember this convention easily by recalling that since benefits *increase* the project account balance their vectors are *up*, and since costs *decrease* the balance their vectors are *down*. End each vector with an arrow pointing away from the timeline. Finally, write down the cashflow amounts by the arrows.

Regarding the convention for vector direction, *always remember*:

Cash Inflows/Receipts/Benefits/Revenues—VECTOR UP
Cash Outflows/Disbursements/Costs/Expenditures—VECTOR DOWN

Let us consider an example to illustrate the above four-step procedure. Assume that you have borrowed $4,000 today and plan to pay back this loan with interest over the next 5 years as annual payments of $1,000 each on the loan anniversaries. Based on what we learned in Section 2.1, your cashflow table will be:

YEAR	0	1	2	3	4	5
AMOUNT	$4,000	−$1,000	−$1,000	−$1,000	−$1,000	−$1,000

Let us go through the four steps to develop the cashflow diagram.

Step 1 *Timeline*
Set out a horizontal line of reasonable length; say 6 inches, as shown in Fig. 2.2(a).

Step 2 *Time Markers*
Beginning at the left end, mark equal lengths on this line to represent all the time periods, as in Fig. 2.2(a). Since there are five time periods, the markers may be an inch apart. Erase the excess length on the right, as in Fig. 2.2(c). Write down the time periods at the markers, as in Fig. 2.2(d). Writing them slightly off to the right allows space for the vectors to be drawn.

Step 3 *Cashflow Scale*
The largest cashflow in the table is $4,000. What length of the vector should represent it? Let us use a 2-inch length for this purpose, so 1 inch = $2,000. A round figure for the scale of the largest cashflow, $2,000 here, yields simpler scaling of the other vectors.

Step 4 *Draw the Vectors*
We now draw the cashflow vectors. Beginning at the left, set out at marker 0 a 2-inch vertically upward line to represent your borrowing of $4,000, as in Fig. 2.2(e). This vector is up since $4,000 is an inflow. Other cashflow vectors are a half-inch long, representing $1,000 each. Since they are all outflows (− sign in the table), these vectors are vertically downward at the appropriate time markers, as in Fig. 2.2(f). Arrow the vector ends and write down the cashflow amounts nearby, as in Fig. 2.2(g) which is the cashflow diagram.

Note that Fig. 2.2 has been drawn step-by-step to explain the procedure. Each sketch is not necessary; it is only the final, composite diagram, Fig. 2.2(g), we are interested in. In other words, Figs. 2.2(a) through 2.2(f) can be bypassed, and only the final diagram (Fig. 2.2(g)) drawn. Once you have practiced these four steps enough, the process will become routine and you will be able to sketch the (final composite) cashflow diagram with ease.

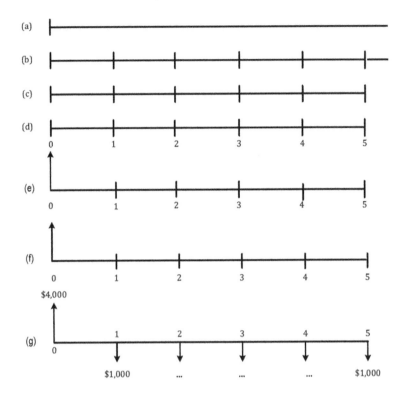

Figure 2.2 The four steps in sketching a cashflow diagram.

Let us now consider the example from lender's viewpoint (the bank). As discussed in Subsection 2.1.1, the signs of the data in the lender's cashflow table are just the opposite of those in the borrower's. This fact reverses the vector directions in the lender's cashflow diagram as shown in Fig. 2.3; note the reversal in comparison to Fig. 2.2(g). In all other respects the two diagrams (borrower's and lender's) are alike.

As discussed in Subsection 2.1.2 and illustrated in Examples 2.1 and 2.2, we often use auxiliary table(s) to comprehend a given problem, summarizing the data finally in a two-column cashflow table. In the same way, one can draw auxiliary cashflow diagram(s) that may have more than one vector at a time marker corresponding to the various cashflows. Such auxiliary diagrams too should be summarized in a (final) cashflow diagram showing at each time marker only one vector representing the *net* cashflow. The vector may altogether be absent at certain time markers if no cashflows there or if the net cashflow is zero.

Do we need to develop the cashflow table before sketching the corresponding cashflow diagram? The answer is both yes and no. If it helps, do it, otherwise don't. Preparing the cashflow table first is usually helpful, especially to beginners. After you have mastered enough skills in sketching the cashflow diagram directly from the problem statement, you can skip the development of cashflow table, unless asked for. In Example 2.3, a direct-to-the-diagram approach is illustrated, while in Example 2.4 the cashflow table is specifically asked for.

Example 2.3: Cashflow Diagram

Monica begins her 4-year program toward a degree in electronics engineering technology. Her parents are interested to help her start on graduation an electronics repair business in her hometown. Their research shows that she will need then a start-up capital of $50,000. They plan to save $300 per month, beginning next month, which will at the end

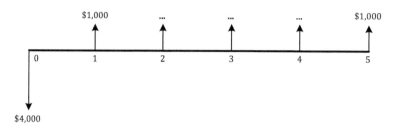

Figure 2.3 Lender cashflow diagram – compare with Figure 2.2 (g).

of 4 years be used as down payment for the bank loan she will need to set up the business. Her parents will, however, withdraw when due the interest earned on the monthly savings for their own use. Draw her parents' cashflow diagram, excluding any consideration of the interest earned.

SOLUTION:

The question does not ask specifically for the cashflow table. We have two choices: (a) prepare the cashflow table as an aid to the development of cashflow diagram, or (b) do not prepare the table, and proceed directly to draw the diagram. What determines the choice? Primarily, it is the level of confidence the engineering economist has for (b). Experienced economists choose (b), bypassing the table. The decision also depends, to a large extent, on the complexity of the problem as perceived by the economist.

This example problem does not seem too complex; let us attempt to sketch the diagram without preparing the cashflow table. We follow the four-step procedure outlined in Subsection 2.2.1.

First, the comprehension of the problem statement. Read the problem once again (more if you need to). Monica will need $50,000 as a start-up capital on graduation 4 years from now. Her parents' monthly savings at maturity will be used as a down payment toward the bank loan. If the savings accumulate to F, then the loan will only be (50,000 – F). Also recall that her parents withdraw the interest earned by their savings. In other words, only the monthly savings of $300 accumulate. Since the savings are monthly over 4 years, there are 4 × 12 = 48 time periods.

With this comprehension, we are ready to draw the cashflow diagram. Let us go through the four steps; keep glancing at Fig. 2.4 as you read:

Step 1: *Timeline*
Set out a horizontal line of reasonable length, say 5 inches.
Step 2: *Time Markers*
Beginning at the left end, mark equal lengths (an inch) on this line to represent the early time periods (months). Since the number of periods is large (forty-eight), they all cannot be shown; so, show discontinuation using dotted lines and mark the last two time periods (see Fig. 2.4). Write down the time periods at the markers, beginning with 0 at the left end and ending with 48 at the right end.
Step 3: *Cashflow Scale*
All cashflows in this problem are the same ($300). We can choose any reasonable vector length, since no further scaling is required. Let us say 1 inch represents the cashflow amount $300.
Step 4: *Draw the Vectors*
Now, sketch the arrowed cashflow vectors. As per the problem statement, there is no cashflow now, i.e., at marker 0. The cashflow begins at period 1, where we draw a 1-inch[2] vector to show the saving of the first month. But, which way? Up or down?

---

[2] Since it is a sketch, the vectors do not have to be exactly 1 inch; draw freehand, simply making them look equal.

The answer lies in determining whether the cashflow is negative or positive. For Monica's parents, the savings are outflows (costs), so the vector should be downward, below the timeline. A smart way to determine whether the vector should be downward or upward is to remember this: If the cashflow brings the account balance *down*, the vector should be *downward*; if it makes the balance go *up*, the vector should be *upward*. In this case, because the cashflow brings down Monica's parents' account balance, the vector is downward. At the other markers we similarly sketch 1-inch downward vectors.

The resulting cashflow diagram is shown[3] in Fig. 2.4.

In problems where the timeline has to be long to accommodate all the time periods, as in the previous example, it is shown broken—a common engineering practice to show continuity. Also, some of the cashflows may be disproportionately large in comparison to the others. Again, the taller cashflow vectors may be shown broken. Fig. 2.5 shows a cashflow diagram in which both the timeline and the tallest vector (5,000 at period 0) are shown broken. Breaks having similar meaning are commonly practiced in engineering drawings. Even when the timeline has a break, its two ends should be shown complete with appropriate point markers. For the vector too, the break should be in the middle rather than at the ends. Another enhancement to a cashflow diagram may be the absence of the currency symbol to add visual clarity, as in Fig. 2.5.

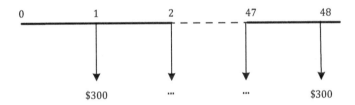

Figure 2.4 Cashflow diagram for Example 2.3.

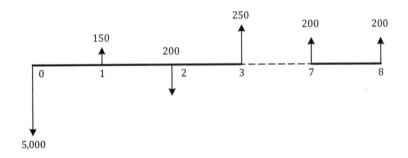

Figure 2.5 Broken lines showing continuation.

---

[3] We considered month as time period in this example as per the data. Since the interest is withdrawn when due, we could have used year—rather than the month—as the time period. In that case, four cashflows of $3,600 (= $300 × 12) each would have been drawn at four time markers 1 through 4. This would have reduced the number of time periods from 48 to 4 and avoided the need to show a broken timeline. Note, however, that if the interests earned were not withdrawn when due, but monthly compounded instead, then adding the $300-monthly cashflows into a $3,600-annual cashflow would have been erroneous.

Example 2.4: Cashflow Table & Diagram

Anjali, a plant supervisor, has been informed that the company has made better-than-expected profit this year and that she can budget funds to replace the robot. She feels that the current robot can be used for another 5 years. The replacement robot to be bought then will cost $50,000 and last for 5 years. The maintenance costs of the new robot will be $1,200 in the first year, increasing thereafter annually by $200 for the remaining 4 years. Considering the time value of money, she budgets this year for $35,750, which is deposited in a 5-year CD to mature to $50,000. Prepare both the cashflow table and the cashflow diagram for the project.

SOLUTION:

We begin with the comprehension of the problem statement. The time period in this example is year. The current robot is expected to last five more years. The new robot will be purchased at the end of this period (5 years from now).

Anjali takes out $35,750 from this year's company profit and deposits this in a 5-year CD. No data is given about the interest rate the CD will earn, but she seems to have worked out that $35,750 will increase to $50,000 at the end of 5 years. This $50,000 sum will be used to buy the new robot that will require maintenance costs during its 5-year life. Thus, this problem involves 10 time periods—5 years of the current robot's life and 5 years of the new one's.

With this comprehension, let us prepare the cashflow table first. The problem seems simple. We can do without an auxiliary table and proceed directly to prepare the two-column cashflow table, whose skeleton will look like:

YEAR	0	1	2	3	4	5	6	7	8	9	10
AMOUNT											

We need to gather the relevant data from the problem statement and post them in the table. For row 0, the cashflow is $35,750; the money Anjali sets aside now (this year). Since this is a cost (paid out) to the company[4], it will have a – sign. For the next 5 years during which the current robot will continue to be used, there are no cashflows. Though the current robot may need maintenance, the associated costs are not given, and hence not a part of the current project. Such costs might have been taken into account while purchasing the current robot. Thus, the cashflows during year 1 through 5 are zero.

At the end of the fifth year, $35,750 would have grown to become $50,000. Depending on how this $50,000 is posted, the cashflow table may have two types of entries for year 5, as explained as follows.

(a) The simplest way to look at the situation is to consider that this $50,000 is paid directly from the CD account to the supplier of the new robot. In that case there is no new cashflow from Anjali's company to the supplier. Looked this way, there should be no entry for year 5.

(b) The approach in (a), which looks simpler, will be objected to by professional accountants. What they will insist on is that the accumulated sum $50,000 be shown as an inflow into company books, and a concurrent payment (outflow) of $50,000 be made out to the supplier. Thus, there will be two entries in the cashflow table corresponding to year 5: One as $50,000 and the other as –$50,000. The former is the inflow (+ sign not shown), while the latter the outflow (payment to the robot supplier). The two entries result in a *net* cashflow of zero for year 5, the same as in (a).

---

[4] Note that we are preparing the company's cashflow table.

The new robot is acquired at the end of year 5 (same as the beginning of year 6) and its use begun. The maintenance cost during the first year of the new robot's use is $1,200. This year is the sixth year of the whole project. Thus, for year 6, the maintenance cost of $1,200 is entered with a –ve sign. Since the maintenance costs for subsequent years increase annually by $200, the cashflows for the subsequent years will be 1,400, 1,600, 1,800, and 2,000, all with –ve signs.

Thus, the cashflow table following approach (a) will be:

YEAR	0	1	2	3	4	5	6	7	8	9	10
AMOUNT	–$33,750	$0	$0	$0	$0	$0	–$1,200	–$1,400	–$1,600	–$1,800	–$2,000

If we follow the strict accounting practice explained in (b) earlier, the cashflow table will be:

YEAR	0	1	2	3	4	5	6	7	8	9	10
AMOUNT	–$33,750	$0	$0	$0	$0	$50,000 (–$50,000)	–$1,200	–$1,400	–$1,600	–$1,800	–$2,000

We can now proceed to draw the cashflow diagram. Since the cashflow table already exists, we do not need to recomprehend the data. Instead we use the cashflow table to develop the diagram. Follow the earlier discussions in Subsection 2.2.1 and the illustrative Example 2.3 on how to develop a diagram.

For case (a), the cashflow diagram[5] is in Fig. 2.6(a), while for (b) it is in Fig. 2.6(b). Note a few *"cosmetic" enhancements* in these diagrams. For example, the time periods have been labeled above the timeline to add clarity. Their usual labeling—below the timeline

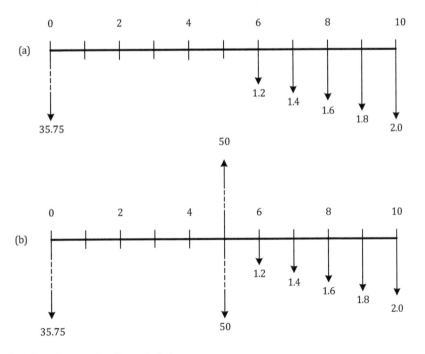

Figure 2.6 Cashflow diagram for Example 2.4.

---

[5] Some engineering economists prefer to show all the vectors at their time markers, for example the two vectors at year 5 in Fig. 2.6(b). Others prefer to show only the net cashflow, which being zero at period 5 does not appear in Fig. 2.6(a).

and slightly off to the right of the time markers—would have been congested. Note also that we did not label each time period, skipping the odd ones 1, 3, ...; when space is limited, this is acceptable. In fact, some analysts do not label the time periods at all, especially if there are fewer periods that can easily be comprehended. Further, the cashflow amounts have been normalized to thousands, reducing the number of digits required to label them. However, a qualifying note to this effect must be made at a prime location in the diagram, as in Fig. 2.6 at the top (Amount in Thousands of Dollars). Some people use the notation k (always lower case), which stands for a thousand. In that case, the cashflow data in Fig. 2.6 might have been labeled as 35.75k, 1.2k, 1.4k, ..., without the qualifying note *Amount in Thousands of Dollars*.

Cosmetic enhancements are meant to add clarity to cashflow diagrams. A clear and unambiguous diagram is likely to lead to an error-free solution. Use your imagination to achieve clarity.

2.3 SUMMARY

Cashflow tables and diagrams are effective tools that aid comprehension of engineering economics problem/project statements. They summarize the given data for further analysis. Their concepts and developments have been discussed in Chapter 2. Engineering economists use them quite often.

A cashflow table is basically a two-column table, in which the first column lists the time periods and the second column lists the corresponding cashflows. Cash outflows (costs, disbursements, or expenditures) are preceded by a negative sign. Cash inflows (benefits, receipts, or revenues) have no sign, meaning that they are positive.

A cashflow diagram is a representation of the cashflows and their timings. It is drawn, in fact sketched, by displaying scaled cashflows as vertical vectors along a timeline showing their time periods. Cash outflows (costs, disbursements, or expenditures) are drawn as downward vectors. Cash inflows (benefits, receipts, or revenues) are drawn as upward vectors. In other words, cashflows with a negative sign in the cashflow table are downward vectors, while those with no sign are upward vectors.

The cashflow table or cashflow diagram, sometimes both, may be helpful in solving a problem. They may be developed independently from the given data. Alternatively, one may be developed from the other.

DISCUSSION QUESTIONS

2.1 Why does a cashflow table usually have only two columns?

2.2 Given the lender's cashflow table, how do you prepare the borrower's cashflow table?

2.3 Explain the role of auxiliary table(s) in the development of a cashflow table.

2.4 Of the two, cashflow table and cashflow diagram, which one offers more visual impact and why?

2.5 Why do the cashflow data displayed in a cashflow diagram not show the sign of the cashflow—whether negative or positive?

2.6 Discuss the four steps involved in developing a cashflow diagram. Which step is most likely to introduce error in the diagram and why?

2.7 Discuss the pros and cons of developing a cashflow diagram directly from the problem statement, without first preparing the cashflow table.

2.8 Explain why all cashflows are treated as if they occurred at the end of the period.

2.9 Discuss the convention of the result in a vector up and vector down cashflow diagram.

2.10 Discuss the correlation between a lender's and a receiver's cashflow.

MULTIPLE-CHOICE QUESTIONS

2.11 A cashflow with no sign represents
   a. receipt of money
   b. disbursement of money
   c. maintenance cost
   d. none of the above

2.12 In a cashflow table, a + sign represents
   a. cash inflow
   b. disbursement
   c. cash outflow
   d. none of the above

2.13 Cash outflow is another term for
   a. disbursement
   b. receipt
   c. revenue
   d. benefit

2.14 Auxiliary table is _____ essential in developing a cashflow table.
   a. always
   b. never
   c. sometimes
   d. none of the above

2.15 An auxiliary table must have _____ columns.
   a. 2
   b. 3
   c. 4
   d. none of the above

2.16 Cashflow vectors are drawn upward when the cashflows are
   a. receipts
   b. disbursements
   c. costs
   d. none of the above

2.17 The cashflow signs in a borrower's and lender's cashflow table are always:
   a. the same
   b. opposite of one another
   c. neither
   d. all of the above

2.18 Which one of the following is not an outflow of cash?
   a. withdrawal
   b. payment for services rendered
   c. payment of operating expenses
   d. receipt of payment for services

2.19 Which of the following is not an inflow of cash?
   a. receipt of payment for services
   b. cash deposit
   c. salaries expense
   d. investment revenue

2.20 Cashflow tables generally have how many columns?
   a. 2
   b. 3
   c. 1
   d. none of these

2.21 On a cashflow diagram, a downward arrow indicates what?
   a. outflow
   b. inflow
   c. revenue
   d. all of the above

2.22 On a cashflow diagram, an upward arrow indicates what?
   a. expense
   b. outflow
   c. inflow
   d. none of the above

2.23 Auxiliary tables are used to do what?
   a. minimize confusion
   b. facilitate calculations
   c. graphical sketch of inflows and outflows
   d. both *a* and *b*

2.24 Once completed an auxiliary table should show what?
   a. timing of transactions
   b. only the principal repaid
   c. amount of transactions
   d. both *a* and *c*

2.25 A cashflow diagram uses what to show magnitude and direction?
   a. pie chart
   b. vectors
   c. expenses
   d. none of the above

2.26 Salvage value of a piece of equipment would show on the cashflow diagram as what?
   a. outflow
   b. expense
   c. revenue
   d. inflow

2.27 Principal repaid to a lender would show as what on the borrower's cashflow table?
   a. outflow
   b. interest expense
   c. revenue
   d. inflow

NUMERICAL PROBLEMS

2.28 Prepare the cashflow table for Example 2.3.

2.29 Perfect Manufacturers decided to finance the company's new car. It paid $5,000 at the time of purchase and agreed to pay $5,000 on each of the next four anniversaries of the purchase. The first year's maintenance is free. The company budgeted for sums of $500, $1,000, $1,500, and $2,000 for the subsequent 4 years' maintenance. It expects to sell the car at the end of the fifth year for an estimated price of $2,500. Prepare the cashflow table for this project.

2.30 Five years ago, a firm purchased a used personal computer (PC) system for $2,500. The PC did not require any repair during the first 2 years. For the subsequent 3 years, the repair costs were $85, $130, and $195 respectively. The computer was sold at the end of the fifth year for $500. Construct the cashflow table and state any assumptions made.

2.31 Engineering Unlimited takes a loan of $100,000 for purchasing a robotic system. The company pays back the loan (capital and interest) through two payments: $50,000 at the end of 6 months, and $65,000 on the first loan anniversary. Sketch the cashflow diagram.

2.32 You have borrowed $1,000 from a bank, which is to be paid back in five equal yearly installments of $235 each, beginning the first loan anniversary. Prepare your cashflow table and draw the cashflow diagram.

2.33 Prepare the bank's cashflow table and the cashflow diagram for the previous problem (2.32).

2.34 Draw the cashflow diagram for Example 2.1 of the text.

2.35 Draw the two cashflow diagrams separately for Example 2.2 of the text.

2.36 Cutting-Edge Services plans to invest $300,000 in an international information service based on the Internet. It expects to earn revenue of $50,000 in the first year, which is likely to increase annually by $4,000 for the next 9 years. After that it will decline annually by $3,000 for the next 5 years. Draw the cashflow diagram for this project.

2.37 Environmentally-Enlightened Farmers is a small co-operative venture based on organic farming. Its cost on gasoline use has been high. It uses 10,000 gallons of gasoline per year. To economize on the fuel cost, it is considering investing in a fuel storage system whose initial cost is $2,500. The system's maintenance cost during the first year is expected to be $50, which will increase annually by $25 for the subsequent 9 years. The system will save 5 cents per gallon. Draw the cashflow diagram for the investment by assuming the system's salvage value to be $450 at the end of its useful life of 10 years.

2.38 An investment of $22,000 in a hydraulic system fetches a benefit of $8,250 in the very first year. The benefits increase annually by 10% during the subsequent 4 years. At the end of the fifth year, the system is sold for $3,000. Sketch the cashflow diagram after preparing the cashflow table.

2.39 A company purchases the neutralization equipment from Example 2.2. With an investment of $105,000 in the equipment they expect a benefit of $10,000 in year 1. The following 6 years the benefits decrease by 15% annually. At the end of 6 years the salvage value is expected to be $0. Sketch the cashflow diagram and cashflow table for this scenario.

2.40 Suppose that in Example 2.4, Anjali invested an additional $15,000 of company profits into another CD in year 3 that will mature in 4 years from the date of investment at $21,000. No information is available on the interest rate. Show this change in the cashflow table and diagram.

2.41 In Example 2.4, show the cashflow diagram of the investment company holding the CD. Be sure to include the information from Question 2.40.

2.42 Suppose that you borrowed $10,000 for a brand-new motorcycle. You agreed to pay back the loan over 5 years. You negotiated an annual interest rate of 7%. Calculate the necessary payments and interest for the loan period. Next, draw the cashflow table and cashflow diagram for the loan.

FE EXAM PREP QUESTIONS

2.43 Annual revenues from a project are expected to grow at a rate of 5% annually from $50,000 from the first year; however, operating and maintenance expenses are expected to remain constant at $15,000. What is the net cashflow in period 4?
 a. $45,775
 b. $60,000
 c. −$ 3,000
 d. −$13,000

2.44 A service contract stipulates that services rendered will meet a minimum of 500 man-hours at market rate. If the market rate is currently $23.00/hour and this cost is expected to rise by 3% per period. What is the amount to be paid in period 3 of the contract?
 a. $13,000
 b. $10,000
 c. $12,200
 d. −$ 4,500

2.45 Production costs are fixed at $350,000 per year and net revenues per unit are $250. What is the minimum amount of units that need to be sold in order to result in a positive net cashflow?
 a. 1400
 b. 1600
 c. 1765
 d.   349

2.46 An investment pays a simple interest rate of 2.5% semiannually. If $1,000 is deposited in period 0, how much would be in the investment account at the end of year 2?
 a. $1,200
 b. $ 450
 c. $1,104
 d. $3,500

2.47 At the end of the useful life of a piece of equipment, the asset accounts for $45,000 worth of production and has an operating cost of $25,000. The equipment has a salvage value of $15,000. The net cashflow in the final year of use is:
 a. −$ 8,000
 b. $15,000
 c. $25,000
 d. $35,000

2.48 $50,000 is deposited into an account paying a simple interest of 4%. The amount in the account after 3 years is closest to which amount?
  a. $54,000
  b. $56,000
  c. $60,000
  d. $45,000

2.49 Operating expenses are expected to rise yearly at 3.5%. If in year 0 the operating expenses are $150,000; what will the operating expenses be in year 6?
  a. $176,946
  b. $189,451
  c. $156,759
  d. $184,389

2.50 A cashflow diagram shows what two things graphically?
  a. magnitude and direction
  b. direction and volume
  c. distance and direction
  d. none of the above

2.51 An inflow vector would point in which direction?
  a. down
  b. right
  c. left
  d. up

2.52 What is the net cashflow of year 1, if the revenues for the period were equal to $100,000 and operating expenses totaled $74,500?
  a. −$ 25,500
  b. $ 25,500
  c. −$174,500
  d. $174,500

Chapter 3

Single Payment

LEARNING OUTCOMES

- Basic equation of compounding
- Law of compound interest
- Interest earned under compounding
- Exponential nature of compounding
- Equation versus first-principles
- How to solve single-payment problems
- Manipulation of basic equation's variables
- Functional notations and their use
- How to interpolate
- Use of compound interest tables
- Nominal and effective interest rates

In Chapter 1 we discussed the concept of interest compounding following the first-principles. We saw there in Section 1.5 how compounding enables a principal to accumulate faster than simple interest does. In this chapter, we extend the discussions on compounding to learn how the first-principles approach becomes cumbersome when n, the number of time periods, is large. Following the first-principles we derive compounding equations that are used to solve engineering economics problems. We develop in Chapter 3 the most *basic equation* of engineering economics—sometimes called the *law of compound interest*. The development and the pertaining discussions *emphasize* the application of the basic equation and its variations in solving problems.

This chapter covers single-payment problems that involve two cashflows. All business transactions comprise at least two cashflows—one by the first party and the other by the second party. When you borrow money from a bank, you as a borrower are the first party, and the bank as a lender is the other party. You may borrow, and pay back, different amounts at different times. In this chapter, however, we consider the simple case where you borrow a lump sum once and pay it back at the end of the loan period along with the interest. So, though the bank and you make a single payment each, there are two[1] cashflows—a pair of cashflows. We discuss a variety of scenarios of this type in this chapter.

In engineering economics, any of the following may be the transacting party: individual investors or lenders, financial institutions such as banks or credit unions, organizations such

---

[1] There may be situations in everyday life where only one party makes a single payment—and that's all—for example, your giving $50 (a single payment) in charity, or receiving a cash gift from grandma on your birthday. The other parties in these examples (the charitable organization in the first and you in the second) do not pay back, at least not *materially*. Such cases do not occur in the business world.

as insurance companies, or federal government. The concept of two parties in a transaction is helpful in comprehending a problem. Consider that as a plant supervisor you spend $1,000 to improve an existing machine in your plant. In here, you representing your company is the first party and the machine (or its vendor) is the second party. The machine is expected to pay back through higher revenues resulting from increased production or better product quality.

In problems involving a pair of cashflows, an initial sum P increases under compounding to become the future sum F. The pairing of cashflows arises from the lending or borrowing of P and its payback later as F. The future sum F is greater than P due to the time value of money, their difference F – P being the total interest earned. Note that symbols other than P and F may also be used to denote present and future.

3.1 THE BASIC EQUATION

Refer to Example 1.2, where interest was compounded following the first-principles. For small values of n, as in that example, the first-principles approach is acceptable in terms of solution effort, especially since no equations are involved. One can, in theory at least, follow the first-principles approach even in problems with higher values of n. But this approach becomes cumbersome, especially lengthy, as the value of n increases. When n is large, it is preferable to use equations that offer more concise solutions than the use of first-principles.

We derive in this section the most basic equation of engineering economics. Consider an initial (or present) sum P invested for n periods under compounding. We use the following notations:

P Present[2] sum (initial principal)
i Interest rate (in fraction)
n Number of interest compounding periods
F Final sum (payable at the end of n periods)

Interest earned during the first period = Initial principal × Interest rate

$$= P_i$$

Due to compounding, this interest is added to the initial principal P. Thus,

Principal at the end of the first period
= Initial principal + Interest during the first period
$$= P + P_i$$
$$= P(1+i)$$

This P(1 + i) becomes the principal at the *beginning of the second period.*

Interest earned during the second period
= Principal for the second period × Interest rate
$$= \{P(1+i)\}i$$

---

[2] P is known by different names, such as *present sum, principal, initial sum, investment*; but they all mean the same thing. Note that P is a lump-sum *single* payment, and so is F.

Thus,

Principal at the end of second period

   = Principal at the beginning of the second period

   + Interest for the second period

   $= P(1+i) + \{P(1+i)\}i$

   $= P(1+i) + iP(1+i)$

Taking P(1+*i*) common from both the terms,

Principal at the end of second period

   $= P(1+i)\{1+i\}$

   $= P(1+i)^2$

Repeating this procedure,

Principal at the end of the third period $= P(1+i)^3$

Principal at the end of the nth period $= P(1+i)^n$

Thus, the (initial) principal P increased under compounding to become $P(1+i)^n$ at the end of the nth period. Noting that this is the final sum F, we can write this in an equation form as

$$F = P(1+i)^n \qquad (3.1)$$

This is the *most basic[3] equation* in engineering economics. All problems of engineering economics rely on this equation for their solutions. To emphasize its importance, Equation (3.1) is often called the *law of compound interest*. Since this equation is fundamentally important, we devote this entire chapter (Chapter 3) to discussions on and about it. We need to thoroughly understand the relationships contained in this equation[4].

3.1.1 Interest Earned

We can use Equation (3.1) to determine the total interest earned from the one-time investment of P for n periods. It is evaluated simply by subtracting the (initial) principal P from the final sum F. Thus,

$$I = F - P$$

where F is given by Equation (3.1). One can simply remember and use this subtraction to determine I, with no need of any equation other than (3.1).

---

[3] A comparison of the derivation procedure for the basic equation and the use of first-principles in Example 1.2 illustrate their equality.

[4] Note that when *i* is zero, Equation (3.1) reduces to F = P, meaning that the investment P remains unchanged. The situation is similar to *burying coins in the backyard* or keeping them under *the pillow*; in other words, not investing the principal or investing it at no interest. In the business world, *i* = 0 makes no sense. But in the personal world, some religious people offer loans to members of their group at no interest. We also hear about some generous, rich people offering interest-free loans to their relatives or charity organizations.

Alternatively, an equation can be derived by substituting for F from Equation (3.1) as:

$$I = F - P$$

$$= P(1+i)^n - P \qquad\qquad (3.2)$$

$$= P\{(1+i)^n - 1\}$$

Note that Equation (3.2) is a *derived* formula. It is in fact redundant as long as one knows the simple fact that the total interest I for the loan period is simply the difference between F and P.

For a given problem, the difference in the values of I obtained from Equations (3.1) and (1.1)[5] yields the additional interest earned due to compounding (over simple interest).

3.1.2 Exponentiality of Compounding

In Equation (3.1) the time period n appears as an exponent. This results in an exponential growth in the function $(1+i)^n$ as n increases, and therefore in the investment P to which this function is a multiplier. Such a growth is illustrated in Table 3.1, as an example for P = $100 and i = 10%.

As seen in this table, a single investment of merely $100 at 10% yearly interest rate, compounded annually, becomes over a million dollars in 100 years. If you want your grandchild's grandchild to be a millionaire, then just invest today a meager sum of $100! (Assuming 25 years as the span of a generation, your grandchild's grandchild will follow you 100 years from now.) Isn't this the easiest way to ensure in future a millionaire in your family!

3.1.3 Compounding Benefits the Lender

Compounding enables the interest earned for a given period to become part of the principal at period-end and, thus, to begin to earn interest for the next period. In other words, under compounding, *interest earns interest*. For a given investment, this results in exponential

Table 3.1 Exponential Growth Due to Compounding (P = $100, *i* = 10%)

YEAR	F (year-end)
0 (now)	100 (= P)
1	110
2	121
3	133
...	...
10	259
...	...
50	11,739
...	...
100	1,378,061

---

[5] See Chapter 1, page ...

growth in the principal than under simple interest. It is this exponentiality that ensures a millionaire great-great-grandchild to a person who is wise enough to invest $100 today.

Since compounding earns additional interest, it is beneficial to the lender or investor, as illustrated in the following example.

Example 3.1: Compounded Interest

After finishing his undergraduate education in construction engineering, John decides to start his own business. His rich, generous uncle has agreed to loan him $80,000 at an annually compounded interest rate of 4% per year. John will pay back his uncle the accumulated sum F on the fifth loan anniversary. How much will John have to pay? How much more will John's uncle earn due to compounding over simple interest?

SOLUTION:

The relevant equation is Equation (3.1). The value of n is 5, since over the loan life compounding takes place five times. The variables' values are:

$P = \$80,000$

$i = 4\% \text{ per year}$

$\quad = 0.04$

$n = 5 \text{ years}$

Substituting these values in Equation (3.1),

$$F = P(1+i)^5$$
$$\quad = \$80,000(1+0.04)^5$$
$$\quad = \$97,332.23$$
$$\quad \approx \$97,332$$

Thus, John pays back his uncle on the fifth loan anniversary a sum of $97,332.

For the second part of the answer, we refer to, and compare with, Example 1.1 where for simple interest was considered. In Example 1.1, John paid back $3,200 four times and $83,200 once as the last payment. So the total payment under simple interest was ($3,200 × 4) + $83,200 = $96,000. Since his payment under compound interest is $97,332, compounding earned his uncle $1,332 more ($97,332 – $96,000). This $1,332 represents the additional interest earned due to compounding. We thus see that compounding benefits the lender (obviously a cost to the borrower!).

3.2 EQUATION VERSUS FIRST-PRINCIPLES

Refer to Example 1.2 in Chapter 1 and compare its solution with that of Example 3.1. The former is based on first-principles and is for a shorter period (n = 3); the latter is based on equation and is for a longer period (n = 5). Even then Example 1.2 is lengthier. Had the value of n been 5 in Example 1.2, the solution would have been much lengthier. Thus, Example 3.1 illustrated the advantage of solution conciseness in using equation over following the first-principles. The golden rule is that if an appropriate equation exists, and you feel confident using it, then use it; equation use normally saves time and effort. Otherwise, follow the first-principles.

3.3 ANALYSIS OF THE BASIC EQUATION

Representing the law of compound interest, Equation (3.1) embodies a very basic relationship in engineering economics. Let us therefore study this equation extensively. Note that it involves four variables, and in its usual form relates P, i, and n with F.

The interest rate i in Equation (3.1) must be expressed as a fraction. For example, if i is given in a problem to be 7.5%, then use its fractional value 0.075 (= 7.5/100) in the equation.

The time period n is an exponent in the basic equation. Its value is a positive integer. In most problems of engineering economics, n is a finite value in the range 3 to 10. The upper limit usually arises due to: (a) limited useful lives of products or services, and (b) uncertainty in the data where time horizons are longer. For high-tech products, n is lower. For example, in the case of personal computers, n is usually 2–4 years due to obsolescence created by feverish on-going innovations in this product. In some projects, especially those involving established technologies, n may be longer than 10 years. In public projects n may be so large that it is almost infinite; examples are bridges, hospitals, schools, and churches, which offer services perpetually.

Most numerical problems dealing with compound interest provide the values of three of the variables, and ask for the evaluation of the fourth (unknown) variable. There are two ways to solve such problems, the choice varying from person to person:

A. *Use the Basic Equation*

Substitute the values of the known variables, rearrange the terms in such a way that the unknown is on the left side of the = sign, and evaluate the unknown.

B. *Use the Derived Formula*. The derived formula[6] will look like:

Unknown = f(known variables),

where f denotes *a "function of."* Substitute the known variables' values in the derived formula to evaluate the unknown.

Given below are four possible analysis situations relating to the basic equation. These include the derived formula for those who prefer the alternative B just mentioned.

1. *Finding F, Given P, i, n.* Use Equation 3.1.
2. *Finding P, Given F, i, n.* Equation (3.1) can be rearranged as

$$P = \frac{F}{(1+i)^n} \tag{3.3}$$

3. *Finding i, given P, F, n.* Equation (3.1) can be rearranged as

$$(1+i)^n = \frac{F}{P}$$

$$1+i = \left(\frac{F}{P}\right)^{1/n} \tag{3.4}$$

$$i = \left(\frac{F}{P}\right)^{1/n} - 1$$

---

[6] Such formulas are derived from the basic equation. The derivation rearranges the variables in such a way that the unknown is on the left of the = sign and the knowns are on the right.

4. *Finding n, given P, F, i.* Equation (3.1) can be rearranged as

$$(1+i)^n = F / P$$

Taking log on both sides,

$$\log\left\{(1+i)^n\right\} = \log\left(\frac{F}{P}\right)$$

$$n \log(1+i) = \log\left(\frac{F}{P}\right)$$

$$n = \frac{\log\left(\dfrac{F}{P}\right)}{\log(1+i)} \tag{3.5}$$

Note that Equations (3.3)–(3.5) are simply rearrangements of the same basic equation. For solving a problem, one can ignore these three formulas altogether, always beginning with Equation (3.1) and rearranging the terms as the solution progresses. However, those who prefer to save the time spent on rearranging can begin the solution with the appropriate derived formula; however, this approach involves having to deal with three more equations.

3.4 SOLUTION METHODS

In this section, we discuss the various ways the basic equation, or its rearrangements as embodied in Equations (3.3)–(3.5), can be used to solve for the unknown in problems involving compounding.

3.4.1 Classical Approach

By classical approach we mean the use of "old-fashioned" paper and pencil and a basic knowledge of algebra; it assumes the absence of a calculator. Any of the four variables can, as an unknown, be evaluated if the values of the other three are given. One begins the solution with the basic equation, or use one of the derived equations ((3.3)-(3.5)). The user must of course be careful about the compatibility of units. Due to the exponential term, evaluation of the unknown may require the use of log tables. However, since n is normally an integer and its value small, the evaluation of $(1+i)^n$ can be done simply by multiplying the value of $(1+i)$ by itself n times. This does away with log tables when i and n are the known variables, i.e., in cases where P or F is the unknown. If i or n happens to be the unknown, the use of log tables becomes necessary.

Since a scientific calculator is ubiquitous today, engineers almost never follow the classical approach. They use instead one of the following three methods (next three subsections) for solving the unknown P, F, i, or n, and for other types of problems discussed in the other chapters.

3.4.2 Scientific Calculator

A scientific calculator is widely available today; its use requires the analyst to:

(a) Select the appropriate equation for the unknown.
(b) Key-in the values of the known variables in the recommended sequence.

The result is displayed almost instantly. However, be careful in using a calculator, since the keys and sequencing of data entry may not be the same in all calculators. Scientific calculators invariably have the exponent key for evaluating the exponential function $(1+i)^n$, where i and n are known. Follow the instructions carefully on how to use your calculator. It is always good practice to confirm the calculator results by approximately working the problem in your head, if you can, or on paper following the classical approach (Subsection 3.4.1).

If you decide to begin the solution with the basic equation, then rearrange the terms carefully to get the unknown on the left of the = sign. The alternative is to use the appropriate derived formula from among Equations (3.3)–(3.5) whose terms are already rearranged.

The following four examples[7] illustrate the use of a scientific calculator in evaluating the unknowns—F in Example 3.2, P in Example 3.3, i in Example 3.4, and n in Example 3.5.

Example 3.2: Compounded Interest

Jenny deposits $300 now in her account and leaves it there for 8 years to earn interest at the rate of 6% per year, compounded annually. How much does she get at the end of the eighth year?

SOLUTION:

The given data are: P = $300, n = 8 years, i = 6% (= 0.06) per year. We need to determine F for which the appropriate formula, wherefrom

$$F = P(1+i)^n$$

$$= 300(1+0.06)^8$$

$$= 300(1.06)^8$$

$$= 300 \times 1.5938$$

$$= 478 \,(\text{ignore the cents})$$

Thus, Jenny gets $478 at the end of the eighth year[8,9].

Example 3.3: Compounded Interest

John will need $5,000 in 4 years as down payment for an apartment he plans to buy after getting a job as an environmental engineer. How much should he save now into a CD that pays 7% yearly interest, compounded annually?

---

[7] These examples, and several others in the text, relate to personal financing. They may leave you wondering whether we are learning about engineering economics. Yes indeed, we are. Many personal financing examples/problems are like engineering economics problems. As an illustration, Example 3.2 has been rephrased as follows to become an engineering economics problem. Note that the rephrasing has made the problem statement unnecessarily lengthy. That is why personal-financing-type examples have been preferred to illustrate a concept. Moreover, they are easier to comprehend since they relate to everyday life. Example 3.2: Jenny, a plant supervisor, has at the end of this fiscal year been left with $300 in her department's expense account. Her company's policy is that a sum under $500 does not have to be returned to the finance and budget office. Such sums may be used by the department heads to enhance productivity in any way deemed appropriate, either in the fiscal year or later. If the money is to be used later it must be invested to earn a reasonable rate of return. She decides to use the $300 left this year to replace a coordinate measuring machine 8 years later, which is estimated to cost $10,000. So, she invests this money in a CD at an interest rate of 6% per year, compounded annually. How much additional capital will be required to replace the machine?

[8] In most problems of engineering economics, cents are normally rounded to the nearest dollar.

[9] The calculator makes it easier to evaluate $(1.06)^8$ in this example. The alternative classical approach would have required multiplying 1.06 by itself eight times.

SOLUTION:

In this, the given data are: F = \$5,000, n = 4 years, i = 7% (= 0.07) per year. If you find yourself confused about any of these data, draw a cashflow diagram, as in Fig. 3.1, to help you comprehend the problem statement. We need to determine P, for which we can adopt one of the following two approaches. Use the one you feel more comfortable with.

1. *Basic Equation*

Substitute the given values in Equation (3.1) and carry out the calculations, rearranging the terms to get the unknown on the left side of the = sign. This leads to the answer, as illustrated as follows.

$$F = P(1+i)^n$$
$$5,000 = P(1+0.07)^4$$
$$= P(1.07)^4$$
$$= P \times 1.3108$$

Therefore,

$$P = \frac{5,000}{1.3108} = 3,814$$

Thus, John needs to save \$3,814 now.

2. *Derived Formula*

For P as the unknown, the appropriate formula is Equation (3.3), wherefrom:

$$P = \frac{F}{(1 + i)^n}$$
$$= \frac{5,000}{(1 + 0.07)^4}$$
$$= \frac{5,000}{(1.07)^4}$$
$$= \frac{5,000}{1.3108}$$
$$= 3,814$$

Thus, John needs to save \$3,814 now.

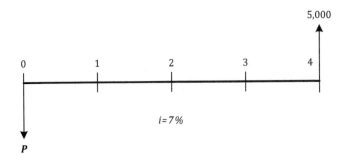

Figure 3.1 Diagram with *P* as the unknown (Example 3.3).

Example 3.4: Compounded Interest

Joe gets back $4,000 for a one-time deposit of $3,200 made 6 years ago in a money market account. What yearly interest rate did he earn, if compounding was annual?

SOLUTION:

Here the given data[10] are: P = $3,200, F = $4,000, n = 6 years. We need to determine i, for which we can adopt one of the following two approaches discussed in Section 3.3. Use the one you feel comfortable with.

1. *Basic Equation*

Substitute the given values in Equation (3.1) and carry out the calculations while rearranging the terms to get i on the left side of the = sign. This leads to the answer, as follows.

$$F = P(1+i)^n$$
$$4,000 = 3,200(1+i)^6$$
$$(1 + i)^6 = \frac{4,000}{3,200} = 1.25$$
$$(1 + i) = 1.25^{1/6} = 1.03789$$
$$i = 1.03789 - 1$$
$$= 0.03789$$
$$= 3.789\%$$
$$\approx 3.8\%$$

Thus, Joe's money market fund earned an annual interest of 3.8%.

2. *Derived Formula*

The appropriate formula for i as the unknown is Equation (3.4), wherefrom:

$$i = (F/P)^{1/n} - 1$$
$$= \left(\frac{4,000}{3,200}\right)^{1/6} - 1$$
$$= 1.25^{1/6} - 1$$
$$= 1.03789 - 1$$
$$= 0.03789$$
$$\approx 3.8\%$$

Thus, Joe earned a yearly interest of 3.8%.

---

[10] We may at times be confused by the meanings of *present* denoted by P and *future* denoted by F, as in this example. We are using P to represent a time in the past—6 years ago—and F to represent the present. Note that the concept of present and future is *logical* rather than *real*, since we are concerned simply with the relativity of timings. Thus, P or F does not have to represent the *real* present or the *real* future. Concentrate instead on the logic of timing. We are concerned primarily with the time gap, in terms of compounding periods, between the investment and its maturity.

Example 3.5: Compounded Interest

Jaya plans to surprise her son with a cash gift of $1,000 that he can use as down payment to buy his first car. She can afford to save $500 now in a bank account that will earn an annual interest of 6%, compounded per year. How long will Jaya have to wait to surprise her son?

SOLUTION:

The given data are: P = $500, F = $1,000, i = 6% (=0.06) per year. We need to determine n, for which we can adopt one of the following two approaches.

1. *Basic Equation*

Substitute the given values in Equation (3.1) and carry out the calculations, rearranging the terms concurrently to get the unknown on the left side of the = sign. This leads to the answer:

$$F = P(1+i)^n$$
$$1,000 = 500(1+0.06)^n$$
$$(1+0.06)^n = \frac{1,000}{500} = 2$$
$$\log\left\{(1+0.06)^n\right\} = \log 2$$
$$n \times \log(1.06) = \log 2$$
$$n = \frac{\log(2)}{\log(1.06)}$$
$$= \frac{0.3010}{0.0253}$$
$$= 11.897 \text{ years}$$

Thus, Jaya will have to wait almost 12 years to be able to surprise her son.

2. *Derived Formula*

For n as the unknown, the appropriate derived formula is Equation (3.5), wherefrom:

$$n = \frac{\log\left(\dfrac{F}{P}\right)}{\log(1+i)}$$
$$= \log\frac{\left(\dfrac{1,000}{500}\right)}{\log(1+0.06)}$$
$$= \frac{\log(2)}{\log(1.06)}$$
$$= \frac{0.3010}{0.0253}$$
$$= 11.897 \text{ years}$$

Thus, Jaya will have to wait almost 12 years to surprise her son.

It may be noted from the previous four examples that derived formulas yield a concise solution. This is because the formula is already in the rearranged form, ready for use. The use of the basic equation, on the other hand, requires manipulation of the terms each time it is used, as solution progresses. However, it does away with the need for derived equations.

The lengthiness of the basic-equation-based solutions in the previous four examples can easily be curtailed by those proficient in algebra. The detailed calculation steps can be collapsed into fewer to yield brevity.

3.4.3 Software

The software-based method assumes that a computer is available along with a suitable engineering economics application program. The program may simply be a spreadsheet. Computer systems, especially PC-based ones, are widely available to students and practicing engineers in the industrialized countries. However, the same may not be true in the developing countries.

If you have access to a computer system and are skilled in using them, the software-based solution method is probably the most suitable for solving engineering economics problems. Most software offer graphics capability that can display the results in an extremely presentable form. This capability is even more desirable in complex problems to enhance result visualization.

In the computer-based method the values of the known variables are input into the system using keys, mouse, or other similar devices. The software evaluates the unknown almost instantly and displays the result on the screen. However, be careful in using the computer for solving problems. Remember GIGO (garbage in, garbage out)! You must have a "feel" of the expected result. It is always good practice to confirm the computer result through sample calculations on a calculator.

3.4.4 Functional Notation

The functional notation method is popular with practicing engineers and technologists, and that is why we discuss it extensively in the next section. In this text, we encourage this method since it is widely used in industry.

As an alternative to evaluating the exponential function $(1+i)^n$ of Equation (3.1) in a problem, their values are made available in a table for different sets of i and n. A portion of such a table is given as follows as an example:

i	n	$(1+i)^n$
5%	3	1.157625
	4	1.215506
	5	1.276282
	8	1.477455
8%	3	1.259712
	4	1.360489
	5	1.469328
	8	1.850930
11%	3	1.367631
	4	1.518070
	5	1.685058
	8	2.304538

Given such a table, the engineering economist simply reads the value of $(1+i)^n$ off the table for the particular values of i and n. Once this value is known from the table, Equation (3.1) reduces to the simple relationship:

$$F = P \times \text{Exponential function's value}$$

Note that the table has freed the analyst from having to calculate the value of $(1+i)^n$. Instead, he or she simply reads the value off the table. Since this approach is based on tables, it may be called the *tabular method*. Also called the *functional notation* method, it is explained in detail in the next section.

3.5 FUNCTIONAL NOTATION METHOD

The functional notation approach, commonly used by practicing engineers and technologists, does away with the need to calculate the value of the exponential factor $(1+i)^n$ for a problem. Rather than calculate, its value is read off tables for the specific interest rate i and the time period n. This and other associated factors are expressed in a special way, called *functional notations*, to minimize the confusion in using them. Tables containing values of functional notations are known as *compound interest tables*. Such tables are sandwiched in between Chapters 8 and 9.

Equation (3.1), given by $F = P (1 + i)^n$, can be rewritten as:

$$F = P\left(\frac{F}{P,i,n} \right) \tag{3.6a}$$

where,

$$\left(\frac{F}{P,i,n} \right) = (1+i)^n \tag{3.6b}$$

Note that the notation (F/P,i,n) is simply another way to express the exponential term $(1 + i)^n$. It too is a function of P, i, and n. It is used for evaluating F by multiplying its value with P. We call it *Compound Amount Factor*, since it actualizes the effect of compounding on a given sum P. The factor (F/P,i,n) is read as: *Find F given P,i,n*. Note that the notation has been so composed that the unknown F is on the left side of the slash, while the knowns P, i, and n are on the right. This is similar to the way we write algebraic equations with the unknown on the left and the knowns on the right.

For conciseness, no algebraic sign has been shown between P and (F/P,i,n) in Equation (3.6a). In reality there is a sign, and it is multiplication—the same as in Equation (3.1) between P and the exponential factor $(1+i)^n$.

The compound amount factor (F/P,i,n) is used to evaluate F, given P, i, and n. In a similar way, it is possible to evaluate P for given values of F, i, and n. We do this by modifying Equation (3.3) as:

$$P = F\left(\frac{P}{F,i,n} \right) \tag{3.7a}$$

where,

$$\left(\frac{P}{F,i,n} \right) = \frac{1}{(1 + i)^n} = (1+i)^{-n} \tag{3.7b}$$

The notation (P/F,*i*,n) is called *Present Worth Factor*, since it is used to evaluate the present worth P of a future sum F.

Refer to the pages of Compound Interest Tables at the end of Chapter 8. There is a page of data for one interest rate. Each page contains several columns. The first column is always for the time period n. For the time being, we concern ourselves with the second and third columns, under *Single Payment*. From these we can read off the value of (F/P,*i*,n) or (P/F,*i*,n) for a given n. As can be seen, the tables contain values for other factors too, which we shall discuss later.

In using the table while solving a problem, simply reach for the page for the particular interest rate, then look for the value of n under the first column, and read off the factor's value under the appropriate column. Can you find the value of F/P factor[11] for *i* = 8% and n = 10 to be 2.159? You can use your calculator to confirm that $(1 + 0.08)^{10}$ is indeed 2.159. Note that some sacrifice is made in the accuracy of a factor's value, and therefore in the final result, due to its limited decimal places.

By the time we would learn about the other factors, there will be eight in number. Since this number is not small, at times confusion may arise as to which column's value to use for a given problem. The confusion can be minimized by looking at the notation closely. Let us consider the expression:

$$F = P\left(\frac{F}{P,i,n}\right)$$

We are interested to check whether the functional notation (F/P,*i*,n) used in this expression is the correct one. Or, should it have been (P/F,*i*,n)? To do this, we compare the two sides of the = sign. On the left we have variable F. So, we should get F on the right side too. On the right, the variable P is logically and visually multiplied with F/P. This multiplication (P × F/P) does indeed yield F, since within the parenthesis numerator P cancels with the denominator P, leaving behind F. Thus, on the right side of the = sign we get F, which is what we have on the left. We have thus checked[12] that the functional notation (F/P,*i*,n) used in the previous expression is the correct one.

In case you are not sure of the functional factor[13] to use in a problem, mentally follow the procedure explained in the previous paragraph.

To illustrate the use of functional notations, we redo the four examples (Examples 3.2–3.5) done earlier. For evaluating F we use Equation (3.6a), and for P Equation (3.7a). There are no direct formulas for evaluating *i* and n using functional notations. Instead, we manipulate the data and rearrange the terms, as illustrated in Examples (3.8) and (3.9) to follow.

Example 3.6: Compounded Interest

Jenny deposits $300 in an account to earn an annually compounded interest rate of 6% per year. How much does she get at the end of the eighth year?

---

[11] F/P is a short form of (F/P,*i*,n); both mean the same.

[12] There is another way to check for this. Mentally consider a = sign in place of / in the parenthesis of the notation. In this case, with = replacing the /, we get F = P within the parenthesis, which is the same as outside; so the functional notation is correct. This comes from the way the functional notations have been composed.

[13] This is an attractive feature of the functional notation approach to solving engineering economics problems. Users do not have to remember any mathematical equation. They only need to be skillful in selecting the correct functional notation. For example, for finding F for a given P, the user simply needs to know that the appropriate format is (F/P,*i*,n,). With this knowledge the user reads off the value in the table and multiplies it with the known P to determine the unknown F. We shall come across other functional notations later. They are all based on the same logic and, therefore, *user-friendly* in comparison to the methods that require calculating the exponential term.

SOLUTION:

We are given P = $300, i = 6% per year, and n = 8 years, and asked to determine F. Select the functional notation formula where the unknown F is on the left side of the = sign. It is Equation (3.6a), as

$$F = P\left(\frac{F}{P, i, n}\right)$$

On substitution of the given values, we get

$$F = 300\left(\frac{F}{P}, 6\%, 8\right)$$

Note that i has been expressed above in %, since the tables are titled in %. However, the factors' values in the table have been calculated using fractional values of i, as required in the basic equation.

From the tables, we need to read off the value of (F/P,6%,8). Turn the pages of the table to locate the page for 6%. Run your finger on the first column (of n) to reach 8. Read off the value of F/P in the second column (Find F Given P) as 1.594. Use this in the previous expression to find F, as

$$F = 300 \times 1.594 = \$478.20$$

Let us compare the notation-based result with that using a calculator to see how much accuracy has been sacrificed by using the tables. We need to evaluate the term $(1+i)^n$, which is what (F/P,i,n) represents. Recalling that i must be in fraction in the exponential form, we find that $(1 + 0.06)^8 = 1.593848$. With this value, F = $300 \times 1.593848 = \$478.15$. The difference in the results for F is five cents ($478.20 – $478.15), which for all practical purposes is negligible.

Example 3.7: Compounded Interest

John will need $5,000 in 4 years as down payment for an apartment he plans to buy after getting a job. How much should he invest now in a CD that pays 7% yearly interest, compounded annually?

SOLUTION:

We are given F, i, and n, and asked to determine P. So select the functional notation formula where the unknown P is on the left side of the = sign. It is Equation (3.7a), namely:

$$P = F\left(\frac{P}{F, i, n}\right)$$

On substitution of the given values, we get:

$$P = 5,000\left(\frac{P}{F}, 7\%, 4\right)$$

Turn the pages of the table to locate the page for 7%. Run your finger on the first column (of n) to reach the entry for 4. Read off the value of P/F in the third column (Find P Given F) as 0.7629. Use this in the above expression to evaluate P, as

$$F = 5,000 \times 0.7629 = \$3,815$$

Thus, John needs to invest $3,815 now.

When the interest rate i or the time period n is the unknown, the functional notation approach requires interpolation, as illustrated in the following two examples.

Example 3.8: Interest Rate

Joe gets back $4,000 for a one-time deposit of $3,200 made 6 years ago in a money market account. What yearly interest rate did he earn, if compounding was annual?

SOLUTION;

In this, the given data are: P = $3,200, F = $4,000, n = 6 years. We need to determine the value of interest rate i. In the functional notation approach there is no formula for evaluating i directly, and therefore we manipulate the basic relationship. We can use either Equation (3.6a) or (3.7a). With the former, we have:

$$F = P\left(\frac{F}{P, i, n}\right)$$

By substituting the values given and rearranging the terms, we find:

$$4,000 = 3,200\left(\frac{F}{P}, i, 6\right)$$

$$\left(\frac{F}{P}, i, 6\right) = \frac{4,000}{3}, 20 = 1.25$$

Our task now is to search in the tables to find the interest rate i, corresponding to n = 6, for which the value of F/P is 1.25.

To do this, turn to any page of the tables. Let us say this happens to be the page for i = 6%. On that page, for n = 6, (F/P,6%,6) is 1.419. This is greater than 1.25 we are searching for. So our search continues. But, shall we search next in the pages of higher interest rates or lower interest rates? We don't know. Let us try the adjoining page of i = 7%, wherefrom (F/P,7%,6) = 1.501. Since 1.501 is greater than 1.419—the F/P value corresponding to 6%—we are heading in the wrong direction. So let us look in the tables for lower interest rates. Let us say we turn the page for i = 5%. For this interest rate, the F/P factor[14] for n = 6, being 1.34 is still greater than 1.25, but we are getting closer. For i = 4%, F/P factor corresponding to n = 6 is 1.265; and for i = 3.5%, it is 1.229. So, the interest rate for which (F/P,i,6) will be 1.25 is somewhere in between these two, namely 3.5% and 4%. To find the exact value of the interest rate, interpolation within this range is essential.

We set out the data to be interpolated into a sketch. Referring to Fig. 3.2, the F/P values are plotted vertically and the interest values horizontally. The unknown i will be found on the horizontal axis. The height ac represents 1.229, the value of F/P corresponding to i = 3.5%, while bd represents 1.265 corresponding to i = 4%. We want to know the value of i corresponding to F/P value of 1.25. Let us say 1.25 is represented[15] by ef. So we need to know how far point f is from point a, i.e., length of the line af. We draw a horizontal line ch, parallel to ab, to generate the two triangles cge and chd. Since these triangles[16] are similar,

$$\frac{eg}{dh} = \frac{cg}{ch}$$

---

[14] F/P factor is another way to express (F/P,i,n). Likewise, we use P/F factor for (P/F,i,n).

[15] Since the value of ef lies in *between* the known boundaries of ac = 1.229 and bd = 1.265, the process of determining the unknown is called *interpolation*.

[16] This interpolation is based on first-principles. Some analysts derive an equation for the purpose of interpolation using x_{min}, y_{min}, etc. I prefer the first-principles approach since it avoids an additional equation. Moreover, the first-principles approach requires nothing to remember—just the understanding of the interpolation concept.

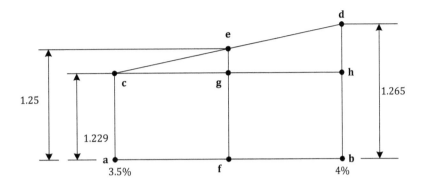

Figure 3.2 Interpolation.

$$\frac{ef - gf}{bd - bh} = \frac{af}{ab} \quad (\text{since } cg = af, \text{ and } ch = ab)$$

$$\frac{1.25 - ac}{1.265 - ac} = \frac{af}{4 - 3.5} \quad \left(\text{since } gf = ac, \text{ and } bh = ac\right)$$

$$\frac{1.25 - 1.229}{1.265 - 1.229} = \frac{af}{0.5}$$

$$\frac{0.021}{0.036} = \frac{af}{0.5}$$

$$af = \frac{0.021 \times 0.5}{0.036}$$

$$\approx 0.292$$

The interest rate represented by point f is the value corresponding to point a plus the value represented by length af, i.e.,

$$i = 3.5\% + 0.292\%$$
$$= 3.792\%$$

Thus, Joe earned a yearly interest of 3.792%. This is slightly higher than the exact value of i (3.789%) in Example (3.4), and is due to the rounding of F/P values in the table (limited to four decimal places).

Example 3.9: Computing n

Jaya is thinking of surprising her son with a cash gift of $1,000 that he can use as down payment to buy his first car. She can afford to save $500 now in a bank account that will earn an annual interest rate of 6%, compounded per year. How long will Jaya have to wait to surprise her son?

SOLUTION:

Here the given data are: P = $500, F = $1,000, $i = 6$ %. We need to determine the value of n. In the functional notation approach there is no formula for evaluating n directly. We use Equation (3.6a) or (3.7a) and manipulate the terms. With the former, we have:

$$F = P\left(\frac{F}{P}, i, n\right)$$

Substituting the values given,

$$1,000 = 500\left(\frac{F}{P}, 6\%, n\right)$$

$$\left(\frac{F}{P}, 6\%, n\right) = \frac{1,000}{500}$$

$$= 2$$

Our task next is to determine from the tables the value of n, for which (F/P, 6%, n) is 2. Turn to the 6% interest rate page. Move your finger down in the F/P column to search for a value of 2. You find that for n = 11 the F/P value is 1.898, and for n = 12, it is 2.012. So the value of n for which the F/P value will be 2 is somewhere in between these two values of n. To determine that value of n we interpolate, similar to that in Example 3.8.

Set out the data to be interpolated into a sketch similar to Fig. 3.2. Refer to Fig. 3.2 for the discussion that follows. Set the F/P values vertically and the n values horizontally. In this case, following the same procedure as in Example 3.8, ac represents 1.898 corresponding to n = 11, while bd represents 2.012 corresponding to n = 12. We want to know the value of n corresponding to F/P value of 2. Let us say 2 is represented by ef. So, we need to know how far point f is from point a, i.e., the length of line af. We draw a line ch to generate the two triangles ceg and cdh. These triangles are similar. Therefore,

$$\frac{eg}{dh} = \frac{cg}{ch}$$

$$\frac{ef - gf}{bd - bh} = \frac{af}{ab} \text{ (Since } cg = af, \text{ and } ch = ab)$$

$$\frac{2 - ac}{2.012 - ac} = \frac{af}{12 - 11} \left(\text{since } gf = ac, \text{ and } bh = ac\right)$$

$$\frac{2 - 1.898}{2.012 - 1.898} = \frac{af}{1}$$

$$\frac{0.102}{0.114} = af$$

$$\approx 0.895$$

The value of n represented by point f is the value corresponding to point a plus the value represented by length af. Thus,

$$n = 11 + 0.895$$

$$= 11.895 \text{ years}$$

Jaya will have to wait 11.895 years to surprise her son. (This value of n is slightly lower than its value of 11.897 years in Example (3.5); the difference is due to rounding of F/P values in the table, limited to only four places of decimal.)

Evaluations of i and n based on interest tables involve search and interpolation[17], as seen in Examples 3.8 and 3.9. The interpolation may have seemed lengthy and cumbersome. But with practice it becomes easier.

---

[17] In the interpolations discussed in this section, and also elsewhere in the text, we make an implicit assumption. We assume that the variation between the two variables being interpolated is *linear*. This linearity assumption introduces some error if the variables have a non-linear relationship, as with the exponential function $(1+i)^n$ or the associated functional notations. As long as the error is not large, it is acceptable for all practical purposes. To minimize the error, keep the boundaries of the variable on the x-axis as close as possible.

A final note on interpolation. In most problems the interpolation process may have to be gone through in the general way explained in Examples 3.8 and 3.9. There are times, however, when it may be possible to cut short this process; for example when the value being searched for is just half-way between the two boundaries. Suppose that in Example 3.8 F is $3,990 instead of $4,000. Then the value of the (F/P,i,n) would be 3,990/3,200 = 1.247. This value happens to be exactly in between 1.229 (for i = 3.5%) and 1.265 (for i = 4%). Therefore, the unknown i would simply be in between 3.5% and 4%, i.e., i = 3.75%. This interpolation was easy. The same way interpolation is easier in problems where the point being searched happens to be at the quarters, thirds, or tenths of the range.

3.6 BASIC EQUATION'S DIAGRAM[18]

We have learned about cashflow diagrams in Chapter 2. How will the basic relationship, Equation (3.1), look as a cashflow diagram where the initial sum P and the final sum F are cashflow vectors?

As discussed in Chapter 2, a cashflow diagram looks different depending on whose viewpoint is considered. From the lender's or investor's viewpoint, the cashflow diagram of the basic equation will look as in Fig. 3.3(a). Here, the initial sum P is invested now (n = 0). Its vector is located at time marker 0 and the direction is *downward* since it is a cost[19] to the investor. At the end of the nth period, a final sum F is received back (a benefit) by the investor. Since F will then increase (move *up*) the investor's account balance, the F-vector is *upward* at time marker n. The timeline is shown broken since n is a variable in the basic equation.

From the borrower's viewpoint, the cashflow diagram will be as shown in Fig. 3.3(b). This diagram is like Fig. 3.3(a) except that the directions of P and F are reversed. The direction convention follows the same rule: P is upward since the borrowed sum P is a benefit, while F is downward since its payment to the lender at the end of the nth period will be a cost.

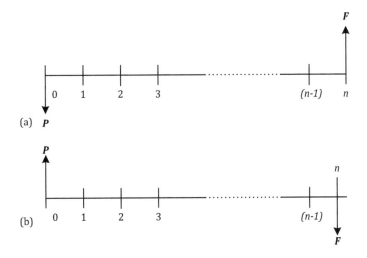

Figure 3.3 Basic equation's cashflow diagram.

---

[18] The cashflow table discussed in Chapter 2 is not of much use in single-payment problems, and hence not used here. The usefulness of cashflow tables becomes evident in problems involving several cashflows, as in later chapters.

[19] Brings *down* his account balance; refer to Section 2.2 (Chapter 2) for the vector convention, if necessary.

Since the final sum F is greater than the initial sum P due to compounding, in both the diagrams of Fig. (3.3) vector F is longer than vector P. The ratio of their lengths is equal to the exponential function $(1+i)^n$, as evident from Equation (3.1). It is this ratio (F/P) that is given in column 2 of the interest tables, while the ratio P/F is in column 3. It may be noticed that any value in column 2 is the reciprocal of the corresponding value in column 3, for all sets of i and n. This is obvious also from a comparison of Equations (3.6b) and (3.7b). Thus,

$$\left(\frac{P}{F}, i, n\right) \times \left(\frac{F}{P}, i, n\right) = 1$$

Problem statements involving P, F, i, and n may at times be difficult to comprehend. In such cases cashflow diagrams should be sketched to facilitate comprehension, as illustrated later through Examples 3.10 and 3.11.

3.7 COMPOUNDING PERIOD

As discussed earlier, compounding is beneficial to the lender. It enables the principal to accumulate faster than under simple interest. Obviously, more frequently the compounding, faster the accumulation, for a given principal and interest rate. In engineering projects, compounding is usually annual. However, it may take place more frequently, such as quarterly, in cases of personal borrowings and investments.

Consider that Rita deposits $2,000 in an account for 2 years at an annually compounded interest rate of 5% per year. At the end of the first year her savings due to compounding becomes $2,100, comprising the initial deposit of $2,000 plus the interest of $100 (= $2,000 × 0.05) for the year. During the second year the interest earned is $2,100 × 0.05 = $105. Thus, at the end of the 2 years Rita's principal is $2,100 + $105 = $2,205. This could have been obtained from Equation (3.1), as well as from the functional notation as:

$$F = 2,000 \left(\frac{F}{P}, 5\%, 2\right)$$

$$= 2,000 \times 1.102$$

$$= 2,204, \text{ which is approximately equal to } \$2,205.$$

Now, consider that the compounding is done every six months. Then the principal at the end of the first six months

$$= \$2,000 + \text{Interest for the first six months}$$

$$= \$2,000 + \left(\$2,000 \times 0.05 \times \frac{6}{12}\right)$$

$$= \$2,050$$

The factor 6/12 in the calculation accounts for the fact that interest is earned only during six of the twelve months of the year.

This $2,050 now earns interest for the second six months. Thus, the principal at the end of the second six months

$$= \$2,050 + \left(\$2,050 \times 0.05 \times \frac{6}{12} \right)$$

$$= \$2,050 + \$51.25$$

$$= \$2,101.25$$

Similarly, the principal at the end of the third six months, i.e., the first six months of the second year

$$= \$2,101.25 + \left(\$2,101.25 \times 0.05 \times \frac{6}{12} \right)$$

$$= \$2,101.25 + \$52.53$$

$$= \$2,153.78$$

And finally, the principal at the end of the last six months, i.e., at the end of the second year

$$= \$2,153.78 + \left(2,153.78 \times 0.05 \times \frac{6}{12} \right)$$

$$= \$2,153.78 + \$53.84$$

$$= \$2,207.62$$

Thus, the six-monthly compounding enables Rita's $2,000 investment to become $2,207.62. Compared to the $2,205 with annual compounding, this is $2.62 more for Rita. While this $2.62 may sound little, for large initial investments or more frequent compoundings, the difference can be quite significant.

Financial institutions out of competition may offer savers compoundings more frequently than yearly. Half-yearly, quarterly, or monthly compoundings are prevalent in the financial marketplace. In some cases, compounding may be daily, and in the extreme it may even be *continuously*—every moment of the time. We do not consider continuous compounding in this text since it is of limited use in engineering projects.

How can we account for more frequent compoundings without having to labor through the first-principles? This is done simply by considering *periods*, rather than *years* or *months* or other units of time, as compounding intervals.

The interest rate usually quoted is yearly; we call it *nominal*[20] interest rate and denote it by r. We use i for the *interest rate per period*. If compounding is annual, then i = r. However, if it is more frequently than annual, then apportion the nominal rate to reflect the interest rate for each period. This will yield i that is used in the formulas. Since for annual compounding i = r, we did not bother to make this distinction earlier between i and r.

In Rita's case, r is 5% per year. Since the compounding is done every six months, i.e., two per year, the interest rate i for the period (6 months) is 5%/2 = 2.5%. Also, over the 2 years of her investment there will be four periods of compounding (four six months in 2 years). So the value of n is 4.

---

[20] Financial institutions use APR (annual percentage rate) or APY (annual percentage yield) for nominal (annual) interest rate r. Sometimes, they quote a daily rate by dividing the APR by 365, the number of days in a year. The daily rate is quoted to convey to the borrower a feeling of lower interest rate.

We can use Equation (3.6a) with P = $2,000, i = 2.5%, and n = 4 to evaluate the final sum F for Rita as:

$$F = P\left(\frac{F}{P}, i, n\right)$$
$$= 2,000\left(\frac{F}{P}, 2.5\%, 4\right)$$
$$= 2,000 \times 1.104$$
$$= \$2,208$$

This answer is approximately the same as that obtained earlier ($2,207.62) from first-principles. The slight difference of 38 cents is due to the limited decimal places in the F/P value 1.104.

Example 3.10: Quarterly Compounded Interest

Sam owns and operates a small electronics repair business that he set up on graduation from the local engineering college. Five years ago he had a very profitable fiscal year. From the profits that year he invested $5,000 in a 10-year CD at 5%, compounded semi-annually. The last few years have been bad. But, this year he has made a decent profit and decided to invest $4,000 for 5 years in another CD at 6%, compounded quarterly. At the end of 5 years, he intends to buy a small CNC drilling machine at an expected cost of $25,000. After using the proceeds of the two CDs, how much more will he have to spend on the machine?

SOLUTION:

This example is slightly more complex. Let us use a cashflow diagram to comprehend the problem. Based on what we learned in Chapter 2, the cashflow diagram is drawn in Fig. 3.4. Referring to the diagram, P_1 ($5,000 deposit) accumulates to F_1, while P_2 ($4,000) to F_2. Thus, the proceeds from the two CDs maturing at the same time will be the total of F_1 and F_2. Sam will have to spend the difference between $25,000 and the total proceeds.

As discussed in Chapter 2, the diagram shows the number of compounding periods (rather than years, quarters, or months). The value of n for the first CD (=n_1) is 20, since two compoundings take place per year over 10 years. For the second CD too, the value of n (=n_2) is 20, since four compoundings take place per year over 5 years.

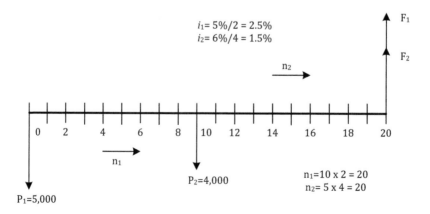

Figure 3.4 Two pairs of cashflows.

Also note in Fig. 3.4 two different interest rates corresponding to the two compounding periods. Since the first investment of $5,000 is compounded twice a year, i_1 is 5%/2 = 2.5%. For the second investment of $4,000 compounded quarterly, i_2 is 6%/4 = 1.5%.

Once the cashflow diagram has served its purpose in problem comprehension, the next step is to solve for the unknown—in this case F_1 and F_2. We can use any of the four methods discussed earlier to evaluate these two future sums. Let us use the (preferable) functional notation method for its simplicity. You should be able to write the following relationship *without referring to any equation whatsoever, if you have mastered the fundamentals presented* in the text so far.

$$\text{Total proceeds} = F_1 + F_2$$

$$= 5,000\left(\frac{F}{P}, 2.5\%, 20\right) + 4,000\left(\frac{F}{P}, 1.5\%, 20\right)$$

$$= 5,000 \times 1.639 + 4,000 \times 1.347$$

$$= 8,195 + 5,388$$

$$= \$13,583$$

Since the drilling machine's cost is $25,000, Sam will have to spend $11,417 more, the difference between $25,000 and the $13,583 proceeds from the two CDs.

Example 3.11: Withdrawal Penalty

Javed, a civil engineer, owns and operates a small consulting firm. He invested $5,000 5 years ago in a 6-year CD at 5.5% annual rate, compounded six-monthly. This investment will mature next year. He had planned at the time of investment to purchase a finite element software for design work. But the software company is offering an attractive deal this year. Future updates of the software will be free if purchased this year. He is considering closing the CD and using the proceeds to purchase the software this year, rather than next year as originally planned. There is, however, a penalty for early withdrawal. Not only does he lose the interest for the current (sixth) year, he also foregoes 1% of the total interest earned. How much will he get if he closes the CD, and what is the cost of early withdrawal?

SOLUTION:

The problem statement is slightly lengthy and may be difficult to comprehend. So, let us draw a cashflow diagram with the hope that it would facilitate comprehension.

Based on what we learned in Chapter 2, the cashflow diagram is sketched in Fig. 3.5. Due to the six-monthly compounding, n = 6 × 2 = 12. For the same reason, i = 5.5%/2 = 2.75% per period. As stipulated, the $5,000 investment is due as F at n = 12. But, since Javed is considering to buy the software now (period 10), a year before the CD is due, he will get T[21]. We need to evaluate T and consider the penalties. The first penalty is that he does not get the interest for the sixth year, i.e., for the last two periods. Thus, T must be evaluated for n = 10. Furthermore, he will lose 1% of the interest earned, which is simply the difference between T and the original $5,000 investment.

---

[21] The use of symbol T for future sum should not be confusing. It is really a future sum with reference to period 0 when the investment was made, but since we are already using F, we had to use another symbol. We chose T; someone would have chosen K or any other symbol. As an alternative to using symbols other than F for future sums, one can use F_1, F_2, etc. This might minimize confusion.

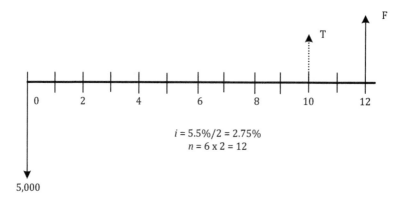

Figure 3.5 Penalty for premature CD closing.

Again we use the functional notation method. The appropriate formula is Equation (3.6a), wherefrom we have

$$T = 5,000 \left(\frac{F}{P}, 2.75\%, 10 \right)$$

Now, we search for the value of the above F/P factor in the tables. Since there is no page for 2.75%, we interpolate the values for 2.5% and 3%. Being exactly in between 2.5% and 3% for which F/P values from the tables are 1.280 and 1.344, the value for 2.75% is the mean of the two; thus,

$$\left(\frac{F}{P}, 2.75\%, 10 \right) = \frac{\left\{ \left(\frac{F}{P}, 2.5\%, 10 \right) + \left(\frac{F}{P}, 3\%, 10 \right) \right\}}{2}$$

$$= \frac{(1.280 + 1.344)}{2}$$

$$= 1.312$$

Thus,

$$T = 5,000 \times 1.312$$

$$= \$6,560$$

Therefore, Javed earns a total interest of

$$I = T - P$$

$$= \$6,560 - \$5,000$$

$$= \$1,560$$

He loses 1% of this, i.e., $15.60 as another penalty. Thus, Javed will get back $6,560 − $15.60 ≈ $6,544 on closing the account now. This is the first part of the answer.

Were the account to run its full course of 6 years, the proceeds would have been F, where

$$F = 5,000 \left(\frac{F}{P}, 2.75\%, 12 \right)$$

Evaluating the same way, we find

$$\left(\frac{F}{P}, 2.75\%, 12\right) = \frac{(1.345 + 1.426)}{2}$$

$$= 1.3855$$

Thus,

$$F = 5,000 \times 1.3855$$

$$= \$6,928$$

Therefore, the cost of early withdrawal

$$= \text{Proceeds if CD not closed} - \text{Proceeds if closed}$$
$$= F - \$6,544$$
$$= \$6,928 - \$6,544$$
$$\approx \$384$$

3.8 EFFECTIVE INTEREST RATE

The interest rates quoted most often are yearly. For a given nominal interest rate, compounding more frequently than yearly yields higher earnings than when compounded annually. This in effect means that the *actual* yearly interest rate is higher than the nominal rate. We distinguish the two rates by labeling the quoted annual rate as *nominal* interest rate denoted by r, and the actual interest rate as *effective* interest rate denoted by i_{eff}. The effective interest rate is also termed *annual yield*.

Let us reconsider Rita's case discussed in the previous section to understand the difference between nominal and effective interest rates. Her 5% yearly interest is the *nominal* interest rate r. Rita's final sum F at the end of 2 years was \$2,205 with annual compounding. With six-monthly compounding it was \$2,208; \$3 more than that with annual compounding. We want to know the *effective* annual rate the six-monthly compounding yielded. In other words, we want to know what annual interest rate would have allowed Rita's \$2,000 investment to become \$2,208, had the compounding been yearly? This rate will be i_{eff}. Let us work it out using functional notation. The solution is very similar to that in Example 3.8. Following the procedure explained there, we have

$$F = P\left(\frac{F}{P}, i, n\right)$$

$$2,208 = 2,000\left(\frac{F}{P}, i, 2\right)$$

$$\left(\frac{F}{P}, i, 2\right) = \frac{2,208}{2,000}$$

$$= 1.104$$

From the tables, the adjoining interest rates for this value of
F/P for n = 2 are 5% with a value of 1.102 and 6% with a value of
1.124. Using the relationship based on similarity of triangles,

as explained in Example 3.8 through Fig. 3.2, we get:

$$\frac{1.104 - 1.102}{1.124 - 1.104} = \frac{af}{6 - 5}$$

$$\frac{0.002}{0.020} = \frac{af}{1}$$

$$af = \frac{0.002}{0.020}.$$

$$= 0.01$$

The interest rate corresponding to the F/P value of 1.104 is thus

$$5 + 0.01 = 5.01$$

This 5.01% is thus the *effective annual* interest rate, i_{eff}. In other words, the growth of $2,000, invested at 5% *nominal* annual rate under six-monthly compounding, in 2 years to $2,208 is equivalent to an *effective* annual rate of 5.01%. Remember that the effective annual interest rate will always be higher than the nominal interest rate when compounding is more frequently than yearly.

Instead of the first-principles, the following formula can be used to evaluate the *effective annual* interest rate i_{eff} for given values of *nominal* interest rate r and the number of compoundings m per year.

$$i_{eff} = \left(1 + \frac{r}{m}\right)^m - 1 \tag{3.8}$$

Recall that when compounding is annual (m=1), i = r. We find this to be true in Equation (3.8); with m = 1, i_{eff} becomes r. Thus, for annual compounding,

$$i = r = i_{eff}.$$

Applying Equation (3.8) to Rita's case where r = 5% (= 0.05) and m = 2, we find:

$$i_{eff} = \left(1 + \frac{0.05}{2}\right)^2 - 1$$

$$= 1.025^2 - 1$$

$$= 0.0506$$

$$= 5.06\%$$

Why use three types of interest rates, which may be confusing? Because each serves a different purpose. Of the three interest rates discussed in this section:

Nominal interest rate r is often the quoted one. It is expressed on an annual basis and is used for comparison by the lay person.

Interest rate i corresponding to the compounding period is the one that is used in calculations and on which the interest tables are based. It is simply nominal interest rate r divided by m, the number of compoundings per year.

Effective annual interest rate i_{eff} is used for comparison of real interest rates offered by financial institutions. The nominal interest rate r does not account for the effect of compounding on the growth of investment; i_{eff} does this.

Lending institutions quote the interest rate they charge their customers. They usually like to quote the nominal interest rate r. To make the interest rate sound low, they sometimes quote r on a monthly or daily basis. For r = 18% per year, the equivalent monthly rate of 1.5% (=18/12) seems so attractive; the daily rate at 0.05% being even more so.

It should be clear by now that what is really important in deciding from where to borrow is the effective annual interest rate i_{eff}; lowest being the best. It is also called *annual percentage yield* (APY) or *annual percentage rate* (APR). Lenders are required by law to quote the APR while lending money to customers. APR allows consumers to compare the cost of borrowing from various lenders. It may take into account the up-front finance charges in securing a loan, combining all the costs of financing into a single value of APR.

Example 3.12: Interest Rates

According to its brochures the local credit union offers to its savers an annual interest rate of 8%, compounded quarterly. What are the values of r, i, and i_{eff}?

SOLUTION:

(a) The given annual interest rate is the nominal interest. Hence r = 8%.
(b) Since there are four compoundings per year, m = 4, the interest rate per compounding period

$$i = \frac{r}{m}$$
$$= \frac{8\%}{4}$$
$$= 2\%$$

(c) We determine i_{eff} from Equation (3.8) for r = 8% (= 0.08), and m = 4 as:

$$i_{eff} = \left(1 + \frac{0.08}{4}\right)^4 - 1$$
$$= 1.02^4 - 1$$
$$= 1.08243 - 1$$
$$= 0.08243$$
$$= 8.24\%$$

3.9 DOUBLING TIME

We have learned by now that an investment increases exponentially under compound interest. We are sometimes interested in knowing how long it takes an investment to double. This doubling time is of interest in other areas too, for example in predicting how long it will take for the population of a country, or the wealth of an individual, to double.

The doubling time of an investment P can be determined from the basic equation:

$$F = P(1 + i)^n$$

For the future sum F to be double of P, we have:

$$(1 + i)^n = \frac{F}{P} = 2$$

To know the value of n that will satisfy the relationship just mentioned, we can take (natural) log of both sides, which yields:

$$\ln\left\{(1+i)^n\right\} = \ln 2$$
$$n\ln(1+i) = 0.693$$
$$= \frac{0.693}{\left\{\ln(1+i)\right\}}$$

(3.9)

Equation (3.9) is used to determine the doubling time for a given interest rate i, which should be expressed in fraction. Note that it involves evaluation of natural logarithm of $(1+i)$. For example, for an interest rate of 5% per year, $\ln(1+i) = \ln(1+0.05) = 0.04879$, and therefore the doubling time will be $0.693/0.04879 = 14.2$ years.

There is a simpler way to determine the doubling time, which does away with the need to use logarithm. The term $\ln(1+i)$ can be expressed in its series as:

$$ln(1+i) = i - \frac{i^2}{2} + \frac{i^3}{2} - \frac{i^4}{2} + ...$$

The value of i is usually small. Even for an interest rate of 10%, the fractional value of i is 0.1 only. Thus, the terms $i^2/2$, $i^3/2$, $i^4/2$, ... on the right side of the relationship just mentioned, being much smaller than the first term i, can be ignored. This reduces $\ln(1+i)$ to i. Thus, from Equation (3.9)

$$n = \frac{0.693}{i}$$

where i is in fraction.

If i is expressed in percent, and 0.693 is approximated by 0.7, then the relationship becomes

$$n = \frac{70}{i}$$

(3.10)

where i is in percent.

Equation (3.10) is "handy" in predicting the doubling time of an investment. For the previous illustration where i was 5%, the doubling time from this equation $= 70/5 = 14$ years, which is very close to what we got (14.2 years) using Equation (3.9). Remember, however, that smaller the value of i, closer the result from Equation (3.10) to that from Equation (3.9). This is because we made such an assumption in ignoring the other terms of the logarithmic series.

Equation (3.10) can be used to determine the doubling time in other areas too. For example, the world human population of 6 billion, under the current growth rate of 1.4%, will take $70/1.4 = 50$ years to double. Consider another case. If you are investing in the stock market and want your investment to double in 5 years, then, from Equation (3.10), your investment must appreciate annually by 14% ($i = 70/n = 70/5 = 14\%$).

3.10 SUMMARY

Compounding of interest is the basis of all investments, loans, and borrowing in the business world. It is embodied in the basic equation $F = P(1+i)^n$, sometimes called the *law of compounding*. Interest earned under compounding is more than that under simple interest. The exponential nature of the basic equation yields faster growth in investment at the latter part of the investment period. For example, a one-time $100 investment for 100 years at 10% interest rate, compounded annually, will bring in over half-a-million dollars as interest during the last 5 years of the investment. Compounding benefits the lender at the cost of the borrower.

In solving single-payment problems, the focus of Chapter 3, one can use either the basic equation or (its) derived formulas. The discussions emphasized the fundamentals of interpolation, as well as of solving the problems. The functional notation method has been discussed, and the compound interest tables have been introduced. The concept of effective annual interest rate has been presented. To determine the doubling time of an investment, use $70/i$, where i is the interest rate in percent.

DISCUSSION QUESTIONS

3.1 Why has Equation (3.1) been given so much prominence in this chapter?

3.2 Explain the law of compound interest to a lay person, using an example from everyday life.

3.3 Of the four methods that can be used to solve problems involving compound interest, which one do you prefer and why?

3.4 Why might practicing engineers prefer the functional notation method?

3.5 Explain the difference between nominal and effective annual interest rates.

3.6 Give an example, not the one in the text, where doubling time relationship, Equation (3.10), can be applied.

3.7 Why does compounding interest benefit the lender or investor?

3.8 Discuss the most basic equation in engineering economics, i.e., the notations and desired end result of computation.

3.9 Distinguish between the *compound amount factor* and *present worth factor*.

3.10 Briefly discuss the concept of interpolation.

MULTIPLE-CHOICE QUESTIONS

3.11 The equation representing the law of compound interest is
   a. $I = P\,i\,n$
   b. $F = P\,(1 + i\,n)$
   c. $F = P\,(1+i)^n$
   d. none of the above

3.12 Compounding is beneficial to
   a. the lender
   b. the borrower
   c. both, depending on interest rate
   d. either, depending on the frequency of compounding

3.13 Present Worth Factor is given by
   a. $(F/P,i,n)$
   b. $(P/F,i,n)$
   c. $(F/P,i,n)/(P/F,i,n)$
   d. $(P/F,i,n)/(F/P,i,n)$

3.14 The use of interest tables is _____ for solving problems involving compound interest.
   a. absolutely essential
   b. essential
   c. somewhat essential
   d. not essential

3.15 You borrow $200 at 6% annual interest rate, compounded quarterly, and plan to pay back the loan and the interest together at the end of fifth year. The sum to be paid can be determined from
a. F = 200(F/P, 6%, 5)
b. F = 200(F/P, 1.5%, 5)
c. F = 200(F/P, 6%, 20)
d. none of the above

3.16 For compoundings more frequently than annual, the effective interest rate
a. is higher than the nominal rate
b. is lower than the nominal rate
c. depends on the amount borrowed
d. equal to the nominal rate

3.17 If a bank charges 1.5% monthly interest for credit card loans, the nominal interest rate is
a. 1.5%
b. 15%
c. 18%
d. 180%

3.18 Find the initial principal of an investment if it was compounded annually at an interest rate of 5% for a period of 8 years. At the end of the loan period the investment totaled $65,000.
a. $35,678
b. $43,995
c. $45,000
d. $63,000

3.19 Find F, given a principal of $34,567 at an interest rate of 12% for 10 years. The amount is compound semiannually.
a. $225,000
b. $450,723
c. $333, 443
d. $237,869

3.20 Find i, given P = $46,050, F = $55,307, and n = 6.
a. 2.75%
b. 4.23%
c. 3.1%
d. 3.345%

3.21 Find n, given i = 4.5%, P = $14,500, and F = $28,061.
a. 9
b. 4
c. 25
d. 15

3.22 Find P, given F = $186,000, i = 7.5%, n = 8.
a. $102,345
b. $104,291
c. $170,890
d. $115,932

3.23 Find F, given P = $90,361, i = 3%, n = 9.
a. $123,349
b. $105,672
c. $117,901
d. $145,902

3.24 Find *i*, given P = $75,000, F = $89,655, and n = 5.
   a. 2%
   b. 3.6%
   c. 4.3%
   d. 5%
3.25 Find n, given P = $105,000, F = $157,884, *i* = 8.5%.
   a. 5
   b. 3
   c. 8
   d. 6
3.26 What is the compound interest factor for F given P at an interest rate of 4.5% for 20 periods?
   a. 2.520
   b. .4146
   c. 2.412
   d. 1.935
3.27 What is the compound interest factor for P given F at an interest rate of 15% for 8 periods?
   a. 3.518
   b. 3.059
   c. .3759
   d. .3269
3.28 Determine n, given a compound interest factor of 7.251 for F/P?
   a. 34
   b. 30
   c. 25
   d. 18
3.29 Determine n, given a compound interest factor of .1827 for P/F?
   a. 12
   b. 15
   c. 18
   d. 23

NUMERICAL PROBLEMS

3.30 Better Mousetrap Incorporated had an excellent fiscal year with profit far more than expected. It decides to invest $12,000 of its profit as a 2-year CD. How much will this investment mature to if the yearly interest rate is 6%, compounded annually?
3.31 If compounding is quarterly in Problem 3.14, what is the answer?
3.32 Determine through interpolation the value of (P/F,8.4%,10).
3.33 How much does one have to invest today for a lump-sum payback of $10,000 in 5 years if the annual interest rate is 4%, compounded quarterly?
3.34 If John is paid back $14,500 for an investment of $12,750 made 3 years ago under compound interest, what was the interest rate, assuming annual compounding?
3.35 Visionary Robotics invests $10,000 in the local credit union that pays back $11,500 at the end of 3 years. What is the nominal interest rate if compounding is quarterly?
3.36 How long will it take for an investment of $4,500 to mature to $6,525 at a yearly interest rate of 5%, compounded annually?

3.37 Monica receives $1,000 as a gift on her tenth birthday. She decides to invest it in a 5-year-CD whose proceeds she plans to use when fifteen as down payment to buy her first car. She has heard that credit unions usually pay higher interests than banks. She can invest with one of the two local credit unions; one affiliated to her father's company and the other to her mother's. The former offers 12.5% interest per year compounded annually, while the latter 12% per year compounded monthly. Which credit union should she invest with?

3.38 Myra received her federal educational loan of $3,000 for the current semester on 31st August. She deposited it the same day with the local bank at 3% annual interest, compounded monthly. For her expenses, she withdrew $800, $500, and $450 on the last days of September, October, and November respectively. How much will her proceeds be if she closes the account on 31st December?

3.39 Savita owns and operates a small construction business she inherited from her parents. Her biggest surprise on the first day at work as construction engineer is a gift of capital from her parents' past savings. Ten years ago her dad invested $8,000 in a CD that compounded interest semiannually. Also, 6 years ago her mom saved $2,000 for her in an account at 6% annual interest compounded monthly. The two savings matured on her first day of work with total proceeds of $15,540. What interest rate did her dad's CD earn?

3.40 How many years will it take a sum to double if it earns annually compounded interest of 9% per year. How much sooner will the doubling occur under monthly compounding? Assume the 9% interest rate to be nominal.

3.41 Engineering Unlimited takes a loan at 16% yearly interest rate, compounded quarterly, for purchasing a robotic system. The company pays back the loan (capital and interest) through two payments: $50,000 at the end of six months, and $65,000 on the first loan anniversary. How much was the loan?

3.42 MegaCorp takes out a loan of $300,000 at a rate of 12% compounded semiannually for new bottling equipment. If MegaCorp pays $60,000 annually:
   a. In how many periods should MegaCorp repay the loan in full?
   b. What is the remaining principal at the end of the second period of compounding?
   c. Draw the cashflow diagram for MegaCorp.

3.43 Conway, President of Twitty Enterprises, decided to invest $10,000 6 years ago in a 10-year CD bearing 5% simple interest. In order to make a move on a business opportunity he decides to prematurely close the investment. However, he will have to forfeit the interest for year 6 and will give back .75% of total interest earned for the early closure. How much will he receive back from the CD closure and what is his premature loss on the investment?

3.44 Dan Glover invested $45,000 in a CD 5 years ago at interest rate of 2.5% compounded semiannually.
   a. Compute F with the given information.
   b. What is the effective interest rate on the investment?

FE EXAM PREP QUESTIONS

3.45 An effective interest rate of 2.5% semiannually is:
   a. 5%
   b. 7.5%
   c. 4%
   d. 1.25%

3.46 An effective interest rate of 1.2% per month is closest to what?
   a. 6%
   b. 10%
   c. 12%
   d. 14%

3.47 A monthly interest rate of 3% is the same as:
   a. 24%
   b. 12%
   c. 36%
   d. 48%

3.48 A nominal interest rate of 15% compounded semiannually is that same as:
   a. 6%
   b. 8%
   c. 7.5%
   d. 12%

3.49 A money market account pays 3% per year compounded monthly, while a bank pays 3.2% per year on a savings account. The money market account is:
   a. equivalent to the bank's rate
   b. lower than the bank's rate
   c. higher than the bank's rate
   d. none of the above

3.50 A nominal rate of 8% per year compounded quarterly is the same as:
   a. 4.5% every six months
   b. 2% every quarter
   c. 2% every six months
   d. 8% per year

3.51 Find n, given F = $33,950, P = $15,000, and i =7%.
   a. 19
   b. 15
   c. 23
   d. 8

3.52 Find i, given F = $36,662, P = $23,000, and n = 8.
   a. 6%
   b. 5%
   c. 10%
   d. 3%

3.53 A nominal interest rate of 6% semiannually is:
   a. 5%
   b. 7.5%
   c. 4%
   d. 12%

3.54 A nominal interest rate of 12.5% compounded semiannually is that same as:
   a. 6.25%
   b. 8.33%
   c. 7.5%
   d. 12%

Chapter 4

Multiple Payments

LEARNING OUTCOMES

- Comprehend problems involving multiple cashflows.
- Further appreciate the importance of the basic equation.
- Solve both regular and irregular cashflow problems.
- Derive equations for uniform, arithmetic, and geometric series cashflows.
- Solve uniform, arithmetic, and geometric series cashflow problems.
- "Tailor" cashflows to the known patterns.

In the previous chapter, we discussed the effect of compounding on a single payment. We saw how an investment P increases under compounding to become future[1] sum F. We learned that both the interest rate i and the time period n affect the relationship between P and F. We analyzed problems involving single payment which, in fact, gives rise to a pair of cashflows—P by one party and F by the other. In this chapter, we extend our discussions to multiple payments involving more than one pair of cashflows.

The pair of cashflows in single-payment problems involves two one-time transactions. They relate mostly to personal financing. Engineering economics problems, on the other hand, usually involve multiple payments. The investment in a resource (machine, equipment, or a similar item) comprises not only the initial (purchase) cost but also recurring costs of maintenance. Besides, there are costs of utilities and insurance that must be paid regularly, such as monthly. In other words, industrial projects encompass several costs and benefits, resulting in multiple-payment problems.

Even in personal financing there are situations that involve multiple payments, for example regular payments for loans on homes, cars, etc. Lease purchases of products and services also require multiple periodic payments.

Multiple-payment problems arise because time has been "sliced" into convenient "chunks" of weeks, months, and years. Wages and salaries are paid weekly or monthly, investment incomes are reported quarterly, and the budgets are drawn annually. These result in engineering projects with cashflows occurring at regular intervals of time.

---

[1] The concept of *present* and *future* in engineering economics is relative, not absolute. Consider that you deposited $500 in a bank account 5 years ago that paid you $575 last year as principal plus the interest. As learned in the previous chapter, we should say that the present sum P is $500, and the future sum F is $575, though both payments have been made in the past. The present, in this case, was logically the time 5 years ago when $500 was deposited, with reference to the future which occurred last year with the $575 payment. Thus, the term *present* does not necessarily mean the present time in the physical sense. It is rather a *reference* point; it could be anywhere on the timeline. The *future*, however, is always after the present. Note that such connotations of *present* and *future* render the word *past* almost meaningless.

Multiple-payment problems fall into one of the following three categories:

1. One cost[2] several benefits
2. Several costs one benefit
3. Several costs several benefits

The first two categories are in fact special cases of the third category. One-time investment on a machine requiring little or no maintenance is a problem of the first category with recurring benefits, while regular savings toward a lump-sum capital outlay is of the second. Problems involving several costs and several benefits are numerous and most frequently encountered.

Multiple-payment cashflows may be *irregular* or *regular*. Irregular cashflows are erratic; they occur inconsistently. Examples are occasional deposits in and withdrawals from a savings account, or repair costs of a machine. We discuss problems of this type in Section 4.1. Cashflows are *regular* if they are repetitive, occurring at each period. Examples are monthly home mortgage payments or annual insurance premiums. We discuss such cases in Section 4.2.

Regular cashflows may have some definite pattern. In one common pattern they are constant, as with monthly home mortgage payments or annual payments on a lease-to-buy machine. Such constant cashflows form a *uniform series*; we discuss them in Section 4.3. In another common pattern, the cashflows increase with time, forming an *arithmetic series*. A machine whose maintenance costs increase each year by a fixed amount is a good example. Such problems are analyzed in Section 4.4. In some cases, the cashflows may follow *geometric* series, increasing each period by a certain percentage. Such cases are covered in Section 4.5. Finally, in Section 4.6 we discuss cashflows that are *approximately* uniform or series-type, and learn how to "tailor" them into one of the three common patterns to simplify the solution.

As we see in the following sections, the basic relationship embodied in Equation (3.1) continues to be the basis of all the formulas in this chapter, as well as elsewhere.

4.1 IRREGULAR CASHFLOWS

In problems involving irregular payments the cashflows can occur at any time period. A bank savings account in which money is deposited when possible and withdrawn when needed comprises irregular cashflows. The cashflow diagram in such a case displays vectors here and there—not at each time period.

Consider a problem whose cashflow diagram is Fig. 4.1. An investment R is made now, none at period 1, S at period 2, T at period 3, and none at period 4. Since the investment has not been made at each period, the cashflows[3] are *irregular*. The payback from these *irregular* investments occurs together at period 5 as Q. Such a several-deposits-one-withdrawal problem is similar to several-costs-one-benefit problems.

Problems like the one in Fig. 4.1 are easily solved by determining the F-value for each deposit. One can use the basic equation or its functional notation to determine the F-values. Using Equation (3.6a), and noting that Q is the total of the three F-values, we get:

$$Q = R\left(\frac{F}{P,i,5}\right) + S\left(\frac{F}{P,i,3}\right) + T\left(\frac{F}{P,i,2}\right)$$

---

[2] As discussed in Chapter 2, the term *cost* is synonymous with *disbursement*, *expenditure*, and *cash outflow*. By the same token, benefit is synonymous with *receipt*, *revenue*, and *cash inflow*.

[3] With reference to the timing of Q, symbols R, S, and T depict "present" sums at their time periods and are later treated as such. Instead of these symbols, we could have used P for present sum by labeling them as P1, P2, P3. Since they are present sums, the appropriate functional notation to determine their future values remains (F/P,*i*,n). Within the parenthesis P is left unchanged to signify "present" what R, S, and T logically are.

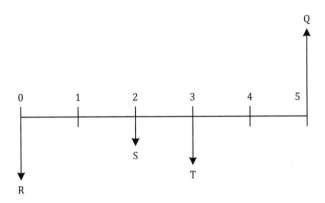

Figure 4.1 Irregular cashflows.

Again, since we used the functional notation approach there was no need to remember any formula. This made it easier to set up the above expression. The basic skill lies in being able to judge the correct value of n for each functional notation, for example 5 for R. The other skill in evaluating Q is to search for the F/P-values in the tables.

Irregular cashflow problems of one-cost-several-benefits or several-costs-several-benefits type are also solved the same way, by accounting for each cashflow separately. The following example illustrates the procedure.

Example 4.1: Irregular Cashflows

A new machine with expected useful life of 5 years is being planned for acquisition next year when it will cost $6,000. This cost includes maintenance for the first 2 years of the purchase. The maintenance for the subsequent 3 years of the useful life is projected to cost $300, $500, and $750 respectively. How much should be budgeted now for the purchase and maintenance of this machine if the yearly interest rate is 6%, compounded annually?

SOLUTION:

Let us sketch a cashflow diagram to help comprehend the problem. Such a diagram is shown in Fig. 4.2. If you have difficulty sketching or understanding this diagram, then reread Chapter 2.

The diagram shows that the machine will be purchased next year for $6,000 (vector down at period 1 since it is a cost). There are no cashflows at periods 2 and 3 since the purchase cost includes the maintenance for the first 2 years. The maintenance costs of the subsequent 3 years are displayed as cashflows at periods 4, 5, and 6. Note that the $300 cost during year 3–4 is posted at period 4 following the period-end assumption. So are the other two maintenance costs.

Altogether there are four costs, which will be paid for from the budget this year, i.e., from P. Thus, P must be *equivalent* to these costs. Note that vector P is shown[4] dotted to highlight that in itself it is *not a cost*, but a budgeted sum which should be enough to pay for the four future costs.

---

[4] Why isn't the vector up? This is because we are setting up the problem in a way where P is the equivalent cost now (t = 0) for the four future costs. Some analysts draw P vertically up, in which case it represents the one-time cash inflow into the machine account to be used for its purchase and maintenance. Thus, P can be vertically up or down depending on the sense of its representation. In either case, the equivalency prevails, yielding the same result.

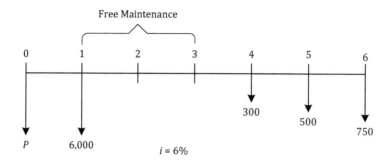

Figure 4.2 Diagram for Example 4.1.

The equivalency between P and the four costs suggests that P can be determined simply by adding the P-value of each future cost. Which equation is appropriate for finding P for a given F? It is Equation (3.7a). Thus, we get

$$P = 6000(P / F, 6\%, 1) + 300(P / F, 6\%, 4) + 500(P / F, 6\%, 5)$$
$$+750(P / F, 6\%, 6)$$

Substituting the values of the functional factors from the interest table for 6%, we have

$$P = 6,000 \times 0.9434 + 300 \times 0.7921 + 500 \times 0.7473 + 750 \times 0.7050$$
$$= 5,660.40 + 237.63 + 373.65 + 528.75$$
$$= \$6,800$$

Thus, the machine should be budgeted this year for $6,800. At the end of the first year, the balance will be $6,800(1 + 0.06) = $7,208, of which $6,000 will be used for the purchase. The remaining $1,208 earning interest on the balance with annual compounding will fully pay for the three maintenance costs.

4.2 REGULAR CASHFLOWS

Several situations arise, both in personal financing and engineering economics, where costs and/or benefits occur regularly at each period. This makes the cashflows[5] *regular*. The cashflow diagram in such a case displays vectors repetitively at each time marker. A savings account in which money is deposited regularly, say on the first day of each month, is a good example of regular cashflows. Another example is the dividend received each quarter from investment in stock markets.

Consider a savings account illustrated by its cashflow diagram in Fig. 4.3. Savings are made regularly: A now, B at period 1, ..., and G at period 6. The balance in the account is withdrawn at period 7 as Q. This is similar to a several-costs-one-benefit problem.

Problems like the one in Fig. 4.3 are solved the way illustrated earlier in Example 4.1. For the value of Q, we have in this case

$$Q = A\left(\frac{F}{P, i, 7}\right) + B\left(\frac{F}{P, i, 6}\right) + \dots + G\left(\frac{F}{P, i, 1}\right)$$

---

[5] Throughout the text, the words *cashflow* and *payment* are used interchangeably.

The solution requires evaluating each term on the right of this expression. Again, in this solution approach—based on functional notation—there is no need to remember any formula.

Regular, multipayment problems may not display any pattern in the cashflow magnitudes, as in Fig. 4.3. However, several such problems possess a certain pattern in their cashflow magnitudes. For example, the payments may be the same at each period, or they may increase or decrease in a certain way. The existence of a pattern may be exploited to avoid treating each payment separately, thus achieving a more efficient solution. We discuss three cashflow patterns commonly encountered in engineering economics, in the next three sections (4.3, 4.4, and 4.5).

4.3 UNIFORM SERIES

Engineering projects often entail uniform series cashflow patterns. The term *uniform series* means that the cashflow exists at each period, and that its magnitude is the same. Referring to Fig. 4.3, it means that A = B = ... = G. Home-mortgage loan and car-lease payments are good examples of uniform series. In these cases, the benefits of owning a home or car are realized by taking a loan now and regularly paying a fixed sum over certain length of time. Such payments generate a *series* of cashflows of constant or *uniform* magnitude[6].

Uniform series problems may involve a loan P now and equal payments A over a predetermined number of periods n. They may also be of the opposite type, i.e., regular deposit of sum A to accumulate capital F at the end of n periods. In a third type of problem a sum is invested now to earn fixed regular income. All these problem types are *conceptually* the same (uniform series), and therefore solved the same way.

We consider next two types of uniform series problems. In one, uniform payments A accumulate to a future sum F at the end of n periods (Fig. 4.4); our aim is to know how A and F are related. In the other, a present loan P is returned through payments of A over n periods (Fig. 4.5(a)). We seek here the relationship between P and A. The opposite case of Fig. 4.5(a) is Fig. 4.5(b), where a sum P invested now (cost) brings uniform dividend (benefit) of A over n periods.

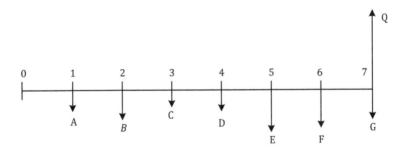

Figure 4.3 Regular cashflows.

---

[6] Two other examples are:Most of us find it difficult to save for retirement once we have received our paychecks. Employers help in circumventing this difficulty by offering pension or retirement plans with deduction at the source. Self-employed people can set up their own retirement plans, usually encouraged by government through tax exemptions. In such situations, a certain sum A is saved regularly over n periods. The savings are invested for the "long haul," earning better than short-term interest. The balance in the retirement account is realized in several ways: in one the retiree receives a lump-sum, and in another a regular sum until death. Small companies raise future capital from within by saving each year out of the profit.

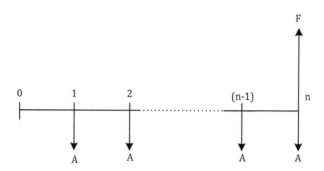

Figure 4.4 Regular saving *A* accumulates to *F*.

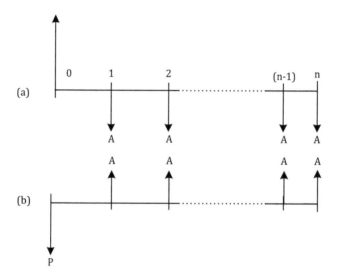

Figure 4.5 (a) Home-mortgage loan; (b) Lump-sum investment as source of regular income.

4.3.1 Relationship between A and F

Consider the cashflow diagram in Fig. 4.4 to represent, for example, a retirement plan with A as the regular contribution. Note that there is no contribution at period 0 (now), which may be considered the beginning of employment when the employee joins the retirement plan by signing-in. The first contribution is posted at period 1 (deducted from the first paycheck). The same amount A is deducted each period, which may be week, month, or year, and invested to earn compound interest. As discussed in Chapter 2, the use of the term *period* avoids any confusion about the unit of time duration, as long as *i* and n are in compatible units. Also note in this diagram that the last deduction is made, from the last paycheck, on the day (period n) the lump sum F is collected.

4.3.1.1 Finding F for Given A, i, n

Increasing under compounding each of the regular saving A contributes toward F. Thus, F is the total of the contributions along with their earned interests. Let us assume that the first contribution A at period 1 accumulates to become F_1 at period n. The same way the second A accumulates to F_2, and so on. Thus, F must equal the total of F_1, F_2, ... and so on. In other words,

$$F = F_1 + F_2 + ... + F_{n-1} + F_n$$

Thus, in Fig. 4.4, the sum F is basically the collection of its components F_1, F_2, ... F_{n-1}, F_n— all bundled together at period n.

We can consider the first contribution A and F_1 as a pair of single-payment cashflows, in which A is like the present sum (P) and F_1 the final sum. We can therefore apply the basic equation, Equation (3.1), to this pair, noting that the value of the time periods for this pair is (n–1). The relationship between the first A and F_1 is thus:

$$F_1 = A(1+i)^{n-1}$$

We can treat the other contributions the same way and write for each a relationship similar to the above, taking care of the value for the exponent. Thus, $F = F_1 + F_2 + ... + F_{n-1} + F_n$ becomes:

$$F = A(1+i)^{n-1} + A(1+i)^{n-2} + ... + A(1+i)^2 + A(1+i) + A$$

Multiplying both sides of this relationship by $(1 + i)$, we get:

$$F(1+i) = A(1+i)^n + A(1+i)^{n-1} + A(1+i)^{n-2} + ... + A(1+i)^2 + A(1+i)$$

Let us now subtract the relationship for F from that for $F(1 + i)$. This results in:

$$F(1+i) - F = A(1+i)^n + A(1+i)^{n-1} + A(1+i)^{n-2} + ... + A(1+i)^2 + A(1+i)$$

$$-A(1+i)^{n-1} - A(1+i)^{n-2} - ... - A(1+i)^2 - A(1+i) + A$$

$$= A(1+i)^n - A$$

Note that the subtraction canceled all other terms on the right side, leaving behind only two terms.

The previous expression can be rearranged as:

$$F(1+i-1) = A\{(1 +i)^n - 1\}$$

$$F = A\frac{(1+i)^n - 1}{i} \qquad (4.1a)$$

Equation (4.1a) is used to analyze uniform series cashflows, specifically to determine F for given values of A, i, and n. In fact, any of the four variables can be evaluated for given values of the other three. Any of the four methods discussed in Chapter 3 can be used for this. But, we continue with the functional notation method, for which Equation (4.1a) can be expressed[7] as:

$$F = A\left(\frac{F}{A,i,n}\right) \qquad (4.1b)$$

---

[7] If you experience difficulty following functional notations, reread Chapter 3 where it has been discussed.

where

$$(F/A, i, n) = \frac{(1+i)^n - 1}{i}$$

The notation (F/A,i,n) is called *Compound Amount Factor* since it is used as a multiplier to find the effect of compounding on uniform series payments A. Note that we used this term earlier too, in Chapter 3 in single-payment analysis. One should therefore be careful with this term, since depending on the cashflow type, compound amount factor can mean either (F/A,i,n) of Equation (4.1b) or (F/P,i,n) in Equation (3.6b).

We prefer in this text to avoid the use of functional notations' names, since they are long and difficult to remember; although each name has a distinct meaning based on what the factor really does. Instead of using the names, we use terms like *F/A-factor*, meaning the factor used to find F for a given A. Sometimes, we also express it as *F-factor for a given A*, which simply means *F/A-factor* or (F/A,i,n).

Similar to that in Chapter 3, one can read the value of F/A-factor off the tables to evaluate F for given values of A, i, and n. Such values are given in the tables under *Uniform Payment Series*. Note that there are four columns under Uniform Payment Series; we have been introduced to one just now. The other three are discussed later in this chapter.

4.3.1.2 Finding A for Given F, i, n

To facilitate the evaluation of A for given values of F, i, and n, Equation (4.1a) can be rearranged as:

$$A = F \frac{i}{(1+i)^{n-1}} \tag{4.2a}$$

Again, any of the four methods discussed in Chapter 3 can be used to determine A using this equation. Any of the four variables of Equation (4.2a) can be evaluated for given values of the other three. Since we prefer the functional notation method, we rewrite this equation as

$$A = F \left(\frac{A}{F, i, n} \right) \tag{4.2b}$$

where

$$(F/A, i, n) = \frac{i}{(1+i)^{n-1}}$$

The notation[8] (A/F,i,n) is called *Sinking Fund Factor* since it eventually *sinks* a given future sum F through uniform series regular[9] payments A. In other words, the future sum F would have *sunk* or vanished altogether with the last payment A. Again, rather than use the long name for the functional factor, we use the terminology *A/F-factor* or *A-factor for a given F*.

---

[8] If you find yourself confused about whether to use A/F or F/A in Equation (4.2b), reread Chapter 3 where this point has been explained. Basically, the factor is correct if A and F follow the same sequence within the parenthesis as outside. In here, A/F has the same sequence as A = F, that is A precedes F; hence the notation used is correct.

[9] There may be cashflows that are regular but not uniform.

Looking closely at what they represent, the two factors (A/F,i,n) and (F/A,i,n) are reciprocal[10] of each other, i.e., (A/F,i,n)(F/A,i,n) = 1. This can also be confirmed from their values for a specific n on any page of the interest tables.

Example 4.2: Retainage

Rani owns and operates a small electronic repair shop. A new instrument, called Rapid Diagnosist, which facilitates faster diagnosis of the items to be repaired, has been marketed recently. Being technologically new, it is however expensive. Based on the past, the costs of such instruments decline fast. Rani predicts that Rapid Diagnosist would sell for $5,000 in the next 5 years. She decides to retain a fixed sum each year from annual profit to raise this capital 5 years hence. The retained sum will be invested regularly in the local bank at 6.5% annually compounded interest rate. How much should be retained each year?

SOLUTION:

First you need to decide whether to draw a cashflow diagram for comprehending the problem statement. This depends on how well you understand the problem. Do you know what is being asked for, what's the unknown, and what variables are known? If you can comprehend all these without the cashflow diagram, you can avoid sketching it. Let us assume that to be the case. We still need to have a mental picture of the problem.

Note that it is a multi-payment problem with regular deposits. It involves A and F, with A being the unknown. The relevant formula is Equation (4.2a) or (4.2b). Use the latter if you favor the functional notation method.

Next, we gather the values of the variables and proceed with the solution. The given data are: F = $5,000, i = 6.5% (per year), and n = 5 (years). Let us say that we decide to use the functional notation method; so from Equation (4.2b), we have:

$$A = 5,000\left(\frac{A}{F,6.5\%,5}\right)$$

Look for the value of A/F-factor in the tables. Since there is no table for 6.5%, look for the values corresponding to 6% and 7% and determine their average (since 6.5% is the average of 6% and 7%). This works out to be:

$$(0.1774+0.1739)/2 = 0.17565, \text{ yielding}$$

$$A = 5,000\times0.17565$$

$$= \$878$$

Thus, Rani must retain $878 each year from the annual profit.

4.3.1.3 Finding i for Given F, A, n

Finding n for Given F, A, i

Any of the four formulas, namely Equation (4.1a), (4.1b), (4.2a), or (4.2b), can be used to evaluate the unknown i for given values of F, A, and n, or the unknown n for given values of F, A, and i. Equations (4.1a) and (4.2a) are formatted for the use of calculators, while

---

[10] The reciprocity is also true between the single-payment factors (F/P,i,n) and (P/F,i,n), as evident in Chapter 3 by what they represent and also by their values in the tables for given i and n.

Equations (4.1b) and (4.2b) of the compound interest tables. As explained in Chapter 3, the use of tables to evaluate i or n involves some interpolation whose basic procedure has been illustrated in Examples 3.8 and 3.9.

4.3.2 Relationship between A and P

Several engineering economics problems involve a present loan P that is paid back through uniform series disbursements A. Though home-mortgage or car-lease loans quickly come to mind as examples, this scenario is common in industry too where capital is borrowed to purchase equipment. As the equipment is used and profits (benefits) from its use realized, certain fixed sum A is paid back to the lender regularly. Let us consider a home-mortgage loan as illustration because of its familiarity in everyday life. But the discussions hold good for engineering economics problems too.

The home-mortgage-loan problem is represented by the cashflow diagram in Fig. 4.5(a), where P is the loan (received now, vector up) and A is the regular payment each period during the loan life. An obvious question is: What is the value of A so that the loan P and the associated interests are fully paid back by the end of the loan life? The lending manager determines A by referring to a special table whose entries are based on the following formula.

4.3.2.1 Finding A for Given P, i, n

There are two ways to develop the relationship between A and P. One is to follow the first-principles, as discussed earlier in Section 4.3.1 for the case of A and F. This derivation is given[11] in Box (4.1) for readers interested in first-principles.

A simpler approach is to combine the expression in Equation (4.2a) with that in Equation (3.1). This approach saves time and is preferred by those who are less interested in theoretical derivations.

On substitution in Equation (4.2a) for F from Equation (3.1), we get:

$$A = P(1+i)^n \frac{i}{(1+i)^{n-1}}$$

Thus,

$$A = P(1+i)^n \frac{i(1+i)^n}{(1+i)^n - 1} \tag{4.3a}$$

Again, any of the four methods discussed in Chapter 3 can be used to determine A using Equation (4.3a). In fact any of the four variables of this equation can be evaluated for given values of the other three. In the functional notation method, we express this equation as

$$A = P\left(\frac{A}{P,i,n}\right)$$

where

$$(A/P, i, n) = \frac{i(1+i)^n}{(1+i)^n - 1} \tag{4.3b}$$

---

[11] Some derivations have been enclosed in boxes to avoid interruption to the discussions. The derivations may be skipped by practicing engineers interested only in the application aspects of equations. Students are encouraged to go through the first-principles-based derivations to "hone" their analytical skills.

The notation (A/P,*i*,n) is called *Capital Recovery Factor* since it is used as a multiplier to recover (get back) the invested[12] capital P through n number of uniform payments A. Rather than use this long name for the functional factor, we often use the terminology *A/P-factor*, or *A-factor for a given P*. The values of A/P-factor are given in the interest tables in one of the columns under Uniform Payment Series.

4.3.2.2 Finding P for Given A, i, n

Equations (4.3a) can be rearranged to find P for given values of A, *i*, and n, as:

$$P = A \frac{(1+i)^n - 1}{i(1+i)^n} \tag{4.4a}$$

Again, any of the four methods discussed in Chapter 3 can be used to find P using this equation. In fact any of the four variables of Equation (4.4a) can be evaluated for given values of the other three. Since we prefer the functional notation method, we rewrite this equation as

$$P = A \left(\frac{P}{A,i,n} \right)$$

where

$$\left(P / A, i, n \right) = \frac{(1+i)^n - 1}{i(1+i)^n} \tag{4.4b}$$

The notation (P/A,*i*,n) is called *Present Worth Factor*[13] since it is used to find the present value of a series of future cashflows A. Rather than use the long name for the functional factor, as mentioned earlier, we use the terminology *P/A-factor* or *P-factor for a given A*.

Again, as obvious from what they represent in terms of *i* and n, the two factors (P/A,*i*,n) and (A/P,*i*,n) are reciprocal of each other. This can also be confirmed from their values for a specific n on any page of the interest tables.

BOX 4.1

Two cashflow diagrams are shown in Fig. 4.5. In (a), benefit P is received now and paid back over n periods as uniform payments A, while in (b) an investment P is made now and the benefits realized over n periods as uniform A. From the analysis viewpoint the two diagrams represent the same problem.

Let us consider the case corresponding to Fig. 4.5(a). In here, sum P is borrowed (vector up since borrowing results in borrower's account balance going up) now at interest rate *i*. This loan along with the interest earned under compounding is paid back (vectors down) over n periods through regular payments A. We are interested in deriving a relationship between P and A for given values of *i* and n.

Each A pays back a portion of P and the accrued interest. We can therefore consider P logically divided in several portions P_1, P_2, ... P_{n-1}, P_n. The payment A at period 1 pays for P_1 and its accrued interest, that at period 2 pays for P_2 and its accrued interest, and so on. Thus, in

---

[12] The home loan is investment capital from the lending institution's viewpoint.
[13] Note the same name under single payment; so be careful in differentiating the two.

Fig. 4.5(a), instead of sum P, we can conceptually consider at period 0 small vectors P_1, P_2, ... P_{n-1}, P_n—all bundled together so as to add to P. Thus, the various portions are related to P as:

$$P = P_1 + P_2 + ... + P_{n-1} + P_n$$

We can consider P_1 and the first payment A at period 1 together as a pair, in which A is like the future sum F for P_1. Using the basic formula, Equation (3.1), where F = A, P = P_1, and n = 1, we can write a relationship between P_1 and the first payment A as:

$$A = P_1 (1 + i)^1$$

Thus,

$$P_1 = A(1 + i)^{-1}$$

We can look at the other portions of P the same way, as a pair with the particular A, and write relationships similar to the above. By adding the resulting relationships, we get:

$$P = A(1 + i)^{-1} + A(1 + i)^{-2} + ... + A(1 + i)^{-(n-1)} + A(1 + i)^{-n}$$

Multiplying both sides of this relationship by $(1 + i)^n$, we get:

$$P(1 + i)^n = A(1 + i)^{n-1} + A(1 + i)^{n-2} + ... + A(1 + i)^2 + A(1 + i) + A$$

Next, multiply both sides of this relationship by $(1 + i)$ to get:

$$P(1 + i)^{n+1} = A(1 + i)^n + A(1 + i)^{n-1} + A(1 + i)^{n-2} + ... + A(1 + i)^2 + A(1 + i)$$

Now, subtract $P(1 + i)^n$ from $P(1 + i)^{n+1}$. Note that this subtraction cancels all the terms on the right-hand side of the above two relationships, except the two terms $A(1 + i)^n$ and A. The net result is

$$P(1 + i)^{n+1} - P(1 + i)^n = A(1 + i)^n - A$$

The relationship can be rearranged as

$$P(1 + i)^n \{(1 + i) - 1\} = A\{(1 + i)^n - 1\}$$

$$Pi(1 + i)^n = A\{(1 + i)^n - 1\}$$

$$P = A\frac{(1 + i)^n - 1}{i(1 + i)^n}$$

This is Equation (4.4a) in the main text, which can be rearranged as Equation (4.3a).

4.3.2.3 Finding i for Given P, A, n

Finding n for Given P, A, i

Any of the four equations, namely (4.3a), (4.3b), (4.4a), or (4.4b), can be used to evaluate the unknown *i* for given values of P, A, and n, or the unknown n for given values of P, A, and *i*. Equations (4.3a) and (4.4a) are in forms suitable for the use of scientific calculators, while Equations (4.3b) and (4.4b) for the use of functional factors. As discussed in Chapter 3, the use of functional factors to evaluate *i* or n involves some interpolation whose basic procedure has been explained in Examples 3.8 and 3.9. The following example illustrates the evaluation of interest rate *i* for known P, A, and n. The procedure for evaluating n is very similar.

Example 4.3: Interpolation

Ray's company borrows $5,000 from the local bank to invest in a personal computer for the quality control department on the term that the company will pay back $1,200 each year for the next 5 years. The first payment is due on the first loan anniversary. What interest rate the bank is charging, if compounding is annual?

SOLUTION:

First, we need to decide whether to sketch a cashflow diagram to facilitate problem comprehension. If you can comprehend without the diagram, there is no need to sketch it, since it is not being specifically asked for. However, if the diagram will help you, then sketch it. Let us say that we can do without the diagram[14].

Note that the problem involves six cashflows; in other words, multi-payment analysis of this chapter is applicable. Next, we need to check whether the payments are irregular or regular. The five $1,200 payments are regular, each being annual. Next, do they follow a pattern? The answer is yes, and the pattern is uniform series since the payments are equal. Thus, the discussions of Section 4.3.2 are applicable, with Equation (4.3b) being the relevant formula for the functional notation method.

We next gather the given values of the parameters and proceed. These are: P = $5,000, A = $1,200, n = 5. With n in years, the unknown *i* will get evaluated as annual. Let us say that we decide to use the functional notation method. Thus, from Equation (4.3b),

$$A = P\left(A/P, i, 5\right)$$

$$1,200 = 5,000\left(A/P, i, 5\right)$$

$$\left(A/P, i, 5\right) = 1,200/5,000$$

$$= 0.24$$

Searching in the tables under the proper column following the procedure explained in Example 3.8, we find that

$$\left(P/A, i, 5\right) = 0.2374, \text{for } i = 6\%, \text{ and}$$

$$= 0.2439, \text{for } i = 7\%$$

Conducting the interpolation as explained in Example 3.8, we get in this case

$$\frac{0.24 - 0.2374}{0.2439 - 0.2374} = \frac{af}{7 - 6}$$

---

[14] The diagram may be physically absent but conceptually present—instead of being on paper, it may be in your head! Get into the habit of developing a mental picture of the diagram before you begin the solution.

$$af = \frac{0.0026}{0.0065}$$

$$= 0.4$$

Therefore, $i = 6\% + 0.4$

$$= 6.4\%$$

Thus, the bank is charging Ray's company an annual interest rate of approximately 6.4%.

4.4 ARITHMETIC SERIES

In the previous section, we discussed problems where cashflows remain constant, forming uniform series. In this and the next section, we consider cases where cashflows increase with time period. When the increase is constant, the cashflows generate an arithmetic *series*. Consider payments of $100 at period 1, $120 at period 2, $140 at period 3, and so on. In here, the increase in successive payments is constant, being $20. Thus, the cashflows $100, $120, $140, $160, ... form an arithmetic series. You must have come across such a series[15] in algebra.

As an industrial example, maintenance costs usually increase with time and may form an arithmetic series. Increasing utility costs, insurance costs, and other similar costs of doing business may also form an arithmetic series.

In most problems involving an arithmetic series, we are interested to know the equivalent present cost of a series of future costs. This helps the engineer in budgeting and also in making decisions based on a comparison of present costs for the alternatives, say machine A and B.

Let us consider the cashflow diagram of Fig. 4.6. In (a), the first cashflow A is at period 1, which may represent the first-year maintenance cost of a machine. The cost increases each year by G, forming the series of Fig. 4.6(a). At period 2 the cost is A+G, at period 3 it is A+2G, and so on; finally at the nth period it is A+(n–1)G. Note that the uniform series discussed in Section 4.3 can be considered a special case of arithmetic series where G = 0.

What we are interested in is to know the present cost implication of the increasing future costs. In other words, how much do we budget now in a maintenance account, i.e., the value of P in Fig. 4.6(b) to pay for the increasing future costs in Fig. 4.6(a). We need to derive an expression that relates P with the arithmetic series payments A, A+G, A+2G, ..., A+(n–1)G. We may also be interested in the equivalent future value R of the costs, as shown in Fig. 4.6(c).

4.4.1 Relationship between P and G

Consider the cashflow diagram combining Fig. 4.6(a) and Fig. 4.6(c). This is shown in Fig. 4.7 as cashflow diagram L, where R is the future equivalent of the arithmetic series payments. We are interested to know the value of R at period n for the payments A, A+G, A+2G, ..., A+(n–2)G, and A+(n–1)G. Since A is common in all the cashflows of this series, we can consider diagram L as composed of two diagrams S and T. Diagram S is based on uniform payments A which are equivalent to Q at period n. Diagram T is based on increasing G, with the series being equivalent to F. Note that there is no cashflow in diagram T at period 1, since the entire cashflow A at this period has been accounted for in diagram S. Since

---

[15] The cashflows $800, $700, $600, ..., $100 also form an arithmetic series with a decrease of $100 (or an increase of –$100) in successive cashflows.

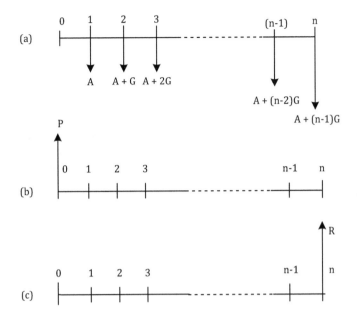

Figure 4.6 Arithmetic series.

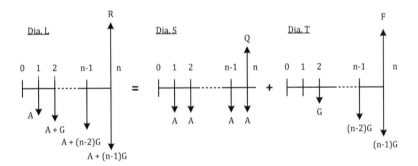

Figure 4.7 Two components of an arithmetic series.

cashflow totals at each period in diagrams S and T equal those in L, the original cashflow diagram, we can say that Q+F must equal R.

4.4.1.1 Finding P for Given G, i, n

Diagram S has been discussed in Section 4.3.1, and for which we have already derived equations. If we can derive now an expression for the diagram T, we should have an equation for diagram L, i.e., for the arithmetic series.

Diagram T requiring derivation is shown on its own in Fig. 4.8. The cashflow begins at period 2 with G, increasing by G and ending at n as (n–1)G. The derivation linking F with G, i, and n is given in Box 4.2, wherefrom

$$F = \frac{G}{i}\left[\frac{(1+i)^n - 1}{i} - n\right]$$

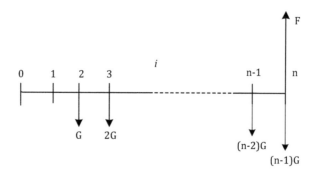

Figure 4.8 Analysis of arithmetic series.

Substituting for F from Equation (3.1), this expression reduces to

$$P(1 + i)^n = G \frac{(1+i)^n - ni - 1}{i^2}$$

$$P = G \frac{(1+i)^n - ni - 1}{i^2(1+i)^n} \tag{4.5a}$$

Any of the four methods discussed in Chapter 3 can be used to find P using this equation. In fact any of the four variables of Equation (4.5a) can be evaluated for given values of the other three. Since we prefer the functional notation method, Equation (4.5a) is expressed as

$$P = G \left(\frac{P}{G, i, n} \right) \tag{4.5b}$$

where

$$(P / G, \ i, \ n) = \frac{(1+i)^n - ni - 1}{i^2(1+i)^n}.$$

The notation (P/G,i,n) is called *Gradient Present Worth Factor* since it is used to find the present value of a series of future payments that increase arithmetically by G. Rather than use the long name for the functional factor, we can use the terminology *P/G-factor* or *P-factor for a given G*. The values of this factor are given in the interest tables under *Arithmetic Gradient*. The following example illustrates the solution procedure involving arithmetic series cashflows.

Example 4.4: Arithmetic Series

Under a government environment incentive plan companies installing air-pollution control equipment are entitled to a 50% grant to offset the purchase and maintenance costs. XYZ decides to avail of this incentive by acquiring equipment costing $50,000. The maintenance costs are estimated to be $4,000 in the first year, rising annually by $300. If the equipment's useful life is 10 years, and annual interest rate is 6% compounded half-yearly, how much will the grant be?

SOLUTION:

If you can comprehend the problem, there is no need to draw the cashflow diagram. Otherwise, draw one. Note that the maintenance cost increases every year by the same amount, so the maintenance costs form an arithmetic series, for which an equation has been derived above. We need to determine the equivalent present cost, 50% of which will be paid by the government as a grant.

The present cost comprises the purchase price and the present value of all the maintenance costs over 10 years. The latter is the equivalent P-value for an arithmetic-series cashflow with A = $4,000 and G = $300. For determining it, we can use Equation (4.5a) or (4.5b). Let us opt for functional-notation-based Equation (4.5b). Since the compounding is half-yearly, i is 3%, and for the same reason n is 20—number of compoundings over 10 years (if you have difficulty with these values of i and n, reread the relevant sections of Chapter 3). Thus, the relevant data are: G = $300, i = 3%, and n = 20, for which the total present cost on the equipment[16]

$$= \text{Purchase cost} + \text{Present value of maintenance costs}$$

$$= 50,000 + \left\{A(P/A,i,n) + G(P/G,i,n)\right\}$$

$$= 50,000 + 4,000(P/A,3\%,20) + 300(P/G,3\%,20)$$

$$= 50,000 + 4,000 \times 14.877 + 300 \times 126.799$$

$$= 50,000 + 59,508 + 38,040$$

$$= \$147,548$$

Being 50% of the total present cost, the grant will be $73,774.

BOX 4.2

Refer to Fig. 4.8 and note that there is no cashflow at period 1. The arithmetic series cashflows increase by a constant amount G. We are interested in determining the future sum F that is equivalent to the cashflows in this series. Note that a payment (n–1)G is made on the day F is due. The derivation should aim at relating F with the series payments: G, 2G, ..., (n–1)G at their respective periods.

Each cashflow of the series contributes toward the final sum F. For example, G at period 2 increases under compounding to become, let us say, F' at period n. Similarly, 2G becomes F", and so on. Thus, F will be equal to F'+ F" + ... We can now pair each series cashflow with its contributing portion and apply the basic expression, Equation (3.1), to each pair. Thus,

$$F' = G (1+i)^{n-2}$$

$$F" = 2G (1+i)^{n-3}$$

...

---

[16] See Fig. 4.7.

Once each payment has been paired, as above, we can write

$$F = F' + F'' + \dots$$

$$F = G(1+i)^{n-2} + 2G(1+i)^{n-3} + \dots + (n-2)G(1+i) + (n-1)G$$

Multiplying both sides by $(1 + i)$, we get:

$$(1+i)F = G(1+i)^{n-1} + 2G(1+i)^{n-2} + \dots + (n-2)G(1+i)^{2} + (n-1)G(1+i).$$

Subtracting the expression for F from that for $(1 + i)F$, we get:

$$(1+i)F - F = G(1+i)^{n-1} + 2G(1+i)^{n-2} + \dots + (n-2)G(1+i)^{2} + (n-1)G(1+i)$$

$$-G(1+i)^{n-2} - \dots - (n-3)G(1+i)^{2} - (n-2)G(1+i) - (n-1)G$$

$$iF = G(1+i)^{n-1} + G(1+i)^{n-2} + \dots + G(1+i)^{2} + G(1+i) - (n-1)G$$

$$= G(1+i)^{n-1} + G(1+i)^{n-2} + \dots + G(1+i)^{2} + G(1+i) + G - nG$$

$$= G\left\{(1+i)^{n-1} + (1+i)^{n-2} + \dots + (1+i)^{2} + (1+i) + 1\right\} - nG$$

The terms within { } add[1] up to $\dfrac{(1+i)^{n} - 1}{i}$, hence

$$iF = \frac{(1+i)^{n} - 1}{i} - nG$$

Dividing both sides by i,

$$F = \frac{G}{i}\left[\frac{(1+i)^{n} - 1}{i} - n\right]$$

Replacing F by Equation (3.1), we get:

$$P(1+i)^{n} = G\frac{(1+i)^{n} - ni - 1}{i^{2}}$$

$$P = G\frac{(1+i)^{n} - ni - 1}{i^{2}(1+i)^{n}}.$$

This relationship appears in the main text as Equation (4.5a).

---

[1] If interested, see any book on college algebra.

4.4.1.2 Finding G for Given P, i, n

This situation rarely arises in practice. But, if it does use Equation (4.5a) or (4.5b) to evaluate G, by substituting the known values and manipulating the terms.

4.4.1.3 Finding i for Given P, G, n

Finding n for Given P, G, i

Equation (4.5a) or (4.5b) can be used to evaluate the unknown *i* for given values of P, G, and n, or unknown n for given values of P, G, and *i*. Equation (4.5a) is suitable for the use of scientific calculators whereas Equation (4.5b) of the functional factors (interest tables). As explained in Chapter 3, the use of functional factors to evaluate *i* or n involves some interpolation, whose basic procedure has been illustrated in Examples 3.8 and 3.9.

4.4.2 Relationship between A and G

Substitution of P in terms of A, *i*, and n from Equation (4.4a) reduces Equation (4.5a) to $i(1 + i)^n$:

$$A\frac{(1+i)^n - 1}{i(1+i)^n} = G\frac{(1+i)^n - ni - 1}{i^2(1+i)^n}$$

$$A = G\frac{(1+i)^n - ni - 1}{i[(1+i)^n - 1]} \tag{4.6a}$$

4.4.2.1 Finding A for Given G, i, n

Again, any of the four methods discussed in Chapter 3 can be used to determine A using this equation. In fact any of the four variables of Equation (4.6a) can be evaluated for given values of the other three. Since we prefer the functional notation method, we rewrite Equation (4.6a) as

$$A = G\left(\frac{A}{G,i,n}\right) \tag{4.6b}$$

where

$$(A/G,i, n) = \frac{(1+i)^n - ni - 1}{i\left[(1+i)^n - 1\right]}$$

The notation (A/G,*i*,n) is called *Gradient Uniform Series Factor* since it is used to determine the equivalent uniform cashflow A of a series of cashflows increasing arithmetically by G. Rather than use the long name for the functional factor, we use the terminology A/G-*factor* or A-*factor for a given G*. The values of this factor are also given in the interest tables under *Arithmetic Gradient*.

4.4.2.2 Finding G for A, i, n

This situation rarely arises in practice. But, if it does use Equation (4.6a) or (4.6b) to evaluate G, by substituting the known values and manipulating the terms.

4.4.2.3 Finding i for Given A, G, n

Finding n for Given A, G, i

Equation (4.6a) or (4.6b) can be used to evaluate the unknown i for given values of A, G, and n, or unknown n for given values of A, G, and i. Equation (4.6a) is suitable for the use of scientific calculators whereas Equation (4.6b) of the functional factors in the tables. As explained in Chapter 3, the use of functional factors to evaluate i or n does involve some interpolation, whose basic procedure has been illustrated in Examples 3.8 and 3.9.

4.5 GEOMETRIC SERIES

In the previous section we discussed cashflows that form arithmetic series. In this section, we consider situations where they form *geometric series*. You must have studied geometric series in algebra. In geometric series, the cashflows vary in such a way that the magnitude at any period is equal to that at the previous period multiplied by a certain factor, which is the same for all cashflows of the series. Consider cashflows of $100 at period 1, $120 at period 2, $144 at period 3, and so on. These cashflows increase by a factor of 1.2 ($120 is 1.2 times the previous cashflow of $100; similarly $144 is 1.2 times of $120). The factor by which cashflows change is called *common ratio* and is denoted by g. As another example, the payments $40, $60, $90, $135, ... form a geometric series with g = 1.5. Costs on maintenance, utilities, and insurance may increase geometrically, especially in an inflationary economy. If g is less than one—rare in engineering economics—the cashflows decrease. Note that the uniform series discussed in Section 4.3 can be considered a special case of geometric series where g = 1.

For geometric series cashflows too, appropriate equations can be derived, as in Box 4.3. Summarized below are the formulas applicable to problems involving geometric series. Note that A_1 is the first cashflow of the series, as shown in Fig. 4.9.

$$P = A_1 \frac{1 - (1 + g)^n (1 + i)^n}{i - g}, \text{ when } i \neq g \tag{4.7}$$

When the interest rate i and the common ratio g are equal, the appropriate relationship is:

$$P = A_1 \frac{n}{1 + i}, \text{ when } i = g \tag{4.8}$$

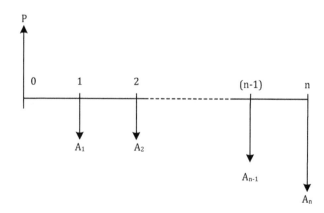

Figure 4.9 Geometric series.

BOX 4.3

Refer to Fig. 4.9 where cash outflows A_1, A_2, ..., A_{n-1}, A_n forming a geometric series. They are related to the first cashflow A_1 through g and the value of n, which is 1, 2, ..., n, as:

$$A_2 = A_1 + A_1 g = A_1(1+g)$$

$$A_3 = A_2 + A_2 g = A_2(1+g) = A_1(1+g)(1+g) = A_1(1+g)^2$$

$$A_n = A_1(1+g)^{n-1}$$

By substituting the value of n, any intermediate cashflow can be related to A_1, the first cashflow. For example, $A_4 = A_1(1+g)^3$ and $A_9 = A_1(1+g)^8$.

Our task is to determine the value of P that is equivalent to the cashflows A_1, A_2, ..., A_{n-1}, A_n. In other words, how much do we budget now to pay for the n future costs increasing geometrically?

Conceptually, each payment of the series is made out of P. The sum P can thus be conceived to comprise portions P', P'', ... such that P equals the total of these portions. In other words,

$$P = P' + P'' + ...$$

We can pair each cashflow with its corresponding portion of P and apply the basic expression, Equation (3.1), to each pair. For A_1 and P', we have

$$A_1 = P'(1+i)$$

Or,

$$P' = A_1(1+i)^{-1}$$

$$= a$$

where

$$a = A_1(1+i)^{-1}$$

Similarly,

$$A_2 = P''(1+i)^2$$

Or,

$$P'' = A_2(1+i)^{-2}$$

Since $A_2 = A_1(1+g)^1$, due to the geometric relationship illustrated above, its substitution gives:

$$P'' = A_1(1+g)^1(1+i)^{-2}$$

$$= A_1(1+i)^{-1}\{(1+g)/(1+i)\}$$

$$= ab$$

where

$$b = \left\{ (1+g)/(1+i) \right\}$$

Similarly, $= A_3(1+i)^{-3}$

$$= A_1(1+g)^2(1+i)^{-3}$$

$$= A_1(1+i)^{-1}\left\{ (1+g)/(1+i) \right\}^2$$

$$= ab^2$$

Once each portion of P has been paired with the corresponding cashflow of the series, as just shown, P = P' + P" + … gives:

$$P = a + ab + ab^2 + \dots + ab^{n-2} + ab^{n-1}$$

Multiplying both sides by b, we get:

$$bP = ab + ab^2 + ab^3 + \dots + ab^{n-1} + ab^n$$

Subtracting the expression for bP from that for P, we get:

$$P - bP = a + ab + ab^2 + \dots + ab^{n-2} + ab^{n-1}$$
$$- ab - ab^2 - \dots - ab^{n-2} - ab^{n-1} - ab^n$$
$$(1-b)P = a - ab^n$$

Rearranging the above,

$$P = a\frac{1-b^n}{1-b}$$

On substituting the values of a and b, the above reduces to

$$P = \frac{A_1}{1+i} \times \frac{1 - \left(\dfrac{1+g}{1+i}\right)^n}{1 - \dfrac{1+g}{1+i}}$$

$$P = A_1\frac{1 - (1+g)^n(1+i)^{-n}}{i-g}$$

This is Equation (4.7) in the main text.

Representation of Equations (4.7) and (4.8) into functional factors so that tables can be used, as done earlier for other equations, becomes clumsy due to the additional variable g. Therefore, we use these two equations as they are to solve geometric series problems. One of these equations, depending on whether $i = g$ or not, is used to evaluate any of the variables P, A_1, i, g, or n provided the values of the others are given. Also note that once A_1 has been evaluated (or is known as given), any of the cashflows in the geometric series can be determined using $A_n = A_1(1+g)^{n-1}$, by substituting the particular value of n.

Example 4.5: Geometric Series

An international company appoints a 30-year Harvard MBA as its new president on a 5-year contract at a salary of $450,000. The salary is to increase each year by 9%. On the first day of the job, a serious past misconduct is discovered about the incumbent. The board of directors therefore decides to fire him the same day. However, as stipulated in the employment contract, he must be paid the present value of the salaries along with a golden handshake in the sum of $500,000. If i = 10%, how much is he paid?

SOLUTION:

The analysis period is 5 years. We can draw a cashflow diagram either from the president's viewpoint or the company's. The answer will be independent of the viewpoint considered. Fig. 4.10 shows the company's diagram with cashflows as costs. The president's initial salary of $450,000 increases at 9% per year. So the value of A_1 is $450,000, and g = 9% = 0.09. The P-value of the salaries plus the $500,000 golden handshake is the total payment to the fired president.

There are two ways to determine the P-value of the salaries. In one, we determine the present value of A_1, A_2, ..., and A_5 individually using the basic relationship $F = P(1 + i)^n$, and add them. In the second approach, we exploit the geometric series the cashflows form, using the equation derived in this section. Let us opt for the equation approach.

With i = 0.10 and g = 0.09, $i \approx$ g, and hence Equation (4.7) is applicable. This equation will yield the P-value of A_1, A_2, ... A_5 at period 0. With n = 5, the P-value of the future salaries is:

$$A_1 \frac{1-(1+g)^5(1+i)^{-5}}{i-g}$$

$$= 450,000 \frac{1-(1+0.09)^5(1+0.10)^{-5}}{0.10-0.09}$$

$$= 450,000 \frac{1-(1.5386 \times 0.6209)}{0.01}$$

$$= 450,000 \times 4.46833$$

$$= \$2,010,749$$

$$\text{Total payment} = \text{P} - \text{value of salaries} + \text{golden handshake}$$

$$= \$2,010,749 + \$500,000$$

$$= \$2,510,749$$

The "sacked" president is thus paid $2,510,749.

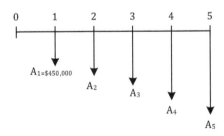

Figure 4.10 Diagram for Example 4.5.

4.6 "TAILORING"

We discussed in the previous sections three common types of cashflow patterns or series: uniform, arithmetic, and geometric. Whenever cashflows follow any of these patterns, the associated equation is invariably used for the solution.

Often we come across cashflows that do not follow the pattern exactly. Such *approximate patterns* can be "tailored" to become one of the three series. Tailoring is done to achieve efficiency of solution offered by the series equations, as illustrated through Examples 4.6 and 4.7.

Example 4.6: Tailoring Cashflow Diagrams

Evaluate P in the cashflow diagram represented by Fig. 4.11(a) for $i = 8\%$.

SOLUTION:

In the given cashflow diagram, P is the present sum, while other cashflows are future sums. The evaluation of P involves converting each future sum in today's value and adding them vectorially. Vectorial addition means taking care of the cashflow direction, i.e., whether the cashflows are above or below the timeline. In this case, the present value of 40 at period 5 will be added to P, the resulting total becoming the present worth of costs. This will be equated with the present worth of benefits by totaling the P-values of the cashflows with upward vectors.

The concept of cost-benefit equivalency is simplified by considering a cashflow diagram as a classical weighing balance. The cashflows may be considered as weights, provided the time value of money is taken into account. The upward vectors, all assembled at any period, may be thought of as being on one pan of the balance, while the downward vectors assembled at the same period on the other pan. The cashflows' equivalency is similar to the balancing of the two pans. This weighing-balance concept makes it easier to set up the cost-benefit relationship. In the present illustration, the cashflow P and P-value of 40 as weights are on one pan (at period 0), while the P-values of the other cashflows are on the other pan (at the same period 0). Thus,

$$P + P - \text{value of } 40 = \text{Total of the P} - \text{values of other cashflows}$$

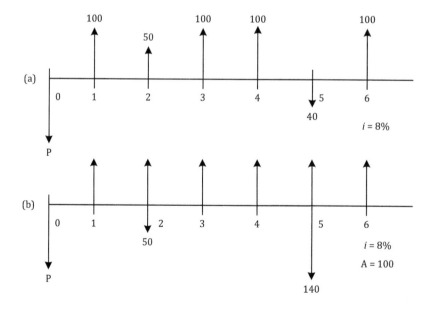

Figure 4.11 Tailoring.

Accounting for the P-values of the cashflows using Equation (3.7a), we get:

$$P + 40(P/F, 8\%, 5) = 100(P/F, 8\%, 1) + 50(P/F, 8\%, 2)$$
$$+ 100(P/F, 8\%, 3) + 100(P/F, 8\%, 4) + 100(P/F, 8\%, 6)$$

Such a solution approach involves evaluation of six functional factors in this example. While this approach is perfectly alright it may be called the *straightforward method* since there is no attempt to achieve solution conciseness offered by the cashflow pattern. Note that the pattern displays a "touch" of uniformity—it does look like a uniform series. It seems that this approximate series can be tailored.

TAILORING THE CASHFLOW DIAGRAM

A closer look at the cashflow pattern in Fig. 4.11(a) suggests that it is approximately of the uniform type discussed in Section 4.3. Of the six future cashflows, four are of the same magnitude, 100 each. If we can modify the cashflows at periods 2 and 5 to be 100 each and vertically up, then we can use the equation for uniform series.

Let us apply the weighing-balance concept mentioned earlier. At period 2 the vector magnitude is 50 up. Let us increase it to 100. Since we are increasing it up by 50, i.e., placing a weight of 50 on one pan of the balance, we need to put a weight of the same magnitude on the other pan, i.e., add a downward vector of 50. In other words, the given upward vector 50 at period 2 can be imagined to be two vectors, one of 100 upward and the other of 50 downward. The modified vectors 100 up and 50 down are equivalent to 50 up, since 100 − 50 = 50.

Applying the same concept at period 5, the downward vector 40 is equivalent to two vectors, one up of 100 and the other down of 140. With these modifications at periods 2 and 5, the *tailored* cashflow diagram looks like as in Fig. 4.11(b). The two cashflow diagrams in Fig. 4.11 are thus equivalent; the solution for P based on Fig. 4.11(b) must therefore yield the same result as that based on Fig. 4.11(a).

To appreciate the benefit of tailoring, let us now base our solution for P on the modified cashflow diagram of Fig. 4.11(b). The uniform series cashflows streamline the conversion of the upward vectors using just one P/A-factor. We still have to convert the two downward cashflows (50 and 140) using P/F-factors. Taking care of the vector directions, we get from cost-benefit equivalency:

$$P + 50(P/F, 8\%, 2) + 140(P/F, 8\%, 5) = 100(P/A, 8\%, 6)$$

Compare this expression with the previous one, by the straightforward method. Due to tailoring we now have to evaluate only three functional factors as compared to six earlier. Thus, tailoring has enhanced solution efficiency.

Reading the values of the factors off the interest tables and substituting them in the tailored expression, we get

$$P + 50 \times 0.8573 + 140 \times 0.6806 = 100 \times 4.623$$

Thus,

$$P = 100 \times 4.623 - 50 \times 0.8573 - 140 \times 0.6806$$
$$= 462.30 - 42.87 - 95.28$$
$$= 324.15$$

In this example, tailoring proved to be a better approach since it saved time and effort. Had the number of future cashflows been fewer (three or less), tailoring would not have been beneficial; the straightforward approach would have been equally efficient. In general, the larger the number of cashflows in the approximate series and fewer the modifications required, the more attractive the tailoring.

Example 4.7: Tailoring Cashflow Diagrams

Evaluate F in Fig. 4.12 for $i = 10\%$.

SOLUTION:

Again, there are two approaches to evaluate F in Fig. 4.12. One is the straightforward approach in which each cashflow is converted to its F-value, using Equation (3.6a). The total of these F-values will yield the answer. Since there are six cashflows this approach will involve six functional factors.

The other approach is to attempt to exploit any pattern in the cashflows for which an equation exists. This approach begins by looking closely at the cashflow diagram to find what patterns exist, if any. In this case, the cashflows at periods 1, 2, and 3 form arithmetic series, while those at periods 4, 5, and 6 form uniform series[17]. It seems that we may be better off tailoring the diagram.

TAILORING THE DIAGRAM

Referring to the diagrams pertaining to arithmetic or uniform series (Fig. 4.6(a) and 4.4), we can notice that

1. An equivalent cashflow located one period left of period 0 can easily be found for the arithmetic-series cashflows at periods 1, 2, and 3 by using P/G factor. Let us call this equivalent cashflow P_1.
2. Also, an equivalent cashflow located at period 3 can easily be determined for the uniform series cashflows at periods 4, 5, and 6 by using P/A factor. Let us call this equivalent cashflow P_2.

Once P_1 and P_2 have been determined, they can then be transferred to period 7 as F-values using F/P factors. The total of these F-values will equal F at period 7.

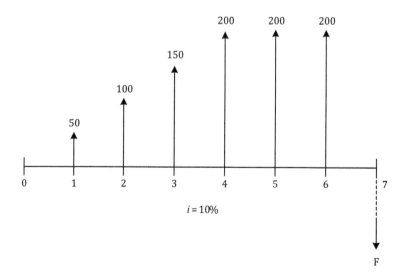

Figure 4.12 Diagram for Example 4.7.

---

[17] One can include the cashflow at period 4 with the arithmetic series.

Following this logic, since G = 50 and n = 4, we get

$$P_1 = 50 \; (P/G, \; 10\%, \; 4)$$
$$= 50 \times 4.378$$
$$= 218.9$$

Note that P_1 is located one period left of period 0.

Next, for the cashflows at periods 4, 5, and 6 forming a uniform series, A = 200. The formula we have derived for P/A factor yields P at just left of the first A, in this case at period 3. Noting that for the series n = 3, we get:

$$P_2 = 200 \; (P/A, \; 10\%, \; 3)$$
$$= 200 \times 2.487$$
$$= 497.4$$

Now we transfer P_1 and P_2 to their equivalent values at period 7 and add the transferred values. For the former n = 8 (since P_1 is at one period left of 0), while for the latter n = 4. Thus,

$$F = 218.9 \; (F/P, 10\%, 8) \; + \; 497.4 \; (F/P, 10\%, 4)$$
$$= 218.9 \times 2.144 \; + \; 497.4 \times 1.464$$
$$= 469.32 \; + \; 728.19$$
$$\approx 1,198$$

Note that the tailoring involved evaluation of four functional factors; the straightforward method would have involved six. Thus, tailoring has been beneficial.

4.7 SUMMARY

Most engineering economics problems involve multiple cashflows. They may be one of the three basic types: one-cost-several-benefits, several-costs-one-benefit, or several-costs-several-benefits. Basic considerations in analyzing such problems have been discussed in Chapter 4. Multiple cashflows may be regular or irregular. They are regular if they occur at each period. Regular cashflows may or may not display patterns, as evidenced in their diagrams or tables.

Problems with no pattern in their cashflows are analyzed by treating each cashflow separately, using the basic formula, Equation (3.1), or its functional notation. Those with patterns are analyzed using the relevant equations.

Most common patterns in cashflows form one of the three series: uniform, arithmetic, or geometric. These three series and their equations have been discussed in Chapter 4. For the uniform and arithmetic series cashflows, the functional-notation method is available to the analyst. For the analysis of geometric series cashflows, the functional notation method is complicated, and hence only formulas are used.

When the cashflows deviate slightly from one of the three standard patterns, there are two choices. One is to ignore their proximity to the pattern and use the basic equation. This is the straightforward method of analysis. The other is to *tailor* the given cashflows so that they get modified to fit in one of the three series patterns, and then use the equation for the pattern. The analyst has to judge which of the two choices makes better sense. Section 4.6 presented discussions and examples on *tailoring* which is a challenging creative task.

May I remind the readers that all the derivations in this chapter were based on Equation (3.1), emphasizing once again that this equation is really the most *basic* formula in engineering economics.

DISCUSSION QUESTIONS

4.1 Give an example from recent personal financing where multiple payments were involved.

4.2 In regular cashflow problems, what is *regular*: cashflow or time period or both? Explain your answer with an example.

4.3 Which pattern—arithmetic or geometric series—is more likely in industrial problems and why? Consider maintenance and utility costs as examples.

4.4 In evaluating i or n in a problem, which one—functional notation or calculator method—do you prefer and why?

4.5 Why is functional notation approach not pursued for geometric series cashflows?

4.6 Compare and contrast uniform series and arithmetic series cashflows.

4.7 What are some possible factors that result in an arithmetic series of cashflows?

4.8 Compare and contrast arithmetic series and geometric series cashflows.

4.9 Why would a cashflow series be tailored? Give an example.

4.10 In your opinion, does tailoring the cashflow series diagram result in an accurate representation of the cashflow data?

MULTIPLE-CHOICE QUESTIONS

4.11 In regular cashflow problems
   a. cashflows must be the same at each time period
   b. a cashflow must occur at each time period
   c. cashflows cannot be unequal
   d. none of the above

4.12 Multi-cashflow problems are always
   a. regular
   b. irregular
   c. in arithmetic series
   d. none of the above

4.13 Referring to Fig. 4.1, the present value of cashflow T is
   a. $T(P/T,i,3)$
   b. $T(F/P,i,3)$
   c. $T(P/F,i,3)$
   d. $T(T/P,i,3)$

4.14 If you set aside 5% of your fixed earnings into a retirement plan, the cashflows will form
   a. a uniform series
   b. an arithmetic series
   c. a geometric series
   d. a logarithmic series

4.15 Which of the following equations is correct?
   a. $F = P(P/F,i,n)$
   b. $P = A(A/P,i,n)$
   c. $F = P(F/A,i,n)$
   d. $P = A(P/A,i,n)$

4.16 For the geometric series with $A_1 = \$100$ and g = 10%, the value of A_6 will approximately be
   a. $ 60
   b. $110
   c. $160
   d. $210

4.17 For the following cashflows the correct relationship is:

Year	0	1	2	3	4	5
Cashflow	−$300	$80	$120	$160	$200	$25

   a. $300 = 80\ (P/A,i,4) + 25\ (P/F,i,5)$
   b. $300 = 80\ (P/A,i,5 - 215\ (P/F,i,5)$
   c. $300 = 80\ (P/A,i,5) + 105\ (P/F,i,5)$
   d. none of the above

4.18 The uniform series is a special case of geometric series with
   a. $g = 0$
   b. $g = 1$
   c. $g = -1$
   d. none of the above

4.19 What is i with the given, $G = \$7,500$, $P = \$254,110$, and $n=10$?
   a. 5%
   b. 4%
   c. 3.25%
   d. 8%

4.20 What is n with the given, $i = 5\%$, $P = \$85,976$, $G = \$450$?
   a. 6
   b. 7.5
   c. 8
   d. 5

4.21 Find i given A = \$1,391, G = \$500, and n = 8.
   a. 4.5%
   b. 13%
   c. 6%
   d. 15%

4.22 Find n given A = \$3,189, G = \$1,200, and $i = 9\%$.
   a. 7
   b. 13
   c. 4
   d. 3.75

4.23 Find P where $A_1 = \$300,000$, $g = 7.5\%$, $i = 12\%$, n = 7, $(i \neq g)$
   a. \$1,300,000
   b. \$2,345,000
   c. −\$ 75,000
   d. \$1,663,536

4.24 What is P when $A_1 = \$60,000$, $g = 5.5\%$, $i = 8\%$, n = 10, $(i = g)$
   a. \$75,000
   b. \$54,786
   c. \$50,117
   d. \$85,956

4.25 What type of cashflow series would have a constant increase in regular cashflows?
   a. geometric series
   b. uniform series
   c. arithmetic series
   d. none of the above

4.26 In what type of series does the cashflow increase vary in such a way that the magnitude of any previous period is equal to the previous period multiplied by a certain factor?
  a. geometric series
  b. uniform series
  c. arithmetic series
  d. all of the above

NUMERICAL PROBLEMS

4.27 Evaluate F in Fig. 4.13 by the straightforward method assuming $i = 10\%$.

4.28 Monica wins one million dollars in a state lottery that cost her one dollar just a week ago for the ticket. She has two options: to receive the million dollars now, or $80,000 per year for the rest of her life. If she expects to live for another 25 years, which option is better for Monica? Assume that the applicable annually compounded interest rate is 10% per year.

4.29 For a mortgage loan of $100,000 Ismayel's quarterly payment for the last 20 years has been $2,000. If he has to continue to pay this amount for another 10 years to fully pay back the loan, what is the nominal interest rate?

4.30 The purchasing department is planning to buy a personal computer system at a cost of $10,000. The system is expected to have a useful life of 5 years during which its updating and maintenance costs will be:

Year	1	2	3	4	5
Cost	$400	$500	$600	$700	$800

How much should it budget for both the purchase and upkeep of the system, if the yearly interest rate is 12%, compounded annually?

4.31 John began a retirement plan by investing $2,000 15 years ago. Each subsequent year's investment was $200 more than the previous year's. Due to the excessive health care costs on his wife, John could not invest in the fifth and seventh year. The investments for years 6 and 8 were $3,000 and $3,400 respectively. How much is there in the account today after this year's investment? Assume an annually compounded interest of 8% per year?

4.32 Twenty-First Century Manufacturers are planning to acquire a new machine that will cost $5,000. The machine's expected life is 10 years. It is expected to be trouble-free

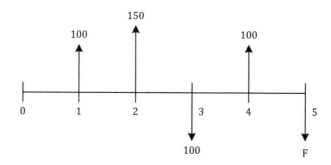

Figure 4.13   Diagram for Problems 4.14 and 4.34.

during the first year; the maintenance costs for the subsequent 4 years are estimated to be $100, $200, $300, and $400 respectively. It will require an overhaul half way through its life at a cost of $2,000. The maintenance costs following the overhaul are expected to be $200, $220, $242, $266.20, $292.82, and $322.10. How much should be budgeted to pay for the purchase and future overhaul and maintenance of the machine? Assume an interest rate of 10% per year, compounded annually.

4.33 If in Fig. 4.1, T = 2S, R = 2.5S, and Q = 8S, what is the interest rate assuming compounding?

4.34 Evaluate F in Fig. 4.13 by tailoring the diagram to exploit the most appropriate cash-flow pattern. Comparing this solution approach with the straightforward approach of Problem 4.14, has tailoring been beneficial? By how much?

4.35 Evaluate D in Fig. 4.14 by tailoring the diagram.

4.36 Evaluate P in Fig. 4.15 by tailoring the diagram.

4.37 Stucco Unlimited signed an 8-year contract with an engineer firm to handle all their projects for $400,000. The contract is to increase by 5% each year. After 2 years,

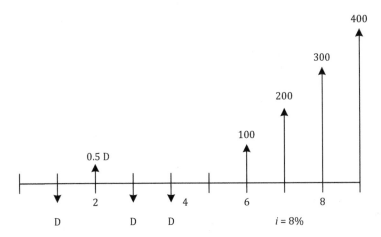

Figure 4.14 Diagram for Problem 4.35.

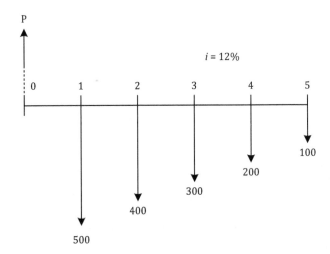

Figure 4.15 Diagram for Problem 4.23.

the construction firm reevaluates the contract and decides to end the relationship. However, as per the contract the construction firm must pay the present value of all future payments with an additional penalty of 7.5% of total present value of future payments. If $i = 8\%$, how much will the engineer firm be paid?

4.38 Due to a new state initiative, manufacturing companies are receiving tax incentives for the purchase of wind turbines to augment plants power usage. Plastics Unlimited plans to take advantage of the incentive by purchasing two wind turbines for $50,000 each. The estimated maintenance expense on the turbines is $2,900 each per year and useful years of service are expected at 15 years. Maintenance expense is projected to increase by 3% each year. Plastics Unlimited is able to write off 25% of the total cost. How much will Plastics Unlimited be able to write off?

4.39 Now assume that the interest rate is unknown for Problem 4.38 and that Plastics Unlimited purchased only one of the wind turbines. Would you be able to calculate the tax incentive?

4.40 Highwaymen Road and Bridge Construction Company decides to take out a loan for $60,000 to buy a piece much needed equipment. The Highwaymen is required to pay $14,000 per year for the next 5 years. The first payment is due on the first loan anniversary. What is the simple interest rate the bank charging?

4.41 Draw the cashflow diagram for Example (4.4), assuming that the air-pollution control equipment has a salvage value of $10,000.

4.42 Lost At Sea boat builders want to purchase a state of the art boat designing software program. The program currently sells for $7,500. However, the builders expect the price to drop 3% each year. The builders plan on purchasing the program in 3 years. The management plans to set aside a fixed sum of money in an account bearing 4.5% interest. How much money should be retained each year in order to make the three year goal?

4.43 A surveying company has ordered a new piece of equipment that will arrive in the next fiscal year. The new item costs $8,000 and has an expected useful life of 7 years. The maintenance costs for the piece of equipment follow in the table. How much should be budgeted in total for the piece of equipment and maintenance costs if the interest rate is 5% compounded annually?

Year	Main. Cost
1–2	$0
3	$400
4	$500
5&6	$550
7	$700

4.44 Sunrise Company wants to save $50,000 over the next 3 years in order to replace several crucial pieces of its distribution fleet. The amount will be retained from annual profits. If the amount is held in an account bearing 4% interest compounded annually, how much should be retained for the initial balance?

4.45 A widget manufacturing plant manager decides to purchase a new drill press with borrowed funds in order to improve efficiency at the plant. The drill press costs $20,000. The operating expense for the press in the first year is projected to be $3,500, increasing annually at a rate of 4% in subsequent years. The drill press has a useful life of 8 years and the yearly interest rate is 7.5% compounded annually, how much will total costs be for the drill press?

FE EXAM PREP QUESTIONS

4.46 What is i with the given, G = \$200, P = \$7,927, and n = 10?

   a. 5%

   b. 4%

   c. 1.75%

   d. 8%

4.47 What is n with the given, i = 3.5%, P = \$3,792, G = \$450?

   a. 6

   b. 7.5

   c. 8

   d. 5

4.48 Find i given A = \$1,917, G = \$600, and n = 8.

   a. 4.5%

   b. 13%

   c. 6%

   d. 15%

4.49 Find n given A = \$3,128, G = \$700, and i = 12%.

   a. 7

   b. 13

   c. 4

   d. 3.75

4.50 Find P where A_1 = \$250,000, g = 5.5%, i = 9%, n = 7, $(i \neq g)$

   a. \$1,250,000

   b. \$ 745,000

   c. -\$ 75,000

   d. \$1,458,862

4.51 What is P when A_1 = \$33,000, g = 3%, i = 6%, n = 10, $(i = g)$

   a. \$274,521

   b. \$165,786

   c. \$373,000

   d. \$256,956

4.52 What type of cashflow series would have a constant increase in regular cashflows?

   a. geometric series

   b. uniform series

   c. arithmetic series

   d. none of the above

4.53 Find P where A_1 = \$20,000, g = 4%, i = 12%, n = 8, $(i \neq g)$

   a. \$300,000

   b. \$111,815

   c. -\$ 75,000

   d. \$450,000

4.54 What is i with the given, G = \$1,525, P = \$31,916, and n = 6?

   a. 5%

   b. 4%

   c. 3.25%

   d. 8%

Part II

Criteria

After having laid the *foundation* of fundamental principles relating to cashflows in Part I, we are ready to *build the structure* in Part II. As the structure depends on its foundation, so do the discussions of this part on those of Part I. In the next seven chapters, we discuss the criteria by which the best alternative is selected from among the many feasible ones. Six different criteria are practiced in industry; we discuss each of them in separate chapters: payback period in Chapter 5, present worth in Chapter 6, future worth in Chapter 6, annual worth in Chapter 6, rate of return in Chapter 7, and benefit-cost ratio in Chapter 8. Finally, we compare the six criteria in Chapter 9.

Chapter 5

Payback Period

LEARNING OBJECTIVES

- Meaning of payback period
- Determination of payback period
- When to apply payback criterion
- Application of payback criterion
- Prevalence in industry
- Limitations of the payback method
- Time-valued payback period analysis

In the four chapters of Part I we discussed the basic principles underlying cashflows and their manipulations. We are now ready to solve engineering economics problems. To do that we need a criterion by which the best alternative will be selected from among the feasible ones. There are six different criteria commonly practiced by industry. In this chapter we learn about one of them, called *payback period*. The other decision criteria are discussed in the next five chapters.

The payback period method of solving engineering economics problems is unique. This is the only one of the six methods that does not require much of what has been discussed in Part I. It differs from the other five, presented in Chapters 6 through 10, in that it ignores the *time value of money* whereas others don't.

5.1 MEANING

Payback period is the duration in which an investment pays for itself. Investments on equipment and other resources are made with the expectation of profit or net income. The time it takes the increased profit to pay back the capital invested is the payback period. Consider a company that invests in an automatic tool changer at a cost of $5,000. The tool changer is expected to increase production, generating $2,000 of additional profit per year. Thus, the $5,000 investment will be paid back in 2.5 years. This time duration of 2.5 years is the payback period of the investment in the tool changer.

Payback period can thus be expressed as:

$$\text{Payback Period} = \frac{\text{Investment Cost}}{\text{Net Income per Unit Time}} \tag{5.1}$$

In this denominator we have used the term *net income*. This is due to the fact that the invested equipment may involve operating costs, for example for maintenance, insurance, and utilities[1]. The net income is the difference between gross income from the invested equipment and its total operating cost. Note also that it is expressed in terms of *per unit time*, usually per year. Being in the denominator, this unit of time becomes the unit of payback period. If the net income is expressed as per month, the payback period will be in months. Instead of the term *investment cost* in the numerator of Equation (5.1), other synonymous terms may also be used. For example, for a new machine the term *first cost* is equally appropriate.

Example 5.1: Payback Period

A hydraulic machine is being planned for acquisition at a cost of $5,000. During its useful life of 7 years, it is expected to generate an annual gross income of $3,400. The cost of space to house the machine and of the utilities is $900 per year. Its maintenance cost is projected to be $300 per year. Determine the payback period for the machine.

SOLUTION:

Here, the first cost (or investment) is $5,000. The recurring expenditure comprises the space and utility cost of $900 and the maintenance cost of $300, both per year. Thus, the total cost in utilizing the machine is $900 + $300 = $1,200 per year. The net income from investment in the machine is obtained by subtracting this total cost from the gross income. This yields a net income of $3,400 – $1,200 = $2,200 per year. Therefore,

$$\text{Payback period} = \frac{\$5,000}{\$2,200\,\text{per year}}$$

$$= 2.27\,\text{years}$$

Note that there was a redundant data in Example 5.1, namely the machine's useful life of 7 years. This data was not needed for the solution. Engineering economics problems may sometimes contain redundant data. In illustrative or exercise problems, redundant data may intentionally be provided to judge the depth of your understanding of the subject matter. In industrial problems, they may creep in during data collection for the project. You must be able to identify any redundant data in a problem and boldly ignore them, rather than ponder on their relevance.

Note that in Example 5.1 *we did not account for the time value of money*. We treated the second year's net revenue at par with that of the first year. This is the major weakness of the payback period method since, as learned in Part I, money does have a time value.

The machine in Example 5.1 has a useful life of 7 years. With a payback period of 2.27 years, it returns the $5,000 investment well before the end of its life. It may continue to be used for another 4.73 years beyond the payback period. The payback period method ignores the benefits and costs beyond the payback period, which is another drawback.

Could the payback period be longer than the useful life of the invested resource? Such a case does not arise normally, since such projects are usually not funded. But if it did, the term *payback period* would lose its meaning since only a portion of the investment would be recovered. The result will be a *partial payback*.

In Example 5.1, the recurring cost of using the machine and the gross income were assumed to be constant. In several problems, these vary from year to year, yielding a variable net income. In that case we track the *cumulative net income*, and payback period is that cut-off duration at which cumulative net income equals the investment. Such problems are discussed next.

---

[1] This term is more prevalent in the United States. It means services such as electricity, gas, telephone, fax, Internet, janitors, etc., that are essential to the operation of the business/industry/plant.

5.2 VARIABLE CASHFLOWS

When gross income and operating cost data for the investment are variable, a tabular or graphical approach is more suitable. The operating cost of the invested resource is likely to vary since costs on maintenance, utilities, insurance, etc., usually increase with time. In such cases the data are summarized preferably in a table, similar to the cashflow table discussed in Chapter 2. Example 5.2 illustrates how problems with variable costs and benefits are solved using the *tabular approach*, while Example 5.3 illustrates the *graphical approach*.

Example 5.2: Payback Period

If for the machine in Example 5.1 expected gross incomes and operating costs are as follows, what is the payback period?

Year	Gross Income	Total Operating Cost
1	$3,500	$1,100
2	$3,350	$1,200
3	$3,600	$1,250
4	$3,200	$1,300
5	$3,450	$1,400
6	$3,400	$1,500
7	$3,350	$1,600

SOLUTION:

The investment remains the same, i.e., $5,000. Both the gross income and the annual operating cost vary[2]. For the net income, subtract the *total operating cost* from the *gross income*. The results are given as follows, in the last column, by extending the data table.

Year	Gross Income	Total Operating Cost	Net Income
1	$3,500	$1,100	$2,400
2	$3,350	$1,200	$2,150
3	$3,600	$1,250	$2,350
4	$3,200	$1,300	$1,900
5	$3,450	$1,400	$2,050
6	$3,400	$1,500	$1,900
7	$3,350	$1,600	$1,750

As evident in this table, net income from the machine is variable. Therefore, Equation (5.1) is not applicable. The solution approach in such a case is to track the *cumulative* value of the net income. The payback period corresponds to the time duration when cumulative net income equals the investment. The calculations can be streamlined by creating another column for the cumulative income[3]. The cumulative income at any point in time

---

[2] In rare cases if they vary by the same amount, their difference, i.e., the net income, will be constant. If so, use Equation (5.1).

[3] Cumulative income means cumulative net income.

is the total of all the previous incomes. As it increases, keep watching its value to check whether it has equaled or exceeded the investment.

Year	Gross Income	Total Operating Cost	Net Income	Cumulative Income
1	$3,500	$1,100	$2,400	$2,400
2	$3,350	$1,200	$2,150	$4,550
3	$3,600	$1,250	$2,350	$6,900
4	$3,200	$1,300	$1,900	
5	$3,450	$1,400	$2,050	
6	$3,400	$1,500	$1,900	
7	$3,350	$1,600	$1,750	

In the last column second row, the cumulative income of $4,550 has been obtained by adding the first year's net income of $2,400 to the second year's $2,150. The third year's cumulative income of $6,900 has been obtained by adding the third-year income $2,350 to the cumulative income at the end of second year ($4,550). Note that we stopped computation as soon as cumulative income exceeded the investment.

This table shows that by the end of the second year the machine recovers $4,550, which is less than the $5,000 investment. But, by the end of the third year, the machine recovers more than the investment.

So, what is the payback period? It is definitely greater than 2 years. Is it 3 years or somewhere in between 2 and 3 years? That depends on how the income is realized. If the profits and costs are accounted for at the end of the year, then the payback period is 3 years. However, if they are realized as "you go," i.e., on continuous basis, then the payback period is under 3 years. In such a case, it can be determined by interpolating between the two limits of 2 and 3 years, as follows:

$$\text{Payback period} = 2 + (5,000 - 4,550) / (6,900 - 4,550)$$

$$= 2 + 0.19$$

$$= 2.19 \text{ years}$$

Thus, the payback period is 2.19 years.

5.3 USE OF TABLE AND DIAGRAM

Are the cashflow table and diagram of much use in the payback period analysis? With some insight we can conceive that the evaluation of the payback period, especially where net income from the investment is variable as in Example 5.2, is based on the manipulation of the cashflow table.

Consider Example 5.1 once again. The problem can be restated through its cashflow table, as follows, where cashflows are the net incomes.

Year	0	1	2	3	4	5	6	7
Cashflow	−$5,000	$2,200	$2,200	$2,200	$2,200	$2,200	$2,200	$2,200

In here, $2,200 is the yearly net income (= $3,400 − $1,200) from the investment. One can use this table and follow the cumulative approach outlined in Example 5.2 to determine the payback period. This first-principles approach does away with the need for any formula, namely Equation (5.1).

The cashflow-table-based approach yields the same result for the payback period, as expected. The choice between this approach and the use of equation is a matter of preference. For investments yielding constant net income, Equation (5.1) is convenient, and therefore preferred. Where incomes from the investment are variable, the standard formula, Equation (5.1), is inapplicable. The tabular approach illustrated in Example 5.2 is the only choice.

Like the cashflow table, cashflow diagrams offer limited use in solving problems under payback criterion. However, they help us enormously in better visualizing the concept underlying the payback period and its limitations.

The cashflow diagrams for Examples 5.1 and 5.2, the two examples discussed so far in this chapter, are given in Fig. 5.1. Example 5.1 corresponds to the diagram in Fig. 5.1(a), while Example 5.2 to Fig. 5.1(b). Note the income vectors' constancy in Fig. 5.1(a), but their variability in Fig. 5.1(b), as dictated by the problem statements. The payback period can be displayed in such diagrams by marking it off the timeline, such as 2.27 years in Fig. 5.1(a), and 2.19 years in Fig. 5.1(b). Note in this figure that the cashflow next to the right of the payback period mark also contributes toward the payback. This is because the cashflows normally occur during the entire period, although we post them at period-ends. For example, in Fig. 5.1(a), the cashflow of $2,200 at period 3 contributes during the initial 0.27 year following period 2. It has been sketched at year 3 only because of the convention of year-end posting of cashflows.

Example 5.3: Graphical Approach

Do Example 5.2 using the graphical approach.

SOLUTION:

The solution for the payback period can also be based on the graphical approach. In that case a graph is drawn between the cumulative income on the y-axis versus time on

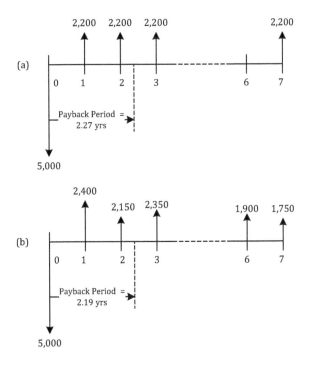

Figure 5.1   Payback period concept.

the x-axis. A horizontal line is drawn to represent the investment. The payback period is read off the x-axis corresponding to the intersection of the cumulative and investment lines.

The graphical approach comprises five steps:

1. Plot the cumulative income data on the y-axis and the time periods on the x-axis.
2. Join the data points by straight lines to draw the cumulative income line.
3. Draw the horizontal investment line.
4. Project the intersection of the investment and cumulative income lines on the x-axis.
5. Read the payback period off the x-axis.

Fig. 5.2 shows the graphical approach to solving Example 5.2, whose cumulative income data have been plotted in this figure. The data points have been joined by straight lines, generating the *cumulative income line*. The intersection of this line and the horizontal investment line yields the payback period on the x-axis as 2.19 years.

Take a moment to ponder that the cumulative line for constant net income as in Example 5.1 will be a single straight line, with one slope. The slope is a measure of the rate at which investment is recovered.

The graphical approach to evaluating payback period, as in Example 5.3, is simply an extension of the tabular approach. In both cases, cumulative income data must be computed. The difference lies in creating an additional column in the data table for posting the cumulative incomes and analyzing them, or in graphically plotting them. The graphical approach offers a superior visualization and is preferred by engineers and technologists, especially if the results are to be communicated to others. Remember however that the accuracy of the graphical method's results is influenced by the accuracy of plotting the data and reading the payback period off the x-axis. This is not a major limitation when graphic calculators or CAD systems—quite a common tool of engineers and technologists today—are used.

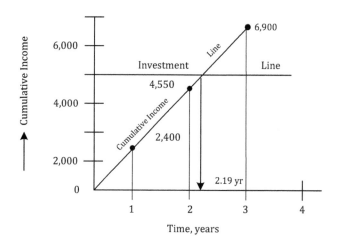

Figure 5.2 Graphical analysis of payback period.

5.4 SINGLE PROJECTS

In Sections 5.1 through 5.3, we learned the procedure of evaluating payback period. We can now discuss decision-making under the payback-period criterion. It is obvious from the previous discussions that this criterion favors projects of shorter payback periods, since the investment is recovered sooner.

The cut-off payback period for funding a project is usually set by the company management after considering various financial and marketing factors. Based on the number of alternatives, projects may be of three types: single, two-alternative, and multi-alternative. Single projects with no alternative are the simplest to analyze. In here, the engineering economist's task is to determine the payback period of the project. If the payback period is shorter than that set up by the management, the project is funded. In Example 5.1, investment in the machine with a 2.27-year payback would be approved if the cut-off period is 3 years. Companies normally finance projects with payback periods under 3 years.

5.5 TWO-ALTERNATIVE PROJECTS

A payback-period-based decision on whether to invest in a resource with no alternative is straightforward, as discussed in the previous section. Two- and multi-alternative problems are comparatively more difficult. We discuss the two-alternative case in this section and the multi-alterative case in the next section.

When engineers are faced with the task of selecting one of the two alternatives based on payback-period criterion, the procedure involves evaluating the payback periods of both the alternatives and choosing the one whose payback period is shorter. The attraction of the shorter-payback-period project is obvious since it will return the investment sooner. Example 5.4 explains the procedure.

Example 5.4: Payback Period

For automated painting of car bodies two models of a robotic system—basic and deluxe—are available in the market. For the following cashflows, which one should be selected on the basis of payback period?

Year	Basic	Deluxe
0	−$50,000	−$70,000
1	$20,000	$30,000
2	$20,000	$28,500
3	$20,000	$27,000
4	$20,000	$25,500
5	$20,000	$24,000

SOLUTION:

Since the cash inflow data are not specific about what they represent, we can assume them to be net incomes. The basic model yields a constant annual income of $20,000. We can therefore find its payback period using the standard formula, Equation (5.1). With an investment of $50,000 and annual income of $20,000, payback period for the basic model is:

$$= \$50,000 / (\$20,000 \text{ per year})$$

$$= 2.5 \text{ years}$$

For the deluxe model, the incomes are variable. So, we use either the tabular or graphical method; both involve the evaluation of cumulative incomes. Following the procedure outlined in Example 5.2 or 5.3, the payback period for the deluxe model is determined to be

$$= 2 + (70,000 - 58,500) / 27,000$$

$$= 2.43 \, \text{years}$$

Based on the desirability of shorter payback period, the deluxe model should be selected.

Two-alternative problems fall into one of the two categories. In one, a comparison of the alternatives' payback periods is made, and the alternative with the shorter period is selected, as in Example 5.4. In this case, it is assumed that the management has not set any cut-off payback period. The alternative with the shorter payback period is selected irrespective of its value.

In the second category of problems, a cut-off payback period has been fixed by the management. The engineer's task is to evaluate the payback periods of both the alternatives and compare each one with the set one. Three situations can arise:

1. Both payback periods are longer than the cut-off payback period. Therefore, neither alternative is selected. However, if the resource is crucial then the compliance to the cut-off payback period may be waived, and the alternative with the shorter payback period is selected.
2. One alternative's payback period is longer than the cut-off period, while the other's is shorter. Obviously, the alternative whose payback period is shorter than the cut-off value is selected.
3. Both alternatives' payback periods are shorter than the cut-off payback period. In this case, the alternative with the shorter payback period is selected.

In Example 5.4, there was no cut-off payback period. The selection was therefore based on comparison between payback periods of the two models. Had the management set the payback period in that example to, say, 2 years, then neither model would have been selected.

5.6 MULTI-ALTERNATIVE PROJECTS

Engineering projects involving more than two alternatives are solved in ways similar to those for two alternatives discussed in Section 5.5. To select the best alternative out of the several feasible ones, the solution procedure comprises: (a) Evaluating the payback period for each alternative, and (b) Selecting the best alternative on the basis of shortest payback period, considering any constraint imposed by the cut-off payback period set by the management. Example 5.5 illustrates the procedure.

Example 5.5: Multi-Alternative Projects

The small drilling machine in the job shop is worn beyond repair. The supervisor has done some research and found that any of the three Makes A, B, or C will suffice as a replacement. The cashflows for these three candidate machines are given as follows. If the company funds projects of payback periods only under 3 years, which one should be selected?

Year	Make A	Make B	Make C
0	−$5,000	−$7,000	−$9,000
1	$2,000	$3,000	$4,000
2	$2,000	$2,500	$3,750
3	$2,000	$2,000	$3,500
4	$2,000	$1,500	$3,250
5	$2,000	$1,000	$3,000
6	$2,000	$ 500	$2,750

SOLUTION:

The cash inflows are assumed to be net incomes. We need to follow the two-step procedure of evaluating the payback period of each make, and selecting the one with the shortest payback period provided it is less than 3 years.

Make A:

Using the standard equation, since incomes are constant, the payback period for Make A is:

$$= \$5,000 / (\$2,000 \text{ per year})$$

$$= 2.5 \text{ years}$$

Make B:

For Make B, the incomes are variable. So, use either the tabular or graphical method; both involve evaluation of cumulative income. Following the procedure outlined in Example 5.2 or 5.3, the payback period for this make is:

$$= 2 + (7,000 - 5,500) / 2,000$$

$$= 2.75 \text{ years}$$

Make C:

For this make too, the incomes are variable. So, use either the tabular or graphical method (Examples 5.2 and 5.3). The payback period for Make C is:

$$= 2 + (9,000 - 7,750) / 3,500$$

$$= 2.36 \text{ years}$$

Since all the three makes satisfy the set cut-off payback period of 3 years, each one is worthy of selection. However, on the basis of shortest payback period, Make C is selected.

Multi-alternative problems can also be categorized into two types. One involves evaluation and comparison of the alternatives' payback periods, and selection of the alternative with the shortest period. In such problems, no cut-off payback period has been set. Such situations may also arise when the resource to be invested in is crucial, and therefore the project is funded irrespective of its payback period.

In the second type of problems, a cut-off payback period is set by the management, as in Example 5.5. The engineer's task is to evaluate the payback periods of all the alternatives and compare them with the cut-off payback period. Projects with payback periods above the cut-off are discarded. Of the remaining, the one with the shortest payback period is selected. In Example 5.5, the three alternatives (Makes A, B, C) had payback periods lower than the set 3-year period. So, the decision was made in favor of the shortest of the three payback periods, resulting in the selection of Make C.

5.7 MERITS AND LIMITATIONS

As a criterion for economic decision-making, the payback period has its supporters and denouncers. The supporters boast about its simplicity, while the denouncers point out its non-consideration of the time value of money. In response to this criticism, the conventional method is sometimes modified to account for the time value of money, as illustrated in the next section.

The payback period method is suited for small investments by, and for, the operating departments. The final decision to invest rests usually with the department supervisor or manager, with no need of approval by the upper management. Normally, the capital involved is small and within the limit set by the management.

Under what circumstances is the payback-period criterion more appropriate? This is an important question. The usual circumstances are:

1. The working capital is low. The company is operating with short-term planning horizon with tight cashflows. It needs to recover the invested capital, as soon as possible, for use elsewhere.
2. The investment is associated with a technology that is changing rapidly[4]. Short payback periods ensure that the investment will be recovered well in time—before the invested resource becomes obsolete. The recovered capital may be "recycled" for upgrading.
3. Investment in innovative technologies should also be governed by shorter payback periods because of the higher risk.

Till recently the payback period has been an extremely popular method in industry for both engineers and non-engineers. It required little knowledge of engineering economics. If the economic climate is stable with low interest rates, ignoring the time value of money may not affect the decision too much, especially for short-term projects. However, the other methods discussed in Chapters 6 through 10 have now become popular, primarily because they account for the time value of money and their analyses are facilitated by computer technology.

The payback period method can lead to decisions that contradict those by other methods, such as the rate of return method (Chapter 9). Engineering economists must be careful about this limitation of the payback period. They may use another method to confirm the decision by the payback period criterion. In fact, the payback-period

---

[4] The result may be obsolescence of the invested resource. How much we repent when the PC bought just last year gets superseded by a newer, more powerful one, and that at a lower cost! Short payback periods minimize the negative impact, through obsolescence, on investment of rapidly changing technologies.

criterion is often used in industry only for initial screening of the projects. Promising projects are then subjected to more rigorous analysis by one of the methods discussed in Chapters 6 through 10.

The merits and limitations of the payback period method are summarized as follows.

5.7.1 Merits

1. Simple to use
2. Suitable for small investments
3. Quicker investment decision since upper management usually not involved
1. Suitable for projects involving rapidly changing technologies
4. Easily understood, even by non-economists

5.7.2 Limitations

1. Time value of money ignored
2. An approximate, less exact method
3. Can lead to erroneous results
4. Benefits beyond the payback period not considered
5. May need confirming of the results by other method(s)
6. Less attractive to companies with "healthy" working capital and/or operating with long-range investment plans

5.8 TIME-VALUED PAYBACK

So far in this chapter, we limited our discussions to the usual procedure of payback period analysis in which the time value of money is ignored. This is done for the sake of simplicity of analysis—the hallmark of the payback period method. Some economists, however, prefer to add realism to this method by considering the time value of money. Such realistic approaches render the method more complex, however. Examples 5.6 and 5.7 illustrate time-valued payback period analysis.

Example 5.6: Time Valued Payback

A hydraulic machine is being planned for acquisition at a cost of $5,000. During its useful life of 7 years, it is expected to generate a gross income of $3,400 per year. Its annual maintenance cost is projected to be $300. The cost of space to house the machine and of the utilities is $900 per year. Determine the payback period for the machine *by considering the time value of money*. Assume annual compounding with i = 10% per year.

SOLUTION:

Note that this is Example 5.1, except that we consider here the time value of money. By subtracting the annual expenditure of $1,200 (= $900 + $300) from the gross income, net income is $3,400 − $1,200 = $2,200 per year. So, the cashflow table is:

Year	0	1	...	...	7
Cashflow	−$5,000	$2,200	...	...	$2,200

Since the $5,000 cost is incurred now, it has no time value with reference to the present (period 0). However, we need to account for the time values of the future cashflows. The first-year income of $2,200 gets modified in terms of today's dollars as:

$$P = 2,200(P/F, 10\%, 1)$$

$$= 2,200 \times 0.9091$$

$$= 2,000$$

Thus, by the end of the first year, $2,000 of the investment is recovered. The investment yet to be recovered = $5,000 – $2,000 = $3,000.

The second-year net income of $2,200 gets modified as:

$$P = 2,200(P/F, 10\%, 2)$$

$$= 2,200 \times 0.8264$$

$$= 1,818$$

Thus, by the end of the second year, the investment yet to be recovered = $3,000 – $1,818 = $1,182.

The third-year net income of $2,200 gets modified as

$$P = 2,200(P/F, 10\%, 3)$$

$$= 2,200 \times 0.7513$$

$$= 1,653$$

Since the third-year time-valued income of $1,653 is more than the remaining investment to be recovered ($1,182), the payback period lies in between years 2 and 3. By interpolating within these two limits we can determine the exact payback period. Following the interpolation procedure explained in Chapter 3,

$$\text{Payback period} = 2 + 1182/1653$$

$$= 2 + 0.71$$

$$= 2.71 \text{ years}$$

The time-valued payback period of 2.71 years is longer than the conventional payback period of 2.27 years determined in Example 5.1. This is expected since consideration of the time value reduced the future benefits, resulting in longer recovery of the investment.

As can be noted in the previous example, consideration of the time value changed Example 5.1 from being uniform-income type to variable-income type. That is why we could not use the standard formula, Equation (5.1), in Example 5.6. In rare cases, the consideration of time value may do the opposite, i.e., change a variable-income problem to a uniform-income type. This occurs when the rate of increase in the future incomes is the same as the interest rate, as illustrated in Example 5.7.

Example 5.7: Time Valued Payback

Considering *the time value of money* and assuming annual compounding with $i = 10\%$ per year, what is the payback period for the following project?

Year	0	1	2	3	4
Cashflow	–$5,000	$2,200	$2,420	$2,662	$2,928

SOLUTION:

Since the $5,000 investment is made now, its time-value with reference to year 0 remains unaffected. As required, the cash inflows should account for the time value of money. The first-year income of $2,200 gets modified as:

$$P = 2,200(P/F,10\%,1)$$

$$= 2,200 \times 0.9091$$

$$= 2,000$$

Thus, by the end of the first year, $2,000 of the investment is recovered. The investment yet to be recovered = $5,000 – $2,000 = $3,000.

For the second year, the $2,420 income gets modified as:

$$P = 2,420(P/F,10\%,2)$$

$$= 2,420 \times 0.8264$$

$$= 2,000$$

Thus, by the end of the second year, the remaining investment yet to be recovered = $3,000 – $2,000 = $1,000.

For the third year, the $2,662 income gets modified as

$$P = 2,662(P/F,10\%,3)$$

$$= 2,662 \times 0.7513$$

$$= 2,000$$

Since the third-year time-valued income of $2,000 is more than the amount to be recovered ($1,000), the payback period is under 3 years. The interpolation between years 2 and 3 is straightforward since the amount to be recovered at the end of second year ($1,000) is exactly half of the recovery during the third year ($2,000). Thus, the payback period is 2.5 years. Alternatively,

$$\text{Payback period} = 2 + 1000/2000$$

$$= 2 + 0.5$$

$$= 2.5\,\text{years}$$

Note that in this example the given incomes were variable (increasing at 10% each year—the same as the interest rate i). The consideration of time value changed these incomes, rendering them uniform at $2,000 each year. We could therefore have used the formula, Equation (5.1), to evaluate the payback period as:

$$\text{Payback period} = \text{Investment}/\text{Income per year}$$

$$= \$5,000/(\$2,000\,\text{per year})$$

$$= 2.5\,\text{years}$$

5.9 SUMMARY

Payback period is the duration of time in which an invested capital is fully recovered through net incomes from the investment. The payback period method usually ignores the time value of money. For being selected under the payback criterion, the project's payback period must be shorter than the cut-off value set by the management. The alternative with the shortest payback period is obviously the best, and hence selected. The payback period method is often used in industry, especially for analyzing projects requiring low capital. Companies operating under short planning horizon, due to tight cashflows, find this method useful. This method is also useful in projects involving rapidly changing technology. It is also used for initial screening of projects competing for funds that are limited. Its simplicity is its hallmark. That the time value of money is not considered is its greatest weakness. The payback criterion should be used carefully, since decision results may at times contradict those by the other methods. Analysts use another method, such as rate of return, to confirm the results of payback period analysis. The conventional analysis can be enhanced by considering the time value of money. However, such an enhancement is achieved at the cost of the payback method's simplicity.

DISCUSSION QUESTIONS

5.1 Why does the payback period method continue to be used in industry in spite of the fact that it ignores the time value of money?

5.2 How would you explain the meaning of the payback period to a lay person?

5.3 Now that you know how to make economic decisions based on the payback period criterion, describe any forthcoming personal project for which you might use it. How will you gather the cashflow data for the project?

5.4 Is the payback period dimensionless? If yes, why? If no, what is its unit?

5.5 Can one apply the payback period criterion to purchasing a car? If yes, how? If no, why not?

5.6 Why would companies favor investments with a payback period under 3 years?

5.7 Do you agree or disagree that the payback period analysis is useful in spite of its limitations? Please explain.

5.8 List examples of projects where the payback period analysis would be appropriate.

MULTIPLE-CHOICE QUESTIONS

5.9 The payback period is expressed usually in
  a. years
  b. quarters
  c. months
  d. days

5.10 In the graphical method for evaluating the payback period the investment line is parallel to the
  a. x-axis
  b. y-axis
  c. z-axis
  d. none of the above

5.11 The payback period method
  a. is exact
  b. is simple
  c. accounts for the time value of money
  d. considers the salvage value in the analysis

5.12 The payback period method is most appropriate for _____ projects.
    a. short-term
    b. long-term
    c. complex
    d. government
5.13 The payback period method is most appropriate for projects involving
    a. established technologies unlikely to change
    b. technology that is changing rapidly
    c. imported technology
    d. appropriate technology
5.14 The payback period method's reliability can be enhanced by
    a. evaluating it in months rather than in years
    b. considering the time value of money
    c. both a and b
    d. neither a nor b
5.15 Which of the following are limitations of the payback period analysis?
    a. simple to use
    b. time value of money ignored
    c. can lead to erroneous results
    d. both b and c
5.16 What is the payback period of a $4,500 piece of equipment if it generates $1,200 of revenue each year?
    a. 3.75
    b. 4
    c. 2
    d. 5.3

For the next questions refer to the table below.

5.17 What is the payback period for Make A?
    a. 3.41
    b. 4.25
    c. 3
    d. 2
5.18 What is the payback period for Make B?
    a. 3.6
    b. 2.75
    c. 4
    d. none of the above

Year	Make A	Make B	Make C
0	−$6,000	−$7,500	−$8,000
1	$1,500	$1,200	$2,000
2	$1,600	$2,300	$2,000
3	$1,700	$2,500	$2,000
4	$3,500	$3,500	$2,000

5.19 What is the payback period for Make C?
  a. 3
  b. 2
  c. 4
  d. all of the above

5.20 Considering the payback period analysis which make would be the best choice for investment?
  a. Make A
  b. Make B
  c. Make C
  d. none of the above

5.21 If a company wants a return on their investment in no more than 3 years, which makes would satisfy the cut-off criteria?
  a. Make A
  b. Make B
  c. Make C
  d. all of the above

NUMERICAL PROBLEMS

(Do Problems 5.22 through 5.27 the usual way, i.e., ignoring the time value of money.)

5.22 Determine the payback period of a project whose cashflows are:

Year	0	1	2	3	4
Cashflow	−$3,500	$1,750	$1,200	$1,050	$950

5.23 Assume that your 4-year education toward the BS degree costs you $40,000 over and above what it would have cost you to live off if you did not attend college. You also lost annual earnings of $15,000 during these 4 years. Assuming that the degree will enhance your future earnings potential by $10,000 per year, what is the payback period for this higher education?

5.24 A new machine for the manufacture of plastic cutlery costs $158,000. The machine will increase the monthly production by 20,000 packs. Each pack costs 50 cents to manufacture and is sold for 75 cents. What is the payback period for the machine?

5.25 Determine the payback period of a PC that costs $9,500 to install and $150 per month to operate. The system will free 50% of a secretary's time whose salary is $15,000 per year and the additional fringe benefits are 60% of the salary.

5.26 Two models of a forklift truck—basic and deluxe—are being planned for acquisition. For the following data which one should be selected on the basis of the payback period? The $3,000 cost for the basic model at year 3 is its required overhauling.

Year	Basic	Deluxe
0	−$50,000	−$60,000
1	$15,000	$17,500
2	$15,000	$17,000
3	$15,000 −$ 3,000	$16,500
4	$15,000	$16,000
5	$15,000	$15,500

5.27 A tool room lathe needs to be replaced, for which there are three choices—A, B, and C—in the market. On the basis of the payback period, which lathe should be procured if the company approves projects of payback periods only less than 2.5 years. The cashflow data are:

Year	A	B	C
0	−$35,000	−$40,000	−$48,000
1	$10,750	$17,500	$17,000
2	$10,750	$17,500	$16,000
3	$10,750	$16,500	$15,000
4	$10,750	$14,500	$15,000
5	$10,750	$13,500	$15,000

5.28 Do Problem 5.27 by considering the time value of money. Assume annual compounding with $i = 12\%$ per year.

5.29 Do Problem 5.28 by considering the time value of money. Assume annual compounding with $i = 8\%$ per year.

5.30 Do Problem 5.29 by considering the time value of money. Assume annual compounding with $i = 10\%$ per year.

5.31 Do Problem 5.30 by considering the time value of money. Assume annual compounding with $i = 15\%$ per year.

5.32 The management of a hydroelectric power plant is evaluating the options on the market to replace one of their outdated turbines. If management wants a return on their investment in no more than 5 years, which turbine would be the best decision considering the payback period analysis? The cashflow data is as follows:

Year	A	B	C
0	−$400,000	−$650,000	−$750,000
1	$ 50,000	$125,000	$160,000
2	$ 50,000	$150,000	$225,000
3	$ 50,000	$200,000	$165,000
4	$ 50,000	$250,000	$100,000
5	$ 50,000	$200,000	$175,000
6	$ 50,000	$175,000	$200,000
7	$ 50,000	$165,000	$215,000
8	$ 50,000	$170,000	$175,000

5.33 Do Problem 5.32 considering the time value of money. Assume annual compounding where $i = 9\%$.

5.34 Do Problem 5.33 considering the time value of money. Assume a semiannual compounding where $i = 5\%$.

5.35 Do Problem 5.34 considering the time value of money. Assume a semiannual compounding where $i = 6\%$.

5.36 Do problem 5.35 considering the time value of money. Assume an annual compounding where $i = 10\%$.

5.37 A manufacturing company is planning on upgrading a conveyor system in their plant. Given the following table, which of the conveyor belt systems would be the best choice if the company wants a return on their investment in 3 years?

Year	Basic	Deluxe
0	−$45,000	−$65,000
1	$15,500	$17,500
2	$15,500	$25,000
3	$15,500	$19,000
4	$15,500	$26,000
5	$15,500	$13,000

5.38 Now do Problem 5.37 considering the time value of money. Assume an annual compounding where $i = 8\%$.

5.39 Draw the cashflow diagram.

5.40 Solve Problem 5.37 using the graphical approach.

FE EXAM PREP

5.41 What is the payback period of a $7,500 piece of equipment if it generates $1,500 of revenue each year?
  a. 5
  b. 4
  c. 2
  d. 5.3

5.42 What is the payback period of a $8,000 piece of equipment if it generates $750 of revenue each year?
  a. 10.3
  b. 9
  c. 7
  d. 3

5.43 What is the payback period of a $15,000 piece of equipment if it generates $1,200 of revenue each year?
  a. 13.75
  b. 14
  c. 2
  d. 12.5

Use the table below to answer the following questions.

Year	Basic	Deluxe
0	−$50,000	−$70,000
1	$20,000	$30,000
2	$20,000	$28,500
3	$20,000	$27,000
4	$20,000	$25,500

YEAR	0	1	2	3	4
AMOUNT	$4,000	–$1,500	–$1,300	–$1,000	–$1,000

5.44 What is the payback period for the basic model?
   a. 3.5
   b. 2.5
   c. 4
   d. all of the above

5.45 What is the payback period for the deluxe model?
   a. 3.5
   b. 2.5
   c. 4
   d. all of the above

5.46 Considering the payback period analysis which model would be the best choice for investment?
   a. basic
   b. deluxe
   c. neither A nor B
   d. both A and B

5.47 What is the payback period for the above table?
   a. 3.3
   b. 2.5
   c. 4
   d. none of the above

5.48 What is the payback period of a $3,750 piece of equipment if it generates $750 of revenue each year?
   a. 10.3
   b. 5
   c. 7
   d. 3

5.49 What is the payback period of an $18,000 piece of equipment if it generates $7,500 of revenue each year?
   a. 10
   b. 19
   c. 24
   d. 30

5.50 What is the payback period of a $25,000 piece of equipment if it generates $11,250 of revenue each year?
   a. 12
   b. 4
   c. 3.1
   d. 2.2

Chapter 6

Time Value of Money

LEARNING OBJECTIVES

- Meaning of present worth (PW)
- Present worth of costs
- Present worth of benefits
- How to determine PW of a project
- Application of PW-criterion
- Input-output concept
- Analysis period
- Perpetual-life projects
- Capitalized cost
- Meaning of *future worth* (FW)
- Evaluation of FW
- Determination of a project's FW
- Application of FW-criterion
- When to use FW-criterion
- Meaning of *annual worth* (AW)
- How to determine a project's AW
- Application of AW-criterion
- How to account for salvage value
- Application of input-output concept in AW method
- Irrelevancy of the analysis period

As discussed in Chapter 5, the greatest weakness of the payback-period method is its nonconsideration of the time value of money. In contrast, the other five analysis methods, to be covered next, account for the time value of money. They are therefore more realistic. In this chapter, we discuss one of the five methods, called present worth (PW), and learn to apply it to engineering projects.

The present worth criterion relies on the sound logic that economic decisions—for that matter all decisions—be made on the basis of costs and benefits pertaining to the time of decision-making. The *past* is over and the *future* is unknown, so relying on the *present* makes sense. It is the present we live in. Most decisions of life are made in the present—looking at the future as far as we can see. Economic decisions should be no exception. In the PW method, the economic data are analyzed through the "window" of the present.

This chapter will also address the future worth (FW) method of analyzing engineering economics problems. It can be considered a proximate extension of the present worth method whose coverage in the previous chapter is significantly pertinent—the only difference being the reference time. The present worth looks at *now* while future worth looks at *then*—a certain time in future. Thus, under the FW-criterion, a decision is made on the basis of a project's economic worth at some time in the future. In the FW-method, the data are analyzed by referring to the end of useful life or analysis period.

Additionally, we will discuss another important criterion, called *annual worth* (AW), and its application to engineering projects. Like the PW- or FW-criterion, the AW-criterion also accounts for the time value of money. Annual worth's greatest attraction is its similitude with the commonly practiced yearly accounting for profit, taxes, and other similar purposes.

AW-based decisions rely on spreading the costs and benefits uniformly over the resource's useful life. The given cashflow data are transformed into their yearly-equivalent whose value dictates the investment worthiness of the project. In the AW-method we look through the "year[1] as the window."

6.1 MEANING

Present worth is the value of something at the current time. Time enters into its meaning because the value of an object is not the same at all times. Consider the worth of an apple now. If you are not hungry, or don't feel like eating fruit, the apple has no worth to you at the present time. In three or four hours, when you will feel like eating something and may not have access to what you would like to eat, the same apple will have worth for you. This of course is a rather simplistic way to explain the worth of something[2] as a function of time.

Decision-making based on the PW-criterion involves evaluation of the present worth. By *present worth* we mean the present value of all the cashflows pertaining to the project. As discussed in Chapter 2, the cashflows may be costs or benefits. The present worth therefore may be of costs or of benefits. Consider the diagram in Fig. 6.1 where the cashflows are in dollars, and assume *i* to be the interest rate per period. Following the convention discussed in Chapter 2, a cost of $500 is incurred now. There are two more costs: $250 at period 3, and $300 at period 5. We also see in the diagram three benefits, one each at time periods 1, 2, and 4. For this diagram, the present worth of costs means the *equivalent cost now* of all the costs, which includes the present $500 cost at period 0. Its determination involves evaluating the present value (P-value) of the two future costs and adding them to the present $500 cost. Thus, we need to convert the $250 cost at period 3 and the $300 cost at period 5 to their equivalents at period 0. These are

$$P - \text{value of the} \$250 - \text{cost} = 250\left(\frac{P}{F,i,3}\right)$$

$$P - \text{value of the} \$300 - \text{cost} = 300\left(\frac{P}{F,i,5}\right)$$

---

[1] The time period in the AW-method is usually year. There is no reason why another time period can't be used in the analysis. If month is used, the decision will be based on *monthly worth*. All discussions of this chapter remain pertinent to monthly-worth analysis, for that matter to analyses based on any time period.

[2] In engineering economics, we consider only those items of a project whose value can be expressed in monetary terms. The items are usually cashflows of first cost, benefits, and maintenance costs, salvage value, etc. However, sometimes we do consider intangible items of benefit or cost that are difficult to express in monetary terms. Usually we ignore the intangibles during the analysis and consider them subjectively at the final stage of the decision-making. If we decide to consider them in the analysis, we guess-estimate their dollar values. Goodwill, whose value may be significant and is usually accounted for while buying or selling an established business, is for example an intangible item.

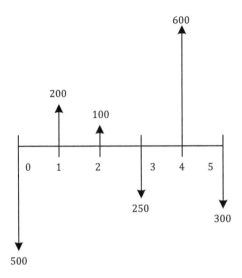

Figure 6.1 Present worth.

Adding these P-values to the $500 cost, the present worth of the costs is

$$PW_{costs} = 500 + 250\left(\frac{P}{F,i,3}\right) + 300\left(\frac{P}{F,i,5}\right)$$

The same way we can determine the present worth of the benefits as

$$PW_{benefits} = 200\left(\frac{P}{F,i,1}\right) + 100\left(\frac{P}{F,i,2}\right) + 600\left(\frac{P}{F,i,4}\right)$$

As can be seen, in determining the present worth of costs, or benefits, all future costs, or benefits, are transferred to their equivalent values at the present time (period 0). This is how time value of money is taken into account in the PW-method.

Future worth is the value of something at a certain time in future. Consider your plan of a car trip lasting several hours. To save time stopping for food, you are thinking about taking two apples to eat while driving. The apples have little value for you when you begin the trip just after lunch. But in a few hours when you will be hungry, they will be valuable. Your decision to take the apples with you is based on their future worth. This of course is a rather simplistic explanation of future worth.

FW-based decision-making involves evaluation of the project's future worth for each alternative and selection of the most favorable alternative. A project's FW is the surplus of its benefits' future worth[3] over that of its costs' future worth. Consider the diagram in Fig. 6.2, in which the given cashflows in dollars are shown by the firm vectors. Following the convention discussed in Chapter 2, a cost of $500 is incurred now. There are two more costs, $250 at period 3 and $300 at period 5. We also see in the diagram three benefits at time periods 1, 2, and 4. For this diagram, the future worth of costs, shown dotted by FW_{costs}, is the *equivalent cost* at period 6 of the three costs. Its determination involves evaluating, and

---

[3] In engineering economics we consider only those elements of a project whose values are expressible in monetary terms. Examples are first cost, maintenance costs, benefits, and salvage value. The *future worth* of a project depends on the values of such elements, i.e., the various cashflows pertaining to the project.

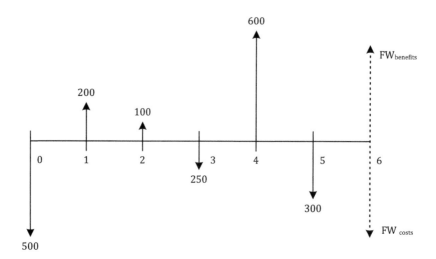

Figure 6.2 Analysis of future worth.

adding together, the F-values at period 6 of the $500 first cost and of the other two costs. In other words, we convert the $500 cost at period 0, the $250 cost at period 3, and the $300 cost at period 5 to their equivalents at period 6, and add these equivalents. For an interest rate i per period, the F-values are:

$$F - value\,of\,the\,\$500 - cost = 500\left(\frac{F}{P,i,6}\right)$$

$$F - value\,of\,the\,\$250 - cost = 250\left(\frac{F}{P,i,3}\right)$$

$$F - value\,of\,the\,\$300 - cost = 300\left(\frac{F}{P,i,1}\right)$$

Adding these F-values, the future worth of costs is:

$$FW_{costs} = 500\left(\frac{F}{P,i,6}\right) + 250\left(\frac{F}{P,i,3}\right) + 300\left(\frac{F}{P,i,1}\right)$$

The same way we determine the future worth at period 6 of all the benefits, shown dotted in Fig. 6.2 by $FW_{benefits}$, as:

$$FW_{benefits} = 200\left(\frac{F}{P,i,5}\right) + 100\left(\frac{F}{P,i,4}\right) + 600\left(\frac{F}{P,i,2}\right)$$

As can be seen, in determining the future worth of costs (or benefits), all the costs (or benefits) of the project are transferred as their equivalents to a specific time marker, using the appropriate functional factors. This way the time value of each cashflow is accounted for. Example (6.11) illustrates the application of the FW-method.

Annual worth is the equivalent net cash inflow each year over the useful life. Consider a project whose cashflow diagram is in Fig. 6.3. There are two cash outflows (costs) M and

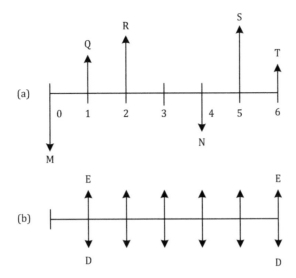

Figure 6.3 Concept of annual worth.

N, and four inflows (benefits) Q, R, S, and T. Let us comprehend the meaning of AW with reference to this diagram.

In the AW-method, we determine the annual equivalents of project costs and benefits, as in Fig. 6.3(a), in terms of uniform-series costs and benefits, as in Fig. 6.3(b). The two costs M and N are shown in Fig. 6.3(b) by their annual equivalent D, while the four benefits Q, R, S, and T by their annual equivalent E. These annual equivalents, D and E, are respectively called *equivalent uniform annual cost* (EUAC) and *equivalent uniform annual benefit* (EUAB). In this text, we prefer to use AW_{costs} for EUAC and $AW_{benefits}$ for EUAB. Since we have used PW_{costs} and $PW_{benefits}$ in Chapter 6, and FW_{costs} and $FW_{benefits}$ in Chapter 7, it will be reader-friendly to use AW_{costs} and $AW_{benefits}$. The excess of $AW_{benefits}$ over AW_{costs}, representing net annual cash inflow, is the annual worth (AW) of a project. In other words,

$$AW = AW_{benefits} - AW_{costs} \tag{6.3}$$

From this equation, AW will be positive when $AW_{benefits}$ is greater than AW_{costs}. So, a positive AW means that the project is investment worthy. Also, the greater the value of +ve AW, the more attractive the project.

It is obvious from this discussion, as well as from Equation (6.3), that AW-analysis involves determining[4] $AW_{benefits}$ and AW_{costs}. This means converting the given cashflows into equivalent annual cashflows. How this conversion can be done efficiently depends primarily on the cashflows' distribution. We have already learned in Chapters 3 and 4 how to carry out conversions.

In the absence of any pattern in the cashflows, we use the straightforward method, treating each given cashflow separately. In case some pattern exists, we utilize any of the equations derived in Chapter 4 for series cashflows. In many cases the given cashflow(s) may have to be converted first to P-value at period 0, which is then converted to its equivalent annual cashflow. Examples 6.1 to 6.3 illustrate the procedure.

---

[4] $AW_{benefits}$ and AW_{costs} are different notations for uniform series vectors A discussed in Chapter 4. $AW_{benefits}$ is the uniform cash inflow vector, while AW_{costs} is the uniform cash outflow vector.

Example 6.1: Present Worth

India England is to begin her 4 years of college to earn a BS degree in environmental engineering. The cost of education during the first year will be $16,000. It is expected to increase annually by 10%. If the money to be spent on education can earn annually compounded 5% interest per year, what is the present worth of costs on her education?

SOLUTION:

The first-year cost is $16,000. Since the cost increases each year by 10%, any year's cost is 10% more than the previous year's. Therefore, the second-year cost will be $17,600 ($16,000 + 10% of $16,000). Similarly, the third- and fourth-year costs will be $19,360 and $21,296.

We can assume the *present* to be the beginning of the college education. Following the convention that a cashflow is posted at the end of the period, the first-year cost of $16,000 is incurred at the end of period 1, and so on. Sketch a cashflow diagram if helpful.

By transferring the future costs to the present and adding them together, the present worth of India's costs on her education is

$$
\begin{aligned}
PW_{costs} &= 16,000\left(\frac{P}{F,5\%,1}\right) + 17,600\left(\frac{P}{F,5\%,2}\right) \\
&\quad + 19,360\left(\frac{P}{F,5\%,3}\right) + 21,296\left(\frac{P}{F,5\%,4}\right) \\
&= 16,000 \times 0.9524 + 17,600 \times 0.9070 + 19,360 \times 0.8638 \\
&\quad + 21,296 \times 0.8227 \\
&= 15,238 + 15,963 + 16,723 + 17,520 \\
&= \$65,444
\end{aligned}
$$

To determine the net effect of the costs and benefits on a project, we determine the project's present worth[5] (PW). The present worth of a *project* is the difference between the present worth of all the benefits from the project and the present worth of all the costs on the project. Thus,

$$
PW = PW_{benefits} - PW_{costs} \tag{6.1}
$$

Note that we subtract the present worth of costs from the present worth of benefits, not the other way around. This ensures that the PW is positive for projects whose present worth of benefits exceeds that of the costs—an obviously desirable aim. Thus, a project is usually funded only if its PW is positive, ensuring a return on investment.

Example 6.2: Present Worth

For $i = 6\%$, determine the PW of the project represented by Fig. 6.1. If the PW is negative, explain its significance.

SOLUTION:

To evaluate the project PW, we first determine the PW of its benefits and costs. Following the earlier discussion on Fig. 6.1, for $i = 6\%$ we have,

---

[5] In some literature, it is called *net present worth* (NPW) or *net present value* (NPV).

$$PW_{\text{benefits}} = 200\left(\frac{P}{F,i,1}\right) + 100\left(\frac{P}{F,i,2}\right) + 600\left(\frac{P}{F,i,4}\right)$$

$$= 200\left(\frac{P}{F,6\%,1}\right) + 100\left(\frac{P}{F,6\%,2}\right) + 600\left(\frac{P}{F,6\%,4}\right)$$

$$= 200 \times 0.9434 + 100 \times 0.8900 + 600 \times 0.7921$$

$$= 189 + 89 + 475$$

$$= \$753$$

$$PW_{\text{costs}} = 500 + 250\left(\frac{P}{F,i,3}\right) + 300\left(\frac{P}{F,i,5}\right)$$

$$= 500 + 250\left(\frac{P}{F,6\%,3}\right) + 300\left(\frac{P}{F,6\%,5}\right)$$

$$= 500 + 200 \times 0.8396 + 300 \times 0.7473$$

$$= 500 + 210 + 224$$

$$= \$934$$

The present worth of the project, therefore, is:

$$PW = PW_{\text{benefits}} - PW_{\text{costs}}$$

$$= \$753 - \$934$$

$$= -\$181$$

The present worth is negative, which means that on the basis of the PW-criterion the project incurs a loss. It should therefore not be funded, unless there are accentuating non-economic reasons.

6.2 PRESENT-WORTH METHOD

The present worth method of solving engineering economics problems involves determination of the present worth of the project or its alternatives. Using the criterion of present worth, the better of the two projects, or the best of the three or more projects, is selected. Thus, the application of this method involves two tasks:

1. Evaluation of the present worth(s).
2. Decision on the basis of the PW-criterion, according to which project with the highest +ve PW is selected.

We discussed the first task in Section 6.1. In this section, we discuss decision-making based on the PW-criterion.

Engineering economics projects can be categorized into one of the two groups: no-alternative projects, or projects with alternatives. In no-alternative projects, there is only one candidate—there is no choice. The data for the sole project are gathered and PW determined, as discussed in Section 6.1. If the PW is positive and capital is available, the project is funded. No-alternative projects therefore do not involve much of the decision-making, beyond the evaluation of PW.

Example 6.3: Present Worth

Determine the present worth of a 10-year project whose first cost is $12,000, net annual benefit is $2,500, and salvage value is $2,000. Assume an annually compounded interest rate of 8% a year.

SOLUTION:

If you can visualize the problem without sketching its cashflow diagram, then proceed directly to the calculations. Otherwise, sketch a diagram, as in Fig. 6.2, to comprehend the problem. Note that the salvage value, recovered at the end of useful life, is a cash inflow to the project. It can therefore be considered a benefit.

Based on Equation (6.1) or the definition of a project's present worth, and the associated functional notations, PW is given[6] by (Figure 6.4)

$$PW = PW_{benefits} - PW_{costs}$$

$$= 2,500\left(\frac{P}{A,8\%,10}\right) + 2,000\left(\frac{P}{F,8\%,10}\right) - 12,000$$

$$= 2,500 \times 6.710 + 2,000 \times 0.4632 - 12,000$$

$$= 16,775 + 926 - 12,000$$

$$= \$5,701$$

In multi-alternative projects, there are several feasible alternatives or candidates. The decision-making involves selecting the alternative whose PW is the most favorable, i.e., the highest of all the positive PWs. For example, if the alternatives' PWs are $300, $650, $268, and $987, then select the one whose PW is $987. If their PWs are all negative, then none should be selected in a normal situation, since negative PW means that the present worth of costs exceeds that of the benefits. If there is an operational necessity so that one of the alternatives must be selected, then select the one whose −ve PW magnitude is the smallest. For example, if the PWs of four alternatives are −$450, −$350, −$896, and −$936, then

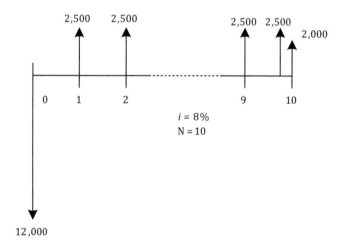

Figure 6.4 Diagram for Example 6.3.

---

[6] If you have difficulty understanding the functional notations used here, reread Chapter 3.

select the one whose PW is –\$350. Being the smallest of the negative PWs, this will minimize the loss, and therefore is the most favorable of the four. In case some of the alternatives yield positive PWs and some negative PWs, then ignore those with negative PWs and select, from among those with +ve PWs, the one with the highest positive PW. The obvious rule is: Decide in favor of financially the most attractive alternative.

Example 6.4: Present Worth

Modern Manufacturers is considering the following five alternative projects as the next year's R&D effort. On the basis of present worth which one should be selected for funding, assuming that the investment must earn an annual return of 12%? All the five projects will take 2 years to develop and earn income during the 5 years beyond development.

Project	Annual Cost (during development)	Annual Income (during profitability)
A	\$2,000	\$3,500
B	\$1,800	\$3,200
C	\$2,200	\$2,300
D	\$1,200	\$2,000
E	\$3,500	\$4,800

SOLUTION:

The R&D costs are incurred during the 2 years of development. The benefits are derived during the 5 years following the development. The decision-making involves two steps: determination of the PW for each project, and decision based on the PW-criterion.

First, we evaluate the PW of each alternative project. Based on the data similarity, the cashflow diagrams of the projects will be alike. Let us analyze project A in details, whose cashflow diagram is given in Fig. 6.3. Its present worth—present being at period 0—can be expressed as

$$PW_A = PW_{benefits} - PW_{costs}$$

To evaluate PW_A, we need to determine project A's $PW_{benefits}$ and PW_{costs}. And to do that we need to look at the diagram more closely. One can determine the present worth of the five benefits easily at period 2 as an upward vector, by treating them as a uniform series. This vector at period 2 must be transferred next to period 0 by treating it as a future sum. These operations on the given cash inflows yield:

$$PW_{benefits} = 3,500 \left(\frac{P}{A,12\%,5} \right) \left(\frac{P}{F,12\%,2} \right)$$

$$= 3,500 \times 3.605 \times 0.7972$$

$$= 10,059$$

The present worth of the costs is easily determined by considering the two cost vectors as a uniform series, yielding (Figure 6.5):

$$PW_{costs} = 2,000 \left(\frac{P}{A,12\%,2} \right)$$

$$= 2,000 \times 1.690$$

$$= 3,380$$

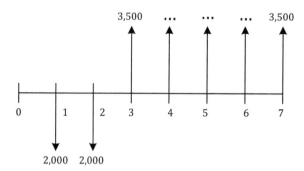

Figure 6.5   Diagram for Example 6.4.

Thus,

$$PW_A = PW_{benefits} - PW_{costs}$$

$$= 10,059 - 3,380$$

$$= 6,679$$

There is another way to determine the present worth of project A. This is based on the concept of tailoring, learned earlier in Chapter 4. The benefits seem to be an approximate uniform series. We can achieve complete uniformity by adding a $3,500 upward vector at period 1 and 2. This will necessitate adding similar vectors downward at these two periods. With these changes in the diagram, the modified benefits form a uniform series over the entire analysis period of 7 years. But the modified costs at periods 2 and 3 are now $2,000 + $3,500 = $5,500 (instead of the given $2,000). Thus, from the tailored cashflow diagram, we get

$$PW_A = 3,500\left(\frac{P}{A,12\%,7}\right) - 5,500\left(\frac{P}{A,12\%,2}\right)$$

$$= 3,500 \times 4.564 - 5,500 \times 1.690$$

$$= 15,974 - 9,295$$

$$= \$6,679$$

Note that the tailoring required evaluation of only two functional notations, in comparison to the three in the straightforward (non-tailored) approach.

Following tailoring, we determine as follows the present worth of the other four alternative projects in a similar way.

$$PW_B = 3,200 \times 4.564 - 5,000 \times 1.690$$

$$= 14,605 - 8,450$$

$$= \$6,155$$

$$PW_C = 2,300 \times 4.564 - 4,500 \times 1.690$$

$$= 10,497 - 7,605$$

$$= \$2,892$$

$$PW_D = 2,000 \times 4.564 - 3,200 \times 1.690$$

$$= 9,128 - 5,408$$

$$= \$3,720$$

$$PW_E = 4,800 \times 4.564 - 8,300 \times 1.690$$

$$= 21,907 - 14,027$$

$$= \$7,880$$

The second step of the analysis involves deciding which project to select. The decision rule is: Select the one that yields the largest positive PW. In this case, with a present worth of $7,880 it is project E, which is therefore selected.

6.3 PRESENT WORTH INPUT-OUTPUT CONCEPT

Engineering economics problems arise in a wide variety of industrial projects. Before an attempt to solve the problem can be made, the problem must be fully comprehended. The *concept of input-output* facilitates problem comprehension. According to this concept, a project can be thought of as an input-output model as in Fig. 6.4. The decision box represents the problem and the solution method, including the decision criterion. The input(s) to the box represents all the costs, including project-associated efforts that can be quantified in monetary terms. There may exist other efforts, essential to project completion that are difficult to quantify in dollars. The output of the decision box is all the benefits from the project (Figure 6.6).

Consider for example research and development (R&D) projects, in which companies invest significant sums to develop new products and services, with the expectation of profits. The expenses for the project are the input and the profits through sales are the output. The basis of decision-making in selecting the best alternative of a project may be one of the following:

1. The output for a given input is the largest.
2. The input for a given output is the smallest.
3. The ratio of the output to input is the largest where both input and output vary from one alternative to another. Such a ratio represents the project's financial or economic efficiency. We discuss this further as *benefit-cost ratio* later in Chapter 8.

On the basis of the input-output concept, engineering economics problems can be grouped into three categories, as discussed in the following subsections. Remember that *input means costs* and *output means benefits*.

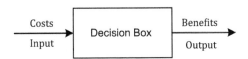

Figure 6.6 Input-output model.

6.3.1 Fixed Output

In many projects the output is fixed, i.e., the benefit from each alternative of the project is the same. Consider for example the need to replace the elevator in a building. There may be several candidate elevators from different suppliers as a possible choice. What is the output in this project? The company does not receive any direct monetary benefit by investing in the elevator. The benefits (output) in this case are indirect, namely the convenience to employees who use the elevator, and the ease with which goods can be moved in and out of the building. In such a project the output is the same irrespective of which elevator is selected to replace the existing one. In other words, the output of this project is *fixed*. Such projects represent *fixed-output-variable-input*, or simply fixed-output, economics problems. The term variable-input signifies the fact that the cost of the candidate elevators may be different for each. The decision-making involved in buying a car is another example of fixed-output problem.

Example 6.5: Input-Output

Siddharth is a plant engineer. One of the machines in his plant is old and needs replacing. Analyze this situation in light of the input-output concept.

SOLUTION:

In applying the input-output concept to Siddharth's decision-making problem of machine replacement, let us first consider the input. The cost of the replacement machine will most likely be budgeted. The purchase cost should include installation and technical training, if any. After receiving quotations from various suppliers, Siddharth will prepare a proposal, for approval by the upper management, which will include justification for replacing the existing machine.

It is not obvious from the problem statement whether the machine is general-purpose or specialized. If specialized, there may not be much choice of suppliers, sometimes none at all. If that is the case, Siddharth's proposal and budgeting will be straightforward. He will enclose with the proposal a copy of the quotation, asking for approval. If, on the other hand, the machine is general-purpose, there may be several choices yielding what we have called multi-alternative project. Siddharth will apply the principles of engineering economics to the various alternatives and select the best based on the company's criterion of investment.

Next, the output. Siddharth may have to consider whether the replacement should just suffice in doing what the existing machine does. Or, should he seize the opportunity to incorporate current technologies by procuring a more modern machine. Such a machine may yield higher productivity or better product quality, or both, using advanced features such as on-line monitoring, predictive maintenance, ports for interfacing with computers, etc.

If the machine is of the general-purpose type, and several alternatives are available at different costs, then the project's input is variable. But all the machines will do the same—replace the current machine. So, the alternatives' output is fixed. Thus, the project is represented by a variable-input-fixed-output problem.

The extent of analysis, depicted by the decision box in Fig. 6.6, will depend on the issues previously mentioned relating to input and output, and on other associated matters. The PW-criterion discussed in this chapter, or any of the other five criteria presented in the other chapters of Part II, can be used to make the final decision.

Fixed-output problems involving multiple alternatives can be solved by comparing only the present worth of alternatives' costs since the benefit from each alternative is the same (fixed). No need to analyze the benefits keeps the solution shorter and simpler. Obviously, the criterion of decision-making in such problems is to *minimize the present worth of costs*, as illustrated in Example 6.6.

Example 6.6: Present Worth

Hari has been using an old car he purchased as a freshman (first year of college). During the 3 years in his current job as an industrial technologist he has saved enough money to buy a new car. He is considering a Honda Accord, Chevy Prism, and Plymouth Acclaim, each with 7-year useful life. Given the following data which one should he buy if the decision criterion is present worth? Assume an annually compounded interest rate of 9% per year.

Car	First Cost	Annual Cost	Salvage Value
Honda Accord	$20,000	$300	$3,000
Chevy Prism	$18,000	$500	$2,000
Plymouth Acclaim	$17,500	$750	$1,500

SOLUTION:

In this problem the benefit—the convenience of owning a car—is the same. The convenience translates into indirect monetary benefits, such as time saved commuting to work, which are intangible. Since the benefit of owning a car is the same, it is a fixed-output problem. The decision can therefore be made solely on the basis of the input, i.e., the present worth of the costs, which should be minimized.

Of the given data, the salvage value represents a cash inflow from the investment. It can therefore be subtracted, after accounting for its time value, from the first cost to determine the net cost on the car. If you can visualize the problem without its cashflow diagram, you should be able to follow the analysis here; else, sketch a diagram to help you.

The present worth of the costs for the three alternatives is:

$$PW_{costs(Accord)} = 20,000 + 300\left(\frac{P}{A,9\%,7}\right) - 3,000\left(\frac{P}{F,9\%,7}\right)$$

$$= 20,000 + 300 \times 5.033 - 3,000 \times 0.5470$$

$$= 20,000 + 1,510 - 1,641$$

$$= \$19,869$$

$$PW_{costs(Prism)} = 18,000 + 500\left(\frac{P}{A,9\%,7}\right) - 2,000\left(\frac{P}{F,9\%,7}\right)$$

$$= 18,000 + 500 \times 5.033 - 2,000 \times 0.5470$$

$$= 18,000 + 2,517 - 1,094$$

$$= \$19,423$$

$$PW_{costs(Acclaim)} = 17,500 + 750\left(\frac{P}{A,9\%,7}\right) - 1,500\left(\frac{P}{F,9\%,7}\right)$$

$$= 17,500 + 750 \times 5.033 - 1,500 \times 0.5470$$

$$= 17,500 + 3,775 - 820$$

$$= \$20,454$$

Comparing the three, Hari should buy a Chevy Prism since its present worth of costs at $19,423 is the lowest.

6.3.2 Fixed Input

The second groups of problems involve projects, whose inputs are fixed, i.e., the alternatives cost the same. The decision maker's task is to select the alternative which maximizes the output (income) from the investment. Such projects are of the *fixed-input-variable-output*, or simply fixed-input type. The term variable-output signifies that the output (benefits) from the various alternatives differ from each other. An example of fixed-input-variable-output project is hiring someone at a given budgeted salary, where the decision maker selects the best[7] from among the applicants.

Since the alternative's input is fixed, the inclusion of costs in the analysis will not affect the final decision; it may however render the analysis lengthy. Thus, in fixed-input projects we select the alternative whose present worth of benefits is the largest. In other words, we maximize $PW_{benefits}$.

6.3.3 Variable Input and Output

Several engineering economics problems relate to projects whose alternatives' costs and benefits vary. Such problems are of the variable-input-variable-output type. The analyses of such problems are relatively lengthier since both the costs and the benefits are to be accounted for. The present worth of each alternative is evaluated, as illustrated in Example 6.2. Once the alternatives' PWs have been determined, they are compared to select the alternative with the largest +ve PW. Example 6.7 offers an illustration.

Example 6.7: Variable Input

The downsizing in her company has finally affected Nina. In spite of 10 years dedicated service to the company, she has to go at the end of this year. Fortunately, she has some savings. After a careful assessment of her financial commitments, Nina decides to venture into the owner-business world. Her father had left her a parcel of land adjacent to the main street in the town she lives. The business possibilities and the related data are:

Business Type	Initial Investment	Net Annual Income	Terminal Value
Farmer's Market	$120,000	$18,000	$45,000
Gas Station	$240,000	$36,000	$60,000
Grocery Store	$140,000	$25,000	$48,500

Nina plans to sell the business at the end of 25 years for the estimated terminal value. Her savings can earn annually compounded interest of 12% per year in a long-term CD. Which business is the best for Nina on the basis of the PW-criterion?

SOLUTION:

There are three alternatives here. Note that both the cost of and the benefit from the three businesses are different. It is thus a variable-input-variable-output problem. We therefore evaluate the PWs of the three businesses and compare them to select the most favorable. The value of i is 12%. If Nina did not have her savings to invest, i would have been the interest rate charged by financial institutions on business loans to customers like her.

Again, a cashflow diagram is not necessary if you can comprehend the problem without it. Using Equation (6.1) and the relevant functional notations, we get:

---

[7] Expected to contribute the most toward company goals.

$$PW_{\text{farmer's market}} = PW_{\text{benefits}} - PW_{\text{costs}}$$

$$= \left\{ 45,000 \left(\frac{P}{F,12\%,25} \right) + 18,000 \left(\frac{P}{A,12\%,25} \right) \right\} - 120,000$$

$$= 45,000 \times 0.0588 + 18,000 \times 7.843 - 120,000$$

$$= 2,646 + 141,174 - 120,000$$

$$= \$23,820.$$

$$PW_{\text{gas station}} = PW_{\text{benefits}} - PW_{\text{costs}}$$

$$= \left\{ 60,000 \left(\frac{P}{F,12\%,25} \right) + 36,000 \left(\frac{P}{A,12\%,25} \right) \right\} - 240,000$$

$$= 60,000 \times 0.0588 + 36,000 \times 7.843 - 240,000$$

$$= 3,528 + 282,348 - 240,000$$

$$= \$45,876$$

$$PW_{\text{grocery store}} = PW_{\text{benefits}} - PW_{\text{costs}}$$

$$= \left\{ 48,500 \left(\frac{P}{F,12\%,25} \right) + 25,000 \left(\frac{P}{A,12\%,25} \right) \right\} - 140,000$$

$$= 48,500 \times 0.0588 + 25,000 \times 7.843 - 140,000$$

$$= 2,852 + 196,076 - 140,000$$

$$= \$58,928$$

A comparison of these present worths suggests that Nina should invest in the grocery store since its PW of \$58,928 is the largest.

6.4 PRESENT-WORTH ANALYSIS PERIOD

While analyzing problems under the PW-criterion, a situation may arise that needs special attention. This happens when the alternatives' useful lives are different. The relevant question is: What analysis period should be used? In other words, for what time horizon should the alternatives be compared with each other?

Consider a machine that needs replacing since it has been a production bottleneck. If it is specialized, there may be only one type in the market, i.e., there is no choice. Such problems are simple; collect the cost and benefit data on the new machine covering the period of its useful life and carry out the required analysis. The analysis period is thus the same as the useful life.

Next consider that the machine is general-purpose with one or more alternatives in the market. The PW-analysis may involve either of the situations discussed in the following two subsections.

6.4.1 Equal Lives

If the alternative machines A and B have the same useful life, there is little difficulty since both can be analyzed over this period. Thus, the analysis period is their useful life. Again, gather all the pertinent cost and benefit data, and evaluate and compare their present worths. Select the one whose present worth is more favorable. The same methodology applies to problems involving more than two alternatives, as illustrated in Example 6.8.

Example 6.8: Variable Input

The loading and unloading of a machine can be enhanced significantly by using a robotic system. Three systems, whose types and data are given as follows, are under consideration. If for the investment the annual interest rate is 15%, which system should be purchased on the basis of the PW-criterion?

Type ,	Initial Cost	Annual Maintenance	Useful Life (in years)
Robota	$65,000	$300	10
Automata	$70,000	$250	10
Robo	$75,000	$ 20	10

SOLUTION:

Note that the useful life for the three types is the same. Therefore, an analysis period equal to their useful life, i.e., 10 years, is decided upon. Based on the input-output concept, the output from each type is the same, i.e., each system will be able to carry out the desired loading and unloading tasks. Thus, the decision may be based solely on the present worth of costs, which are:

$$PW_{costs(Robota)} = 65,000 + 300\left(\frac{P}{A,15\%,10}\right)$$

$$= 65,000 + 300 \times 5.019$$

$$= 65,000 + 1,506$$

$$= \$66,506$$

$$PW_{costs(Automata)} = 70,000 + 250\left(\frac{P}{A,15\%,10}\right)$$

$$= 70,000 + 250 \times 5.019$$

$$= 70,000 + 1,255$$

$$= \$71,255$$

$$PW_{costs(Robo)} = 75,000 + 20\left(\frac{P}{A,15\%,10}\right)$$

$$= 75,000 + 20 \times 5.019$$

$$= 75,000 + 100$$

$$= \$75,100$$

Comparing these, Robota is selected since its present worth of costs at $66,506 is the minimum.

6.4.2 Unequal Lives

If the useful lives of the alternatives are different, then what analysis period to use for the problem? Let us say that machine A has a useful life of 3 years and B of 5 years. For what time horizon do we compare A with B? Three years or 5 years?

At the end of 3 years, machine A will need to be replaced again, while B will still be useful. If we replace A and keep it for another 3 years, accounting for the associated costs and benefits, then A will be in use when B will need to be replaced at the end of the fifth year. Thus, we seem to be chasing a time "mirage." This is the difficulty unequal lives create in deciding an appropriate analysis period.

The difficulty is resolved by extending the analysis period to an extent that both A and B terminate together. Such an analysis period[8] corresponds to the *least common multiple* (LCM) of their useful lives. In the case under discussion, it is 15 years (LCM of 3 and 5 years). Thus, the comparison between alternative A and B should be made by looking through a 15-year time "window." During this period, machine A will be replaced four times (4 × 3 = 12 years of life due to replacement, plus 3 years of the first acquisition), while machine B twice. At the end of 15 years both would be ready for simultaneous replacement for further use.

When an analysis period longer than the alternative's useful life is considered in a problem, three major assumptions are implicitly made. These are: (i) the suppliers will continue to be in business, (ii) they will offer the alternative at the same price, and (iii) benefits will remain unchanged throughout the analysis period.

Example 6.9: Useful Life

Abraham's personal computer (PC) is slow and lacks features for the current application programs. There are three choices: upgrade the existing system, buy a new PC and software, or buy a one-year-old "fully-loaded" PC. For the following data, which one should he opt for if the decision criterion is PW and the applicable yearly interest rate is 9%, compounded annually?

Choice	First Cost	Annual Cost	Useful Life (in years)	Salvage Value
Upgrade	$3,000	$200	2	$250
New PC	$8,000	$100	4	$500
Old PC	$5,000	$150	3	$300

SOLUTION:

Since PW is the criterion and the useful lives of the three choices are different, the analysis period should be the LCM of their lives of 2, 3, and 4 years, i.e., 12 years. Fig. 6.7 shows the cashflow diagrams[9] for the choices. Note that some vectors have not been labeled since they are repetitive.

The present worth[10] of the choices are found as

---

[8] If the LCM-based analysis period becomes too long, then (a) Use an analysis period equal to the useful life of one of the alternatives along with its costs and benefits, but use the prorated costs and benefits of the other alternatives, or (b) Use any of the other analysis methods, such as annual worth (Chapter 8) or rate of return (Chapter 9).

[9] If you have difficulty comprehending the diagrams, reread Chapter 2.

[10] Following the standard convention that costs (vectors down) are negative, while benefits (vectors up) are positive.

$$PW_{upgrade} = -3,000 - 200\left(\frac{P}{A,9\%,12}\right) - (3,000-250)\left\{\left(\frac{P}{F,9\%,2}\right)\right.$$

$$+\left(\frac{P}{F,9\%,4}\right)+\left(\frac{P}{F,9\%,6}\right)+\left(\frac{P}{F,9\%,8}\right)+\left.\left(\frac{P}{F,9\%,10}\right)\right\}$$

$$+250\left(\frac{P}{F,9\%,12}\right)$$

$$= -3,000 - 200\times7.161 - 2,750(0.8417+0.7084+0.5963$$

$$+0.5019+0.4224) + 250\times0.3555$$

$$= -3,000 - 1,432 - 2,750\times3.0707 + 89$$

$$= -3,000 - 1,432 - 8,444 + 89$$

$$= -\$12,788$$

$$PW_{new} = -8,000 - 100\left(\frac{P}{A,9\%,12}\right) - (8,000-500)\left\{\left(\frac{P}{F,9\%,4}\right)\right.$$

$$+(P/F,9\%,8)\left.\right\} + 500\left(\frac{P}{F,9\%,12}\right)$$

$$= -8,000 - 100\times7.161 - 7,500(0.7084+0.5019)$$

$$+500\times0.3555$$

$$= -8,000 - 716 - 7,500\times1.2103 + 178$$

$$= -8,716 - 9,077 + 178$$

$$= -\$17,615$$

$$PW_{old} = -5,000 - 150\left(\frac{P}{A,9\%,12}\right) - (5,000-300)\left\{\left(\frac{P}{F,9\%,3}\right)\right.$$

$$+\left(\frac{P}{F,9\%,6}\right)+\left.\left(\frac{P}{F,9\%,9}\right)\right\} + 300\left(\frac{P}{F,9\%,12}\right)$$

$$= -5,000 - 150\times7.161 - 4,700(.7722+0.5963+0.4604)$$

$$+ 300\times0.3555$$

$$= -5,000 - 1,074 - 4,700\times1.8289 + 107$$

$$= -6,074 - 8,596 + 107$$

$$= -\$14,563$$

Comparing the three present worths which are all negative[11], Abraham decides to upgrade the existing system since it is the most favorable one (smallest −ve PW, i.e., lowest net cost).

---

[11] One could have used the *present worth of costs* as the decision variable since it is a fixed-output problem. In that case, all would have been positive. The decision would nevertheless have been the same.

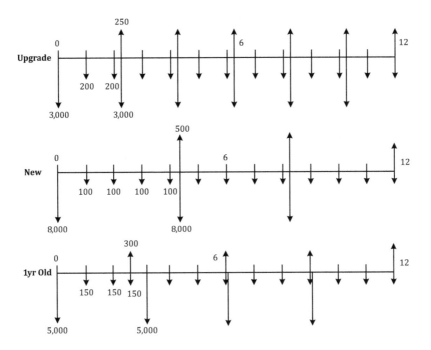

Figure 6.7 Alternative with unequal lives.

6.5 PRESENT WORTH CAPITALIZED COST

Engineers sometimes develop and implement facilities that are expected to provide use or service forever. Highways, bridges, dams, and school buildings are some of the examples. Their useful life is perpetual. Such facilities are usually constructed by governments with the taxpayers' money.

Since perpetual-life projects last forever, their economic analysis must cover a long period; in fact, the analysis period is infinite. Besides the initial cost, they may require continual maintenance and renovation, as and when necessary. The on-going costs must be budgeted for, or means to raise them in future planned for, at the time of initial construction.

If the funds for maintenance are budgeted at the time of initial construction, they are kept in an interest-bearing account, and the earned interest used as planned. Such a fund is called *capitalized cost*, since it continues to exist as capital—as *seed corn*. City and other local governments often make provision in their budgets for capitalized costs of various projects. If P is the fund invested as capitalized cost at an annual interest rate *i*, then the year-end interest is *i*P. This *i*P can be disbursed (used up) annually without depleting P.

Thus, a perpetual-life project often has two components: initial cost and capitalized cost. In such cases, no funds will have to be raised in future for the project. As an alternative to capitalized cost, annual taxation and bonds may be used to raise the maintenance funds as and when necessary.

An example of the capitalized-cost concept is found in higher education. Colleges and universities often raise funds as endowments to offer perpetual scholarships. To offer an annual scholarship A perpetually, the required endowment gift is P so that *i*P = A. In other words,

$$P = \frac{A}{i}$$

where P is the endowment (capitalized cost).

The donor's gift P is invested at an annual interest rate i, generating an annual income of A perpetually. The same logic is valid also for maintaining perpetual projects as illustrated in Example 6.10.

Example 6.10: Capitalized Cost

The City of Hattiesburg finds Hardy Street congested during peak traffic hours and is planning to build a parallel street to reduce congestion. The new street is to be 10 miles long and the initial cost of construction is $500,000 per mile. What is the total project cost if a provision of $10,000 for annual repairs is also made? Assume $i = 5\%$ per year, compounded annually.

SOLUTION:

$$\text{Initial cost} = \$500,000 \text{ per mile} \times 10 \text{ miles}$$

$$= \$5,000,000$$

$$\text{Capitalized cost for annual repairs}, P = \frac{A}{i}$$

$$= \frac{10,000}{0.05}$$

$$= \$200,000$$

Therefore,

$$\text{Total project cost} = \$5,000,000 + 200,000 = \$5.2 \text{ million}$$

Example 6.11: Future Worth

Khalda begins her 4 years of college to earn a BS degree in computer science. The annual cost of education is expected to be $15,000. The BS degree is likely to enable her to earn $10,000 more (in relation to if she did not have the degree) per year during her expected professional career of 30 years. What is the worth of additional earnings at career-end if $i = 8\%$ per year[12]?

SOLUTION:

The annual cost of the BS degree is $15,000 for 4 years. The financial return from the degree is the $10,000 additional annual income over 30 years. Following the convention that a cashflow is posted at period-end, the complete diagram is shown in Fig. 6.8. Note that the question specifically asks about the future worth of only the additional income. Thus, the cash outflows at periods 1 through 4 are irrelevant to the solution. The question is also explicit about the time period at which future worth is to be determined. It is at the end of the 30-year career, i.e., at period 34 (30 years of professional career following 4 years of education[13]).

---

[12] In the absence of any specific mention, always assume the compounding to be annual. Also, if unspecified, the interest rate should be considered annual.

[13] For the purpose of this example, one can sketch another diagram that does not show periods 1 through 4. Such a diagram will begin at period 4 of Fig. 7.1 as period 0, and end with n = 30.

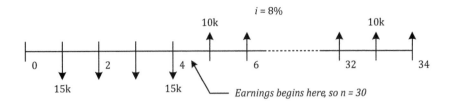

Figure 6.8 Diagram for Example 6.11.

With the uniformity in the cash inflows it is obvious whose exploitation will save time and effort. For the given data: A = $10,000, n = 30, and *i* = 8%,

$$FW_{benefits} = F = A\left(\frac{F}{A, i, n}\right)$$

$$= 10,000\left(\frac{F}{A, 8\%, 30}\right)$$

$$= 10,000 \times 113.283$$

$$= \$1,132,830$$

Thus, the future worth of additional earnings from the BS degree is more than a million dollars.

A project's future worth (FW) accounts for the net effect of benefits over costs, and is the basic parameter for decision-making under the FW-criterion. It is the difference between the future worth of benefits and future worth of costs. Thus,

$$FW = FW_{benefits} - FW_{costs} \tag{6.2}$$

Note that we subtract the future worth of costs from the future worth of benefits. This renders the project's FW positive in cases where future worth of benefits exceeds future worth of costs—a desirable aim. Thus, a positive value for FW means that the project is financially attractive, and hence normally funded.

Example 6.12: Future Worth

Determine the FW of Khalda's higher education in Example 6.11 and comment on its value.

SOLUTION:

From Example 6.11, the future worth of benefits is already known to be $1,132,830.

As a continuation of the discussions therein, we next determine the future worth of the costs. Referring to Fig. 6.2, the four costs—$15,000 each at periods 1 through 4—need to be transferred to period 34, which we should do as efficiently as possible. One[14] of the most efficient ways is to transfer them first to period 0 as their present value P, and then

---

[14] The other way is to treat them as A and transfer first to period 4 as F. This F is then treated as P and transferred to period 34 as F-value.

transfer P to period 34 as the final sum F. Again, let us exploit the uniform-series pattern in these cash outflows. Thus,

$$P = A\left(\frac{P}{A,i,n}\right)$$

$$= 15{,}000(P/A,8\%,4)$$

$$= 15{,}000 \times 3.312$$

$$= \$49{,}680$$

Next,

$$FW_{costs} = F = P\left(\frac{F}{P,i,n}\right)$$

$$= 49{,}680\left(\frac{F}{P,8\%,34}\right)$$

$$= 49{,}680 \times 13.690$$

$$= \$680{,}119$$

The future worth of the project is thus:

$$FW = FW_{benefits} - FW_{costs}$$

$$= \$1{,}132{,}830 - \$680{,}119$$

$$= \$452{,}711$$

Since the project's FW is +ve, the BS degree is a good investment[15] on the basis of the FW-criterion.

6.6 FUTURE WORTH METHOD

The FW-method of solving engineering economics problems involves determination of the future worth of the project or future worths of project alternatives. Using the FW-criterion, the better of the two alternatives, or the best of all the alternatives, is selected. Thus, the application of this method involves two tasks:

1. Evaluation of the future worth(s)
2. Decision on the basis of the FW-criterion

We discussed the first task in Section 6.1. In this section we discuss decision-making based on the FW-criterion.

From an engineering economics viewpoint, projects can be categorized in two groups: no-alternative projects, or projects with alternative. In no-alternative projects, there is only one candidate for analysis; there is no choice. The data for the sole project are collected and analyzed to determine FW, as discussed in Section 6.1. No-alternative projects do not involve

---

[15] Note that we have not accounted for—because it is difficult—other non-monetary benefits higher education may bring to enhance Khalda's quality of life (directly) and of the community she lives in (indirectly).

much of the decision-making, beyond the evaluation of FW, as illustrated in Examples 6.11 and 6.12, and their approval if the FW is +ve.

Example 6.13: Future Worth

Determine the future worth of a 10-year project whose first cost is $12,000, net annual benefit is $2,500, and salvage value is $2,000. Assume an annually compounded interest rate of 8% per year.

SOLUTION:

If you can visualize the problem without its cashflow diagram, then proceed directly to the calculations. Otherwise, sketch a diagram first to comprehend the problem. Note that the salvage value is recovered at the end of useful life, and as cash inflow it is a benefit to the project. Since it is not obvious from the problem statement when the benefits begin to accrue, we need to make an assumption. A reasonable assumption is that the benefits begin to be realized immediately following the investment.

Using Equation (6.2) and the associated functional factors, FW is determined[16] as:

$$FW = FW_{benefits} - FW_{costs}$$

$$= \left\{ 2,500 \left(\frac{F}{A,8\%,10} \right) + 2,000 \right\} - 12,000 \left(\frac{F}{P,8\%,10} \right)$$

$$= 2,500 \times 14.487 + 2,000 - 12,000 \times 2.159$$

$$= 36,218 + 2,000 - 25,908$$

$$= \$12,310$$

This example has the same data as Example 6.3, and therefore, the present worth answer in Example 6.3 must be equivalent to the future worth answer in Example 6.13. The value of PW (or P) in Example 6.3 was $5,701, which indeed is equal to the value of FW (or F) in Example 6.13, since:

$$F = P \left(\frac{F}{P,i,n} \right)$$

$$= 5,701 \left(\frac{F}{P,8\%,10} \right)$$

$$= 5,701 \times 2.159$$

$$= \$12,308$$

The $2 difference between $12,308 based on Example 6.3's PW and $12,310 of Example 6.13 is due to the error introduced by the rounding-off of functional factors' values in the interest tables.

In multi-alternative projects, there are several feasible alternatives (candidates) to the same project. For example, as a replacement of an old machine there may be three types of machines—A, B, and C—in the market. In such a case, the process of selecting the best machine is the project[17], while the three candidate machines (A, B, and C) are the alternatives.

---

[16] If you have difficulty understanding the functional factors used here, reread Chapter 4.

[17] The alternatives themselves may sometimes be called *projects*. In multi-alternative cases, therefore, the two terms *project* and *alternative* are used synonymously.

The decision-making in multi-alternative projects involves selecting the alternative whose FW is the most favorable, i.e., the largest of all the +ve FWs. For example, if the FWs of the alternatives are $300, $650, $268, and $987, then select the one whose FW is $987. If the FWs of all the alternatives are negative, then none should be selected in a normal situation, since –ve FW means that the investment will incur a loss. If there are compelling business reasons that demand that a selection must be made, then select the alternative whose –ve FW is the smallest. For example, if the FWs of four alternatives are –$450, –$350, –$896, and –$936, then select the one whose FW is –$350. Being the smallest of the –ve FWs, this will minimize the loss in investment, and is thus the most attractive of the four alternatives. Where some of the alternatives yield positive FWs and others yield negative FWS, discard those with –ve FWs and select the one with the highest +ve FW. In short, the decision must eventually be in favor of financially the most *attractive* alternative.

Example 6.14: Future Worth

World's Best Manufacturer is considering five new projects[18] as part of next year's R&D efforts. On the basis of future worth which one should be selected for funding, assuming that the investment must earn an annual return of 12%? All five projects will take 2 years to complete and will be profitable during the subsequent 5 years.

Project	Annual Cost (during development)	Annual Income (during profitability)
A	$2,000	$3,500
B	$1,800	$3,200
C	$2,200	$2,300
D	$1,200	$2,000
E	$3,500	$4,800

SOLUTION:

The R&D costs are incurred during the 2 years of the development. The benefits are realized during the 5 years following development. The decision-making involves two steps: determination of FW for each project, and decision based on the FW-criterion.

Let us first evaluate the FWs. Since the projects are similar except the data, their cash-flow diagrams will be alike. Referring to the one for project A in Fig. 6.5, its future worth can be determined as

$$FW_A = FW_{benefits} - FW_{costs}$$

$$= 3,500\left(\frac{F}{A,12\%,5}\right) - \left\{2,000\left(\frac{F}{P,12\%,5}\right) + 2,000\left(\frac{F}{P,12\%,6}\right)\right\}$$

$$= 3,500 \times 6.353 - \{2,000 \times 1.762 - 2,000 \times 1.974\}$$

$$= 3,500 \times 6.353 - 2,000(1.762 + 1.974)$$

$$= 3,500 \times 6.353 - 2,000 \times 3.736$$

$$= 22,236 - 7,472$$

$$= \$14,764$$

---

[18] In here, the term *projects* is more appropriate than *alternatives* since each may be technically different from the others. From the analysis viewpoint, however, they are in fact alternatives.

The calculations for the other projects will be similar except for the last three lines. Thus,

$$FW_B = 3,200 \times 6.353 - 1,800 \times 3.736$$

$$= 20,330 - 6,725$$

$$= \$13,605$$

$$FW_C = 2,300 \times 6.353 - 2,200 \times 3.736$$

$$= 14,612 - 8,219$$

$$= \$6,393$$

$$FW_D = 2,000 \times 6.353 - 1,200 \times 3.736$$

$$= 12,706 - 4,483$$

$$= \$8,223$$

$$FW_E = 4,800 \times 6.353 - 3,500 \times 3.736$$

$$= 30,494 - 13,076$$

$$= \$17,418$$

The second step involves selecting the project[19]. As discussed earlier, the most attractive project should be selected. In this case, the future worth of all five projects are positive, so select the one with the largest +ve FW[20], i.e., project E.

6.7 WHEN TO USE FUTURE WORTH

The FW-criterion and its application to engineering economics problems are very similar to the PW-criterion. An obvious question is when to use one, and not the other. The simple rule is: If the *present* is deemed to be important in decision-making, use the PW-method; by the same token, if the *future* is important, then use the FW-criterion. The time period to which *future* refers should be the one of *relevance*.

Consider financial planning by an investor who desires to build retirement equity. In here, the *future*—rather than present—is relevant. And the future occurs at the time the investor would like to retire, which becomes the pertinent time period in decision-making. The obvious criterion in this case is future worth. In contrast, for capital investment projects the appropriate criterion is present worth. In here, the aim is to invest in a resource now for generating profits later. In general, engineering companies are not in the business of financial investment, but in engineering investment. So, the PW-criterion is more suited to the engineering industry, while the FW-criterion to the financial industry.

---

[19] Being the same as Example 6.4 except the criterion, this example offers an opportunity to compare its FW-result with the corresponding PW-result in Example 6.4. For instance, FW_B in Example 6.14 can be compared with the FW_B, obtained from PW_B, of Example 6.4, as $FW_B = PW_B$ (F/P,12%,7) = 6,155 × 2.211= \$13,608 The \$3 difference in this value (\$13,608) of FW_B and that (\$13,605) in Example 6.14 is due to the rounding-off error in the functional factor's tabular value.

[20] As an alternative approach the FWs in Example 6.14 could have been determined directly from the corresponding PWs in Example 6.4.

Example 6.15: Annual Worth

Judy has just won $400,000 in a lottery. Compare this windfall with her husband's annual salary of $40,000; her husband expects to continue in his job for the next 25 years. Assume i = 6% per year.

SOLUTION:

Judy's windfall of $400,000 is a benefit which she receives now. Her husband, on the other hand, expects to earn $40,000 annually for the next 25 years. Their cash inflows can be compared by looking at the data through the same time "window." There are two ways to compare. We can determine the present worth (PW) of her husband's future salaries and compare it with her $400,000 windfall. Alternatively, we can determine her windfall's equivalent uniform-series annual worth AW and compare it with her husband's $40,000 annual salary. Following the latter, the AW of Judy's windfall for n = 25 is:

$$AW = 400,000\left(\frac{A}{P,6\%,25}\right)$$

$$= 400,000 \times 0.0782$$

$$= \$31,280$$

Thus, Judy's windfall is equivalent to an annual salary of $31,280 over the next 25 years. Her husband will thus continue to earn $40,000 – $31,280 = $8,720 more per year.

Example 6.16: Annual Worth

What is the annual worth of the cashflows in Fig. 6.9(a) if M = $1,200, N = $400, Q = $300, R = $600, S = M, and T = Q? Assume i = 6% per year.

SOLUTION:

The annual worth AW is the difference between $AW_{benefits}$ and AW_{costs}. To determine it, we therefore evaluate $AW_{benefits}$ and AW_{costs} which, referring to Fig. 6.9(b), are E and D.

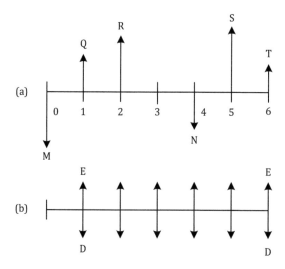

Figure 6.9 Analysis of future worth.

Since there exists no "strong" pattern in the given cashflows, we will follow the straight-forward method. But how do we transfer the given cashflows into E and D? There is no direct way to do this. Instead, the given irregular benefits and costs are converted first to their P-values at period 0. These P-values are then converted to their A-values (E and D).

Note that S = M = \$1,200, and T = Q = \$300. For the P-value of the cash inflows (benefits) we have:

$$PW_{benefits} = Q\left(\frac{P}{F,6\%,1}\right) + R\left(\frac{P}{F,6\%,2}\right) + S\left(\frac{P}{F,6\%,5}\right) + T\left(\frac{P}{F,6\%,6}\right)$$

$$= 300 \times 0.9434 + 600 \times 0.8900 + 1,200 \times 0.7473 + 300 \times 0.7050$$

$$= 283.0 + 534.0 + 896.8 + 211.5$$

$$= \$1,925.3$$

For the P-value of the cash outflows (costs) we have:

$$PW_{costs} = M + N\left(\frac{P}{F,6\%,4}\right)$$

$$= 1,200 + 400 \times 0.7921$$

$$= 1,200 + 316.8$$

$$= \$1,516.8$$

We next convert these P-values to their annual equivalents yielding us:
$AW_{benefits}$ and AW_{costs} respectively.
Thus,

$$AW_{benefits} = PW_{benefits}\left(\frac{A}{P,6\%,6}\right)$$

$$= 1,925.3 \times 0.2034$$

$$= \$391.6$$

$$AW_{costs} = PW_{costs}\left(\frac{A}{P,6\%,6}\right)$$

$$= 1,516.8 \times 0.2034$$

$$= \$308.5$$

As a difference between $AW_{benefits}$ and AW_{costs} (same as using Equation (6.3)), the annual worth is:

$$AW = AW_{benefits} - AW_{costs}$$

$$= 391.6 - 308.5$$

$$= \$83.1$$

Example 6.17: Annual Worth

For the project whose cashflow diagram is in Fig. 6.10, what is the value of $AW_{benefits}$? Assume $i = 8\%$ per year.

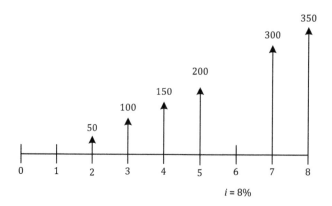

Figure 6.10 Application of tailoring.

SOLUTION:

The value of $AW_{benefits}$ depends on the six cash inflows. These inflows are observed to follow an arithmetic series pattern, except that there is no vector at period 6. We can tailor the cash inflows by inserting a $250 upward vector at period 6. This will yield an arithmetic series with G = 50. To compensate for the inserted $250 upward vector, a downward vector of the same magnitude is applied at the same period, i.e., at 6.

Now, how do we convert the arithmetic series data to their equivalent uniform annual cash inflow? Have we discussed any relationship earlier somewhere that can do it? Yes indeed, we have, and it is Equation 4.6(b) in Chapter 4. Thus, the equivalent uniform annual cash inflow R_1 for the tailored arithmetic series is:

$$R_1 = G\left(\frac{A}{G, i, n}\right)$$

$$= 50\left(\frac{A}{G, 8\%, 8}\right)$$

$$= 50 \times 3.099$$

$$= \$154.95$$

But we must also account for the effect of the compensating downward vector at period 6 on $AW_{benefits}$. This requires that we first transfer the $250 vector from period 6 to its P-value at period 0, and then convert this P-value to its equivalent A-value, say R_2, as follows.

$$R_2 = \left\{250\left(\frac{P}{F, 8\%, 6}\right)\right\}\left(\frac{A}{P, 8\%, 8}\right)$$

$$= 250 \times 0.6320 \times 0.1740$$

$$= \$27.49$$

Note that since R_1 is upward and R_2 is downward, it is their difference that yields the value of $AW_{benefits}$. Thus,

$$AW_{benefits} = R_1 - R_2$$

$$= \$154.95 - \$27.49$$

$$= \$127.46$$

6.8 ANNUAL WORTH METHOD

The application of the AW-method involves two tasks:

1. Evaluation of the annual worth(s)
2. Decision-making under the AW-criterion (selection of the most favorable project)

We discussed the first task in Section 6.1. In this section we discuss the AW-criterion and how to apply it.

Engineering economics projects can be categorized into one of the two groups: no-alternative projects, or multi-alternative projects. In no-alternative projects, there is only one candidate, with no choice. The data for the sole project is collected and AW determined, as discussed in Section 6.1. If the AW is +ve, the project is funded. No-alternative projects therefore do not involve much of the decision-making, beyond the evaluation of the AW, as illustrated earlier in Example 6.11 and in Example 6.13.

In multi-alternative projects, there are several feasible alternatives to the project. The decision-making involves selecting the alternative whose AW is the most favorable, i.e., the largest of all the positive AWs. For example, if the alternatives' AWs are $300, $650, $268, and $987, then select the one whose AW is $987. If the AWs of all the alternatives[21] are negative, then none should be selected in a normal situation, since –ve AW means that the annual cost is greater than the annual benefit. However, if there is a business compulsion that demands that one of the alternatives must be selected (or there are significant intangible benefits), then select the one whose negative AW is the smallest. For example, if the AWs of four project alternatives are –$450, –$350, –$896, and –$936, then select the one whose AW is –$350. Being the smallest of the –ve AWs, this minimizes the loss, and therefore is the most favorable. In case some of the alternatives yield positive AWs while others yield negative AWs, then discard all the alternatives with –ve AWs, and select from the remaining the one with the largest +ve AW. In short, the decision must be in favor of financially the most *attractive* project. Example 6.14 illustrates the application of AW method to a multi-alternative project.

6.9 SALVAGE VALUE

The salvage value is recovered at the end of useful life. It should be converted, by treating it as a future sum, to its equivalent uniform annual cashflow. This cash inflow can be thought of as a benefit and is generally added to $AW_{benefits}$, as in Example 6.13. It can, on the other hand, be thought of as partial recovery of the first (initial) cost and therefore subtracted from the project's AW_{costs}. In AW-analysis, the final decision in a multi-alternative project is not affected by how we treat the salvage value—as part of the benefit or as a recovery of the first cost—as long as all the alternatives account for the salvage value the same way.

Example 6.18: Annual Worth

Determine the annual worth of a 10-year project whose first cost is $12,000, annual benefit is $2,500, and salvage value is $2,000. Assume an annually compounded interest rate of 8% per year.

---

[21] The terms *project* and *alternative* are used synonymously.

SOLUTION:

If you can visualize the problem without its cashflow diagram, then proceed directly to the calculations. Otherwise, sketch a diagram to comprehend the problem. Note that the salvage value is recovered at the end of useful life as a benefit (inflow to the project).

Using Equation (6.3) and the associated functional notations, AW is given[22] by:

$$AW = AW_{benefits} - AW_{costs}$$

$$= \left\{ 2,500 + 2,000 \left(\frac{A}{F,8\%,10} \right) \right\} - 12,000 \left(\frac{A}{P,8\%,10} \right)$$

$$= 2,500 + 2,000 \times 0.0690 - 12,000 \times 0.1490$$

$$= 2,500 + 138 - 1,788$$

$$= \$850$$

Example 6.19: Annual Worth Comparison

Best Manufacturers is considering the following five new projects as next year's R&D initiatives. On the basis of annual worth which one should be selected for funding, assuming that the investment must earn an annual return of 12%? All five projects will take 2 years to develop and will be profitable during the subsequent 5 years.

Project	Annual Cost (during development)	Annual Income (during 5-year life)
A	$2,000	$3,500
B	$1,800	$3,200
C	$2,200	$2,300
D	$1,200	$2,000
E	$3,500	$4,800

SOLUTION:

The R&D costs are incurred during the first 2 years of the development. The benefits are realized during the 5 years following the development. The decision-making involves the twin-steps: determination of each project's AW, and selection based on the AW-criterion.

Let us go through the first step and evaluate the AW of each project. Since the cashflow pattern of the projects are similar, their cashflow diagrams will be alike. Consider project A first, whose cashflow diagram is shown in **Fig. 6.3**. To determine the annual worth from this diagram, some tailoring of the cash inflows (benefit vectors) is required to exploit the uniform-series pattern. The tailoring involves adding two 3,500 vectors *upward*, one at period 1 and the other at period 2, so that the cash inflows $3,500 can become $AW_{benefits}$. The added vectors are compensated for by having two 3,500 vectors *downward* at the same time periods, i.e., at periods 1 and 2. This compensation increases the two cash outflows at these periods to $5,500 from the original $2,000.

The tailored diagram is next analyzed to determine AW_{costs}. The two downward $5,500 vectors are transferred to period 0 as P-value which is then converted to A-value spread over 7 years, i.e., to AW_{costs}. This yields:

---

[22] If you have difficulty understanding the functional notations used here, reread Chapter 4.

$$AW_A = AW_{benefits} - AW_{costs}$$

$$= 3,500 - \left\{ 5,500 \left(\frac{P}{A,12\%,2} \right) \right\} \left(\frac{A}{P,12\%,7} \right)$$

$$= 3,500 - 5,500 \times 1.690 \times 0.2191$$

$$= 3,500 - 5,500 \times 0.3703$$

$$= 3,500 - 2,037$$

$$= \$1,463$$

In a similar way,

$$AW_B = 3,200 - 5,000 \times 0.3703 \qquad AW_C = 2,300 - 4,500 \times 0.3703$$

$$= 3,200 - 1,852 \qquad\qquad\qquad = 2,300 - 1,666$$

$$= \$1,348 \qquad\qquad\qquad\qquad = \$634$$

$$AW_D = 2,000 - 3,200 \times 0.3703 \qquad AW_E = 4,800 - 8,300 \times 0.3703$$

$$= 2,000 - 1,185 \qquad\qquad\qquad = 4,800 - 3,073$$

$$= \$815 \qquad\qquad\qquad\qquad = \$1,727$$

The second step involves deciding which project to select. Since AWs of all the five projects are +ve, the most favorable project is the one whose AW is the largest. So, project E is selected.

6.10 ANNUAL WORTH INPUT-OUTPUT CONCEPT

The input-output concept discussed in Chapter 6 is equally useful in annual worth analysis. Read the particular section in that chapter to grasp this concept. On the basis of the input-output concept, engineering economics projects can be grouped into three categories:

6.10.1 Fixed Output

In many projects, the output is fixed, i.e., the benefit from each alternative is the same. Such multi-alternative projects are solved by comparing the alternatives' annual worths of costs, since the annual worth of benefit(s) is the same (fixed) for each alternative. Exclusion of benefits from the analysis keeps the calculations simpler. Obviously, the criterion of decision-making in such problems[23] is to *minimize the annual worth of the costs*, as illustrated in Example 6.15.

Example 6.20: Annual Worth Input-Output

Hari has been using an old car since his first year at the college. During the last 3 years as an industrial technologist he has saved enough money to buy a new car. He is considering a Honda Accord, Chevy Prism, and Plymouth Acclaim. Given the following data which

---

[23] The terms *problem* and *project* have been used synonymously.

one should he buy if the decision criterion is annual worth? Assume each car's useful life as 7 years, and an annually compounded interest rate of 9% per year.

Car	First Cost	Annual Cost	Salvage Value
Honda Accord	$20,000	$300	$3,000
Chevy Prism	$18,000	$500	$2,000
Plymouth Acclaim	$17,500	$750	$1,500

SOLUTION:

In this problem, the output, which is the convenience of owning a car, is fixed. The convenience translates into indirect benefits which are difficult to quantify in monetary terms. Of the given data, the salvage value represents a cash inflow and is treated as a benefit. Since it is a fixed-output problem, the selection could be based on the annual worth of the costs. If you can visualize the problem without its cashflow diagram, you should be able to follow the following analysis; else, sketch a diagram to help you.

The annual worth[24] for the three choices, from Equation (6.3), are:

$$AW_{Accord} = 3,000\left(\frac{A}{F,9\%,7}\right) - \left\{20,000\left(\frac{A}{P,9\%,7}\right) + 300\right\}$$

$$= 3,000 \times 0.1087 - (20,000 \times 0.1987 + 300)$$

$$= 326 - (3,974 + 300)$$

$$= -\$3,948$$

$$AW_{Prism} = 2,000\left(\frac{A}{F,9\%,7}\right) - \left\{18,000\left(\frac{A}{P,9\%,7}\right) + 500\right\}$$

$$= 2,000 \times 0.1087 - (18,000 \times 0.1987 + 500)$$

$$= 217 - (3,577 + 500)$$

$$= -\$3,860$$

$$AW_{Acclaim} = 1,500\left(\frac{A}{F,9\%,7}\right) - \left\{17,500\left(\frac{A}{P,9\%,7}\right) + 750\right\}$$

$$= 1,500 \times 0.1087 - (17,500 \times 0.1987 + 750)$$

$$= 163 - (3,477 + 750)$$

$$= -\$4,064$$

All the three AWs[25] are −ve. Comparing them, Hari should buy a Chevy Prism since, with its lowest −ve annual worth, it is the most favorable. In other words, the Chevy Prism is selected because the equivalent annual cost of purchasing and operating is the least of all.

---

[24] This problem can be treated differently by subtracting the salvage value's equivalent annual worth from that of the initial cost's. This, for example, will yield for Honda Accord $AW_{costs\ (Accord)}$ = 300 + {20,000 (A/P,9%,7) − 3,000 (A/F,9%,7)}This approach too will lead to the same decision. Try doing it.

[25] Note that each car's annual worth is −ve. Then, why invest in a car? The explanation is that we did not, perhaps could not, account for the benefits of owning a car; only the salvage value contributed toward the benefit. Such a situation arises in many projects where the benefits are difficult to quantify. If we account for all the benefits of owning a car, for example, savings on bus or taxi fares, cost of time saved from driving rather than walking to work, and so forth, we are likely to find that a car's AW is actually +ve. Moreover, there is the intangible beneficial feeling of prestige in owning a car, which can at times be egoistic, especially with expensive cars.

6.10.2 Fixed Input

In the second group of projects, the input is fixed, i.e., the alternatives cost the same. Such projects are of the *fixed-input-variable-output* type. By variable-output we mean that the benefits (output) from the alternatives differ from each other. The decision maker's task is to select that alternative which maximizes the output (benefit). The decision criterion in such projects is the largest +ve annual worth of benefits. In other words, maximize $AW_{benefits}$. Since the input is fixed, its inclusion in calculations for alternatives' annual worths would not affect the decision, but unnecessarily complicate the analysis. An example of fixed-input project is the hiring of an employee at a fixed budgeted salary, where the decision maker's task is to select the best from among the applicants. Another example is lump sum investment in a CD after "shopping" around for the highest effective interest rate to maximize the return.

6.10.3 Input and Output Vary

Several engineering economics projects do not involve alternatives in which neither input nor output is fixed. The analyses of such variable-input-variable-output problems are relatively lengthier. The annual worth of each project alternative is evaluated using Equation (6.3), as illustrated in Example 6.11 and elsewhere. Once AWs are known, selection is made in favor of the most attractive project, usually the one with the largest +ve AW (smallest –ve AW if alternatives' AWs are all –ve).

6.11 ANNUAL WORTH ANALYSIS PERIOD

In Chapter 5 we discussed the situation where alternatives' useful lives are different. There we answered the question: What analysis period should be used? This question does not arise at all in AW-analysis, because AW translates the given cashflows in an equivalent uniform annual cashflow. As long as the alternatives of different useful lives can be replaced for the same cost and yield on replacement the same benefits—usually a sensible assumption—AW is unaffected by the analysis period. Thus, in AW-analysis of alternatives with unequal lives, each alternative's analysis period is its own useful life[26]; there is no need for an LCM-based analysis period as in PW-analysis. This renders the AW-analysis simpler, as illustrated in Example 6.21.

Example 6.21: Analysis Period

Abraham's personal computer (PC) is slow and lacks features essential for running current application programs. There are three choices: upgrade the existing system, buy a new PC, or buy a 1-year-old PC. For the following data, which one should he select if the yearly interest rate is 9%, compounded annually? Use the AW-criterion for decision-making. Salvage values are $250, $500, and $300 for Upgrade, New PC, and Old PC, respectively.

Choice	First Cost	Annual Cost	Useful Life (in years)
Upgrade	$3,000	$200	2
New PC	$8,000	$100	4
Old PC	$5,000	$150	3

---

[26] Also means that alternatives are analyzed over different analysis periods.

SOLUTION:

Since the AW-criterion is the basis of decision-making, it does not matter that the useful life of the three choices are different, especially if we assume that the replacement cost and the resulting benefit data will remain the same.

We need to evaluate the annual worths of the three choices and select the most favorable one. Considering the salvage value as a benefit, their annual worths[27] are[28]:

$$AW_{upgrade} = AW_{benefits} - AW_{costs}$$

$$= 250\left(\frac{A}{F,9\%,2}\right) - \left\{3,000\left(\frac{A}{P,9\%,2}\right) + 200\right\}$$

$$= 250 \times 0.4785 - (3,000 \times 0.5685 + 200)$$

$$= 120 - (1,706 + 200)$$

$$= -\$1,786$$

$$AW_{new} = AW_{benefits} - AW_{costs}$$

$$= 500\left(\frac{A}{F,9\%,4}\right) - \left\{8,000\left(\frac{A}{P,9\%,4}\right) + 100\right\}$$

$$= 500 \times 0.2187 - (8,000 \times 0.3087 + 100)$$

$$= 109 - (2,470 + 100)$$

$$= -\$2,461$$

$$AW_{old} = AW_{benefits} - AW_{costs}$$

$$= 300\left(\frac{A}{F,9\%,3}\right) - \left\{5,000\left(\frac{A}{P,9\%,3}\right) + 150\right\}$$

$$= 300 \times 0.3051 - (5,000 \times 0.3951 + 150)$$

$$= 92 - (1,976 + 150)$$

$$= -\$2,034$$

Comparing[29] the three annual worths which are all −ve, Abraham selects to upgrade the existing system since, with its smallest −ve AW, it is the most favorable.

---

[27] Follow the standard convention that costs (vectors down) are −ve, while benefits (vectors up) are +ve.

[28] The convention for subscripts followed in this text is the same everywhere. If there is no subscript the notation AW (or PW) relates to the project as a whole, used normally for no-alternative projects. With *benefits* or *costs* as subscript, the AW (or PW) is of the benefits or costs of the project. For multi-alternative projects, a key term is used as a subscript, as in $AW_{upgrade}$, to represent the AW (or PW) of the project alternative. An additional subscript is used to denote benefits or costs, for example $AW_{costs\ (Accord)}$ in the footnote of Example 6.20.

[29] Note the brevity of analysis in Example 6.21 based on the AW-method in comparison to that in Example 6.9 based on the PW-method. This is because PW analysis is conducted over a long analysis period based on LCM. The AW-method does not require a common analysis period, instead the alternatives are analyzed over their own useful lives.

6.12 SUMMARY

The present worth (PW) method is commonly practiced by the engineering industry. The application of the PW-criterion to engineering projects involves two tasks: determination of the present worth(s), and decision-making based on the criterion. A project's PW is the difference between its present worths of benefits and costs. It is usually positive for a project to be funded. In a multi-alternative project, the alternative with the most favorable PW is selected.

The input-output concept, under which project costs are considered input and benefits the output, streamlines the analysis. Based on this concept problems may be categorized in one of the three types: fixed output, fixed input, or variable-input-variable-output. In fixed-output (same benefit) problems, the alternative with the lowest input (cost) is selected. In fixed-input (same cost) problems, the alternative with the highest output (benefit) is selected. Irrespective of the problem type, the aim is always to maximize the desirable (benefit) and/ or minimize the undesirable (cost).

Present worth analysis requires that the alternatives be compared within the same time window (analysis period). This occurs automatically in a project whose alternatives are of equal useful lives. If the alternatives' useful lives are unequal, an analysis period equal to the least-common-multiple (LCM) of their lives is selected. If the LCM-based analysis period is unreasonably long, there are two choices: (a) Use an analysis period equal to the useful life of one alternative along with its costs and benefits, but use the prorated costs and benefits for the other alternatives, or (b) Use any of the other analysis methods, such as annual worth or rate of return.

The maintenance and upkeep of projects of perpetual life can be provided for by capitalizing the necessary fund. A project's capitalized cost is the initial lump sum whose interest income pays for the operation and maintenance cost. The concept of capitalized cost is applicable to non-engineering projects as well, for example in setting up perpetual scholarships or professorships in higher education.

The application of the future worth (FW) criterion consists of two steps: evaluation of FW, and decision-making to select the best alternative. A project's FW is defined as the surplus of its benefits' future worth over its costs' future worth. It is evaluated by analyzing the given cashflows through first-principles or using Equation (6.2). Alternatively, it can be evaluated by transferring the project's PW, if known, to the specific future time. In multi-alternative projects, the criterion of decision-making is the largest +ve FW. If all the FWs are −ve, the best choice is the smallest −ve FW. The FW-criterion is most suited for projects where savings accumulate for disbursement in the future. Compared to the PW-criterion, the FW-criterion is less frequently applied to the analysis of engineering projects. It is more appealing to the financial sector of the economy.

A project's annual worth (AW) is the surplus of its equivalent uniform annual benefit, $AW_{benefits}$, over its equivalent uniform annual cost, AW_{costs}. Its determination therefore involves evaluation of both $AW_{benefits}$ and AW_{costs} for the project. The application of the AW-criterion to engineering economics projects is a two-step process: determination of AW(s), and selection of financially the most attractive alternative. The input-output concept categorizes problems into fixed output, fixed input, and variable-input-variable-output types. By input we mean cost(s) and by output we mean benefit(s). In fixed-output problems, the alternatives yield the same benefit; therefore, the decision is based on minimization of AW_{costs}. In fixed-input problems, on the other hand, maximum $AW_{benefits}$ is the decision criterion. In variable-input-variable-output problems, the alternative with the largest +ve AW is usually selected. If the alternatives' AWs are all −ve, then select the alternative whose −ve AW is the smallest. The AW-method is most suitable for analyzing projects with alternatives of different useful lives. It takes care of unequal lives by itself, since AW normalizes the cashflows on a yearly basis.

DISCUSSION QUESTIONS

6.1 Give an example from your own experience where the present worth criterion would be appropriate for decision-making.

6.2 Explain the meaning of *present* in the present worth method.

6.3 How does the present worth criterion account for the time value of money?

6.4 How does the input-output concept help in analyzing engineering economics problems?

6.5 How useful is the input-output concept in analyzing engineering economics problems?

6.6 Discuss when a project of zero PW can be funded.

6.7 Give an example from your own experience where the future worth criterion would be most appropriate for decision-making.

6.8 How does the future worth criterion account for the time value of money?

6.9 Explain the significance of *future* in the future worth method.

6.10 Can the FW-criterion lead to a decision different from that based on the PW-criterion? If yes, under what circumstances? If no, why not?

6.11 Of the two steps involved in FW-analysis, which one is more difficult and why?

6.12 Explain the significance of *annual* in the annual worth method.

6.13 Give an example from personal financing where the annual worth criterion would be most appropriate for making a decision.

6.14 How does the annual worth criterion account for the time value of money?

6.15 How is input-output concept helpful in AW-analysis?

6.16 Why do engineering economists not worry about alternatives' unequal useful lives while analyzing problems by the AW-method?

MULTIPLE-CHOICE QUESTIONS

6.17 If the present worth of costs exceeds that of the benefits, PW for the project is
   a. positive
   b. negative
   c. equal to zero
   d. always an integer

6.18 Your manager has received approval from the upper management to upgrade your personal computer at a cost of $1,000. This represents a _____ economics problem.
   a. variable-input-variable-output
   b. fixed-output
   c. fixed-input
   d. fixed-input-fixed-output

6.19 While comparing the PW of alternatives of unequal useful lives, the following assumptions are made EXCEPT
   a. The suppliers of the resources will continue in business.
   b. The alternatives will cost the same.
   c. The benefits will remain unchanged.
   d. The alternatives are perpetual.

6.20 To attend college, you need to purchase a car. This represents a project of
   a. fixed input
   b. fixed output
   c. fixed-input-fixed-output
   d. variable-input-variable-output

6.21 John buys a video game for $300. After using it for 2 years, he expects to sell it for $20. If $i = 10\%$, the present worth in dollars is
   a. 300 (F/P,10%,2) – 20
   b. 280
   c. 20 (P/F,10%,2) – 300
   d. 300 – 20 (P/F,10%,2)

6.22 There are two choices for replacing a punch press. The basic model has a useful life of 8 years while the deluxe model of 12 years. The most appropriate analysis period for this problem is
   a. 20 years
   b. 96 years
   c. 1.5 years
   d. 24 years

6.23 The local city government has an offer of a gift from a rich, generous Hollywood actress who grew up in the town and remembers her numerous visits to the city park as a child. She wants to donate towards perpetual upkeep of the newly built children's park. If the upkeep cost is $50,000 per year, and her gift as a long-term investment can earn 10% annually, the gift should be
   a. $ 50,000/12
   b. $ 50,000
   c. $500,000
   d. $500,000/12

6.24 If the future worth of costs exceeds future worth of benefits, the project FW will always be
   a. positive
   b. negative
   c. in fraction
   d. an integer

6.25 John buys a video game for $300. After using it for 2 years, he expects to sell it for $20. If $i = 10\%$ per year, the future worth in dollars is
   a. 300(F/P,10%,2) – 20
   b. 280 + 10% of 280
   c. 20 – 300(F/P,10%,2)
   d. 300 – 20(P/F,10%,2)

6.26 For most projects, FW is likely to be
   a. smaller than PW
   b. equal to PW
   c. greater than PW
   d. any of the above depending on the cashflow pattern

6.27 Bob is worried about the harm he might be doing to his lungs from cigarette smoking. Besides, he has begun to see the benefit of quitting since he enrolled in engineering economics. He smokes $60 worth of cigarettes a month. If he quits and saves this money each month in a 20-year CD that pays 12% annual interest, compounded monthly, the future worth of the economic benefit from quitting is
   a. 60(F/A,12%,240)
   b. 60(F/A,12%,20)
   c. 60(F/A,1%,240)
   d. 60(F/P,1%,240)

6.28 With help from her financial advisor, Gita is planning to retire in 10 years. The criterion most suitable in this planning is
    a. payback period
    b. present worth
    c. future worth
    d. benefit-cost ratio

6.29 AW-analysis of engineering economics problems may be preferred over other methods because
    a. it is the simplest
    b. LCM-based analysis period becomes irrelevant
    c. financial institutions require it
    d. it is easier to budget for

6.30 In fixed-output problems the objective should be to
    a. maximize AW
    b. maximize $AW_{benefits}$
    c. minimize AW_{costs}
    d. minimize AW

6.31 In fixed-input problems the objective should be to
    a. maximize AW
    b. maximize $AW_{benefits}$
    c. minimize AW_{costs}
    d. minimize AW

6.32 John buys a video game for $300. After using it for 2 years, he expects to sell it for $20. If $i = 10\%$ per year, the annual worth in dollars is
    a. 20 (F/A,10%,2) – 300 (P/A,10%,2)
    b. 280 + 10% of 300
    c. 20 (F/A,10%,2) + 300 (P/A,10%,2)
    d. 300 (P/A,10%,2) – 20 (F/A,10%,2)

6.33 There are two choices for replacing a punch press. The basic model has a useful life of 8 years while the deluxe model of 12 years. The analysis period for this problem under the AW-criterion is
    a. 24 years for both
    b. 4 years for basic and 6 years for deluxe
    c. 8 years for basic and 12 years for deluxe
    d. 20 years

6.34 In analyzing a variable-input-variable-output problem we need to determine
    a. $AW_{benefits}$ only
    b. AW_{costs} only
    c. a or b depending on which one is +ve
    d. both a and b

6.35 In analyzing a project whose alternatives have unequal lives, the best method is
    a. present worth
    b. future worth
    c. annual worth
    d. a, b, or c, depending on the number of alternatives

NUMERICAL PROBLEMS

6.36 The initial cost of a new photocopier is $12,000. It is expected to generate a net annual income of $4,000 in the first year, decreasing annually by $300. At the end of its 5-year useful life, the photocopier is expected to be sold for $500. Assuming $i = 10\%$ per year with annual compounding, determine the photocopier's present worth.

6.37 Within 2 years of graduation, Shyam has established a significant market for non-CFC Styrofoam cups. He has decided to purchase a new machine rather than continue to operate with the current one. The maintenance cost of the machine is expected to be $500 in the first year, $1,000 in the second year, $1,500 in the third year, and so on. Assuming an annually compounded interest rate of 12% per year, how much should be included in the current budget to pay for the maintenance if the machine's useful life is 10 years?

6.38 A rental fax machine can be replaced with one of the two models. The basic model with a life of 4 years costs $6,000 and is expected to bring in net $2,000 annually. For the deluxe model, the data are 8 years, $10,500, and $2,500 respectively. Assuming no salvage value and an annually compounded interest rate of 8% per year, which model is a better choice on the basis of the PW-criterion?

6.39 Two candidate machines are under consideration. For machine A, the initial cost is $100,000, operating benefits are $40,000 per year, and useful life is 6 years. The respective data for machine B are $150,000, $50,000, and 9 years. Neither machine has any salvage value. On the basis of present worth which machine would you select? Assume $i = 8\%$.

6.40 It costs Hattiesburg City $50,000 a year to maintain Hardy Street. If this street is resurfaced at a cost of $150,000, the annual maintenance cost will reduce to $20,000 for the first 5 years and to $30,000 for the subsequent 5 years. At the end of 10 years, the maintenance cost is expected to revert back to $50,000 a year. Should the city resurface this street if the annually compounded interest rate is 5% per year? Use the PW-criterion for the decision.

6.41 Isaac asks his brother-in-law to lend him $16,000 so that he can buy a new car. The car will help Isaac sell more merchandise. Isaac promises to pay back $3,000 at the end of first year, $6,000 at the end of second year, and $9,000 at the end of third year. If his brother-in-law can earn 10% per year on his $16,000 from the local bank, will he do Isaac a favor by lending? Use the present worth criterion.

6.42 Three *mutually exclusive*[30] alternatives X, Y, and Z are being considered as a permanent solution to an assembly-line bottleneck. The data are:

	X	Y	Z
Initial Cost[a]	$10k	$14k	$12k
Annual Benefit	$2,600	$5,500	$1,500
Useful Life, years	5	3	15

a k as a prefix in SI units is an acronym for kilo, meaning 1,000.

Assuming no salvage value, annually compounded interest rate of 8% per year, and unchanged data at replacement times, which alternative should be selected on the basis of the PW-criterion?

---

[30] The term *mutually exclusive* means that the selection of one of the alternatives precludes the others from being selected. In other words, the alternatives are independent of each other. This is one of the assumptions in multi-alternative decision-making. The term may be explicitly expressed, as in this problem, but in most problems we make an implicit assumption to this effect.

6.43 For equipment being considered for acquisition at a cost of $80,000 the following data have been compiled.

Year	Cost	Income
1	$4,000	$20,000
2	$5,000	$24,000
3	$6,000	$29,000
4	$7,000	$32,000
5	$8,000	$34,000

Assuming a 7% annual interest rate, should the equipment be purchased under the PW-criterion?

6.44 ZeroDefect Manufacturing Company needs to build additional space for inventory storage. There are three choices:

- Spend $12 million to build space that will meet the needs for the next 10 years,
- Spend $8 million now to meet the needs of the first 6 years and another $5 million during the sixth year to meet the needs of the next 4 years, or
- Rent a nearby space for 10 years at a yearly cost of $2 million.

With reference to the input-output concept, which type of project is this? Which is the best choice on the basis of the present worth criterion? Assume no salvage values, and $i = 8\%$ per year.

6.45 Within a few years of graduation from the Engineering

   Department of The International University, Tom has amassed enormous wealth from his patent on a microorganism-based solid waste disposal system. He wants to endow a scholarship at his alma mater in his mother's name. The endowment will be placed in an interest-bearing long-term financial instrument earning 12% annual interest. The yearly proceeds should generate a $5,000 income for the annual scholarship. How much should the endowment fund be?

6.46 The initial cost of a new photocopier is $6,000. It is expected to generate a net annual income of $2,000 in the first year, decreasing thereafter by $300 each year. At the end of its 7-year useful life the photocopier is expected to be sold for $750. Assuming $i = 10\%$ with annual compounding, determine the photocopier's future worth.

6.47 Determine the FW of the project whose cashflow diagram is depicted in Fig. 7.1. Assume $i = 10\%$.

6.48 There are two models of a piece of metrology equipment in the market. The basic model with a life of 4 years costs $6,000 and is expected to bring in $2,000 net annually. For the deluxe model, these data are 8 years, $10,500, and $2,500 respectively. Assuming no salvage value and an annually compounded interest rate of 8% per year, which model is a better choice on the basis of the FW-criterion? (Figure 6.11)

6.49 Saira Banu turns fifteen today. Her elder sister who is enrolled in engineering economics has helped her appreciate the power of compounding in building up equity. Saira decides to set aside the cash gift from her grandparents' trust at each birthday, beginning with the sixteenth and ending with the fifty-fifth. The trust provides a cash gift ten times her age. If her savings can earn an annually compounded interest rate of 10% per year, how much will have accumulated on her 60th birthday? Though she does not save beyond the fifty-fifth birthday, the savings continue to earn compounded interest.

6.50 Two candidate machines are under consideration. Machine A's initial cost is $100,000, operating cost is $4,000 per year, and useful life is 6 years. The respective data for

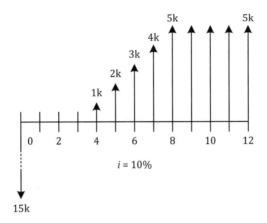

Figure 6.11   Diagram for Problem 7.12.

machine B are $150,000, $5,000, and 9 years. Neither machine has any salvage value. On the basis of future worth which machine would you select if $i = 7\%$?

6.51 Madhuri begins her career as an electronics engineer at an annual salary of $40,000. She expects a yearly increase of $2,000 in her salary during a 30-year professional career. If she invests 5% of her salary each year in an IRA (individual retirement account) at 10% annually compounded interest, how much will she get at the end of her career?

6.52 Isaac asks his sister-in-law to lend him $16,000 so that he can buy a car. The car will help him sell more merchandise. Isaac promises to pay back $3,000 at the end of first year, $6,000 at the end of the second year, and $9,000 at the end of the third year. If his sister-in-law's $16,000 can earn 10% annual interest from the local bank, is she doing him a favor by lending? Use the FW-criterion for your decision.

6.53 ZeroDefect Manufacturing Company needs to build additional space for inventory storage. There are three choices:

1. Spend $12 million to build space that will meet the needs for the next 10 years.
2. Spend $8 million now to meet the needs of the first 6 years and another $5 million during the sixth year to meet the needs of the next 4 years.
3. Rent a nearby space for 10 years at a yearly cost of $2 million.

Which is the best choice on the basis of the future worth criterion? Assume zero salvage value, and $i = 8\%$.

6.54 For a piece of equipment being considered for acquisition the following data have been compiled.

Year	Cost	Income
0	$70,000	None
1	$ 4,000	$20,000
2	$ 5,000	$24,000
3	$ 6,000	$29,000
4	$ 7,000	$32,000
5	$ 8,000	$34,000

Assuming that money is worth 7%, should the equipment be acquired if the decision criterion is future worth?

6.55 For the following cashflows, determine FW for an annually compounded interest rate of 9% per year.

Year	0	1	2	3	4	5
Cashflow	−$6,000	$3,000	$5,000	−$1,500	$500	$2,500

What will the percentage change be in FW if compounding is monthly?

6.56 Amanda has bought a second-hand car for $4,000. She plans to keep it for 4 years, at the end of which it is likely to fetch $500. If the maintenance cost is $350 per year, and $i = 8\%$ per year, what is the car's net annual cost?

6.57 Urmila is interested in a mink coat she is likely to wear thrice a year while partying. The choices are: buy for $5,000, or rent each time when needed for $120. If she owns the coat, she insures it for an annual premium of $50. If she rents it, she spends $20 on dry cleaning each time she uses it. Based on the AW-criterion, should she buy or rent? Assume $i = 5\%$ per year, n = 10 years, and zero salvage value if bought.

6.58 A fax machine needs replacing. There are two models on the market. The basic model with a life of 4 years costs $6,000 and is expected to generate annually a net income of $2,000. For the deluxe model, these data are 8 years, $10,500, and $2,500 respectively. Assuming zero salvage value and an annually compounded interest rate of 8% per year, which model is a better choice on the basis of the AW-criterion?

6.59 Two candidate machines are under consideration. Machine A has an initial cost of $100,000, operating costs of $4,000 per year, and useful life of 6 years. The respective data for machine B are $150,000, $5,000, and 9 years. Neither machine has any salvage value. On the basis of annual worth which machine would you select? Assume $i = 8\%$ per year.

6.60 A CNC lathe costs $50,000 to purchase. Its expected useful life is 10 years and salvage value is likely to be 10% of the initial cost. The annual operating cost will be $800. The first-year maintenance cost is estimated to be $700, increasing each year by $300. Determine AW_{costs} for an annual interest rate of 12% compounded twice a year.

6.61 Three *mutually exclusive*[31] alternatives X, Y, and Z are being considered as a permanent solution to an assembly-line bottleneck. The data are:

	X	Y	Z
Initial Cost[a]	$10k	$15k	$20k
Annual Benefit	1.6k	1.5k	1.9k
Useful Life (years)	5	10	15

[a] As a prefix in SI units, k is an acronym for kilo, meaning 1,000.

[31] The term mutually exclusive means that the selection of one of the alternatives precludes the others from being selected. In other words, the alternatives (choices) are independent of each other. This is a basic assumption in multi-alternative projects, which may be explicitly expressed, as in this problem. In most problems, however, such an assumption is made implicitly.

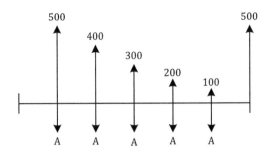

Figure 6.12 Diagram for Problem 6.12.

Assuming zero salvage value and an annually compounded interest rate of 8% per year, which alternative should be selected based on the AW-criterion?

6.62 For an equipment being considered for acquisition the following data have been compiled. Assuming a yearly interest rate of 7%, should the equipment be procured if the decision criterion is annual worth?

Year	Cost	Income
0	$70,000	None
1	$ 4,000	$20,000
2	$ 5,000	$24,000
3	$ 6,000	$29,000
4	$ 7,000	$29,000
5	$ 8,000	$34,000

6.63 An automation system comprising two robots can do the job of an operator who annually costs $30,000. The robots cost $40,000 each, are expected to remain useful for 10 years, and will have no salvage value. The annual maintenance and operating cost for the system is estimated to be $3,000. If $i = 10\%$ per year, what is AW?

6.64 Evaluate A in Fig. 6.12.

6.65 Tony is considering setting up a faxing service business. The equipment and accessories will cost $8,000 and last for 4 years with no resale value. Rent, labor, insurance, and maintenance costs add up to $10,000 a year. If he expects an annual benefit of $25,000, should he set up the business on the basis of the AW-criterion? Assume $i = 15\%$ per year.

FE EXAM PREP QUESTIONS

6.66 There are two choices for replacing a punch press. The basic model has a useful life of 8 years while the deluxe model of 12 years. The most appropriate analysis period for this problem is
   a. 20 years
   b. 96 years

    c. 1.5 years

    d. 24 years

6.67 If the present worth of costs exceeds that of the benefits, PW for the project is

    a. positive

    b. negative

    c. equal to zero

    d. always an integer

6.68 If the future worth of costs exceeds future worth of benefits, the project FW will always be

    a. positive

    b. negative

    c. in fraction

    d. an integer

6.69 If $100 is deposited each month into a 25-year CD that pays 8% annual interest, compounded monthly, the future worth of the economic benefit from quitting is

    a. $100(F/A,12%,240)

    b. $100(F/A,12%,20)

    c. $100(F/A,1%,240)

    d. $100(F/P,1%,240)

6.70 Richard buys a used truck for $1,500. After using it for 3 years, he expects to sell it for $800. If $i = 7.5\%$ per year, the future worth in dollars is

    a. $1,500(F/P,7.5%,3) − 800

    b. $ 700 + 10% of $700

    c. $ 700 − $1,500(F/P,7.5%,3)

    d. $1,500 − $700(P/F,7.5%,3)

6.71 In fixed-input problems the objective should be to

    a. maximize AW

    b. maximize $AW_{benefits}$

    c. minimize AW_{costs}

    d. minimize AW

6.72 What is the future worth of an asset if the initial cost was $23,000, the net benefit is $7,000 per year, the yearly interest rate is 8%, and the useful life is 5 years?

    a. $30,000

    b. $12,000

    c. $ 3,000

    d. $ 7,282

6.73 A manufacturing plant manager has purchased a new bologna slicing machine for $24,500. The new machine is expected to produce an extra $8,000 of revenue and have an annual operating expense of $3,000. If the useful life is expected at 6 years and $i = 7\%$ per year, what is the PW?

    a. $25,000

    b. $35,000

    c. $23,835

    d. $50,000

6.74 A new drill press costs $12,000. It would potentially produce an additional $4,000 of revenue per year and have an operating expense of $1,200. What is the PW of the drill press if the new equipment is expected to last 8 years and $i = 10\%$?

    a. $14,938

    b. $15,534

    c. $21,735

    d. $19,345

6.75 The Wheel of Fortune Entertainment Company is considering replacing the wheel crucial to their operations. If the new tricked out wheel costs $50,000 and generates $21,000 worth of extra revenue and accumulates operating expenses of $7,000 annually. If $i = 12\%$, what is the PW of the crucial component? Assume 4 years life.

   a. $70,169

   b. $65,000

   c. $89,001

   d. $42,518

Chapter 7

Rate of Return

LEARNING OBJECTIVES

- Meanings of rate of return (ROR)
- How to calculate a project's ROR
- Application of ROR-criterion
- Difficulties in evaluating ROR
- Concept of "pure" investment
- Hybrid cashflows
- Internal and external rates of return
- Cashflow sign change rule
- Incremental method of analysis
- Analysis period for ROR method

So far in Part II we have discussed four analysis methods, namely payback period, PW, FW, and AW in Chapters 5, and 6. In this chapter, another method called rate of return (ROR) is presented. An obvious question is why so many methods. This question is answered in Chapter 9 where we compare all the methods for their merits and limitations, and discuss when one is preferred over the other.

In industry, ROR is the most widely used method. It is easier to comprehend, but its analysis is relatively complex, and the pertaining calculations are lengthy. In PW-, FW-, or AW-analysis, the interest rate i is known; in the ROR-analysis, i is the unknown.

We evaluate a project's ROR to ensure that it is higher than the cost of capital, as expressed by *minimum acceptable rate of return* (MARR). MARR is a cut-off value, determined on the basis of financial market conditions and a company's business "health," below which investment is undesirable. In general[1], MARR is significantly higher than what financial institutions charge for lending capital. The term *significantly* is befitting since investing in engineering projects is risky. If MARR is higher only marginally, doing engineering business does not make much of a business sense.

7.1 MEANING

Of the three commonly used measures—PW, AW, and ROR—of investment worthiness, rate of return is the most comprehensive and easily understood. This is primarily because ROR is expressed as a percentage which is an *absolute* measure, being independent of the amount of investment. Besides, we are accustomed to the percentage measure in our daily lives in so many

---

[1] For marketing and other technical reasons, a project may sometimes be funded even if its ROR is less than MARR.

different ways. The PW- and AW-measures of a project, on the other hand, being in monetary units are *relative*; they must be judged in conjunction with the amount of investment.

Economists explain the meaning of ROR in several ways. According to one meaning, ROR is the interest earned on remaining balance of investment. By another, ROR is *the value of i that renders PW zero*. In other words, ROR is that interest rate for which cash inflows (benefits) equal cash outflows (costs). Of course, the time value of money must be taken into account in this equivalency.

7.2 EVALUATION

ROR is basically the interest rate for which project costs are fully recovered through its benefits. As explained earlier, the costs and benefits of a project can be considered as weights on the two pans of a balance. As per this analogy, ROR is that interest rate which renders the balance in balance. We illustrate this meaning of ROR through Example 7.1.

Example 7.1: ROR

An investment of $5,000 is paid back after 5 years as a lump sum of $7,345. What is the ROR assuming annual compounding?

SOLUTION:

By equating (or balancing) the single cash outflow P ($5,000) with the single cash inflow F ($7,345), from F = P(F/P,$i$,n) for n = 5,

$$7,345 = 5,000 \left(F/P, i, \ 5 \right)$$

Transferring the unknown to the left side of the = sign,

$$\left(F/P, i, \ 5 \right) = 7,345/5,000$$
$$= 1.469$$

Thumbing through the interest tables, (F/P,i,5) is 1.469 when i is 8%. Thus, i = 8% per year. For this interest rate, the $7,345 payback (benefit) equals the investment (cost); hence ROR = i = 8% per year.

Example 7.1 was set up intentionally to make the evaluation of ROR easier. The value of i was simply read off the tables. ROR-evaluation is rarely so simple. In fact, it is usually more involved even for simple problems, as illustrated in Example 7.2.

Example 7.2: ROR

Evaluate ROR for the project whose cashflows are diagramed in Fig. 7.1.

SOLUTION:

From the diagram, an investment (or cost) of $500 now brings in benefits of $200 at period 4 and $500 at period 6. We can easily express the cost-benefit equivalency[2] by transferring the two F-values to period 0, as

$$500 = 200 \left(P/F, i, 4 \right) + 500 \left(P/F, i, 6 \right)$$

---

[2] Note that equivalency is valid only if all the costs and benefits pertain to the same time period, for example to period 0 in this case.

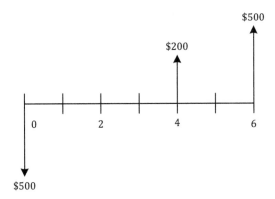

Figure 7.1 Evaluation of ROR.

Our task is to evaluate i in this expression. There are two approaches[3] to do this. In one, we substitute the equations for functional notations and solve for i. The substitution (Equation 3.7b) yields:

$$500 = 200 \left(1 + i\right)^{-4} + 500 \left(1 + i\right)^{-6}$$

Multiplying throughout by $(1 + i)^6$ and dividing by 500, and rearranging the terms, we get:

$$\left(1 + i\right)^6 - 0.4 \left(1 + i\right)^2 - 1 = 0$$

This relationship can be solved for i in several ways, including iteration[4] methods. However, we skip this solution approach and use instead the table-based functional notations emphasized throughout the text. This is the second approach of evaluating i in the original expression: $500 = 200\ (P/F,i,4) + 500\ (P/F,i,6)$.

Rearranging the terms, we get:

$$500 \left(P/F,i,6\right) + 200 \left(P/F,i,4\right) - 500 = 0$$

This expression[5] can be solved for the unknown i through iteration and interpolation, as explained as follows. The iteration effort can be reduced by shrinking the large numbers 200 and 500 by dividing the expression throughout by 500. This yields:

---

[3] Readers will notice by the end of this section that the two approaches are basically the same.

[4] In some simple cases we may be able to evaluate i directly. As an illustration, if there was only one benefit in this example, say of $750 at period 6, then we would have

$500 = 750\ (1 + i) - 6$

$(1 + i)6 = 750/500 = 1.5$

Using natural logarithms, this gives

$6\ \ell n\ (1 + i) = \ell n\ 1.5$ $1 + i = e0.06758$

$6\ \ell n\ (1 + i) = 0.405465$ $1 + i \approx 1.07$

$\ell n\ (1 + i) = 0.06758 =$ $0.07 = 7\%$

However, problems where such a direct evaluation of i is encountered are rare, and hence, in ROR analysis the equation-based solution approach is limited in scope.

[5] Many analysts prefer the format

$500\ (P/F,i,6) + 200\ (P/F,i,4) = 500$

instead of the one used in the text. Either format is acceptable since the result will remain the same. So use the format you feel comfortable with.

$$(P/F,i, 6) + 0.4 (P/F,i, 4) - 1 = 0$$

One could have divided by 200 instead to get

$$2.5 (P/F,i,6) + (P/F,i,4) - 2.5 = 0$$

The extent of shrinking to achieve smaller multipliers is your own preference, since the evaluated i will not be affected by such manipulations.

Let us consider the first expression for further analysis, namely

$$(P/F,i, 6) + 0.4 (P/F,i, 4) - 1 = 0$$

Note that the left side of this expression is related to the PW of the cashflows. We therefore call it PW-function. There is no simple way to solve the PW-function for i, except trial and error in which different values of i are tried to see which one satisfies it. For each i tried, the functional factors' values are read off the interest tables and the PW-function evaluated. The trials and errors[6] in this case are summarized as follows.

i (%)	$(P/F,i,6) + 0.4(P/F,i,4) - 1$
4	$0.7903 + 0.4 \times 0.8548 - 1 = 0.132$
5	$0.7462 + 0.4 \times 0.8227 - 1 = 0.075$
6	$0.7050 + 0.4 \times 0.7921 - 1 = 0.022$
7	$0.6663 + 0.4 \times 0.7621 - 1 = -0.028$
8	$0.6302 + 0.4 \times 0.7350 - 1 = -0.076$

We notice that the function changes sign when i changes from 6% to 7%. Thus, i that will render the function zero (the right side of the expression) must lie in between 6% and 7%. We therefore interpolate[7] between these two interest rates[8]. From the interpolation[9],

$$ROR = i = 6 + 0.022/(0.022 + 0.028)$$

$$= 6 + 0.44$$

$$= 6.44\%$$

As seen in Example 7.2, evaluation of ROR even for a simple problem can be lengthy. This is the major drawback of the ROR method. Programmable calculators, spreadsheets, and canned software make the task of ROR evaluation easier. Their advent is one reason why the ROR method has recently gained popularity in industry.

---

[6] Error is the difference in the value of expression, for the tried i, on the left and that on the right which is zero. For example, for i = 5%, the error is 0.075 − 0 = 0.075.

[7] Interpolation assumes a linear relationship between the PW-function and i, which is not true as illustrated in Fig. 7.2. Thus some sacrifice in the accuracy of results is made. For highest accuracy, keep the interest boundaries within which interpolation is done as close to each other as the interest tables will allow.

[8] See Chapter 3 to recall the interpolation procedure.

[9] As an alternative to algebraic interpolation, one can draw a graph between the values of PW-function and i, and read i (= ROR) off the graph where the function is zero. Fig. 7.3 shows the graph for this case; such a graph is called PW-profile. Note that additional points had to be calculated to get a smooth curve. A programmable calculator is very useful in evaluating ROR this way. With graphic capability such a calculator can also display the PW-profile.

7.2.1 Where to Begin

In evaluating ROR in Example 7.2 we began the trial with i as 4%. Why did we begin with 4%? Why not 15% or 35%? Also why did we increment i by 1%? Answers to these questions are important for efficient analysis. Unfortunately, there are no simple answers except that the analyst should use intelligent guesses about where to start and how much to increment or decrement i.

Start with a *reasonable* value of i. If the value of the PW-function is getting closer to zero[10], increment i by as little as the interest tables would allow. Should you increment or decrement? Here the rule is: Do whichever will lead toward zero. Notice this in the trial results of Example 7.2, where PW-function's value of 0.132 for $i = 4$% decreased to 0.075 as i was incremented to 5%. This decrease confirmed that incrementing was helping (leading toward zero) in the search. Thus, keep watching the function's value as the search for i continues, to ensure that you are heading in the right direction, and also to decide whether to slow down the change in i, say from 1% to 0.5% (Fig. 7.2).

The value of i with which the trial is begun can be approximated by analyzing the diagram mentally. Let us do this for Example 7.2 with reference to Fig. 7.1. Neglecting the time-value of money for this approximation, a payback of $700 ($200 plus $500) for an investment of $500 means that the total interest is $700 – $500 = $200. Assuming period 5 as the approximate time of payback since it lies in between periods 4 and 6, the $200 total interest is equivalent to 200/5 = $40 interest per year. For a $500 investment, this $40 annual interest works[11] out as a rate of 8% per year. So, one could have begun in this case with $i = 8$%. The resulting –ve value for PW-function from trying $i = 8$% would have suggested decrementing i for the next trial.

One could have approximated differently[12], getting another value of i to begin the trial. But it would have been in the same range, not 35–50%. To guess the value of i with which to begin the trial,

A. Estimate the approximate total interest and the number of periods over which it is earned, and

B. From the above, work out a rate for the investment. (Figure 7.3).

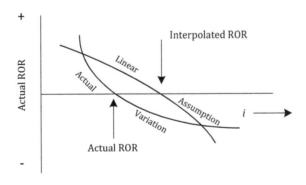

Figure 7.2 Approximation of nonlinear variation.

---

[10] Right-hand side of the evaluated expression.

[11] This 8% is higher than the actual 6.44% because, in the approximation, we assumed all $700 to be located at period 5, and also assumed simple interest for the ease of mental arithmetic.

[12] For example, one can apply the Rule of 70 (see Section 3.9) according to which a sum doubles every 70/i periods, where i is in percentage per period. For example, if $i = 9$% per year, then a sum will double in 72/9 ≈ 7.8 years. So if it takes N periods for a sum to double, then i can be approximated by 70/N.

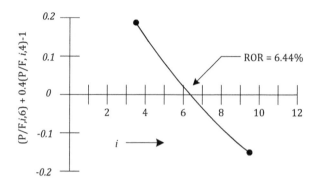

Figure 7.3 Graphical evaluation of ROR.

Given below are some tips on determining the beginning value of *i*, on varying *i* during interpolation, and on calculating ROR.

1. Approximate the cashflows and their timings in a way that facilitates estimation of *i*'s beginning value.
2. If it is difficult to approximate the beginning *i*, then use a *reasonable* value. Values like 5% or 10% are more reasonable as annual interest rate than 50% or 60%.
3. In the beginning of the trials, you may increment or decrement *i* by more than 1%, say 2% or 5%, to reduce the number of iterations.
4. You do not have to only increment or decrement. Thesearch may be either way—forward with higher *i* or backward with lower *i*.
5. Use a programmable calculator to save time and effort in evaluating ROR. One with graphic capability can plot the PW-profile and let you read ROR off the display.

7.2.2 ROR and *i*

The interest rate *i* is a generic term applicable to all situations. ROR, on the other hand, is specific to a project. As a project-specific parameter, it is used to assess the project's investment worthiness. It answers the question: At what interest rate the benefits from the project pay for its costs? For a project to be attractive its ROR should be greater than *i*—the interest rate *i* prevailing in the financial marketplace, also called *cost of capital*. If the mousetrap you intend to design, manufacture, and market using your own capital is expected to yield 8% ROR while the local bank pays its savers 9% interest, shouldn't you hand over your capital to the bank and go fishing!

7.2.3 MARR as *i*

Rather than the value of its ROR, we are often interested in simply knowing whether the project is worthy of investment. To judge this, companies set a cut-off limit on ROR below which projects are not funded. This limiting "bottom-line" ROR is the *minimum acceptable rate of return*, MARR, which should obviously be greater than the cost of borrowed capital. It is also used for rationing capital for competing projects.

If MARR is given and the task is to determine whether or not the project should be funded, then simply determine the project PW with MARR as interest rate. If this PW is +ve (benefits outweigh the costs at MARR), the project[13] is investment worthy; otherwise not. Example 7.3 illustrates the point.

Example 7.3: ROR with MARR

A milling machine's first cost is $25,000. Its likely benefits are $5,000 in the first year, $4,750 in the second year, and so on decreasing annually by $250. The machine's useful life is 10 years and salvage[14] value is $1,000. Based on ROR-criterion should the machine be approved for purchase if MARR is 15%?

SOLUTION:

MARR is given, and our task is assessing the machine's investment worthiness on the basis of ROR. Note that the value of ROR is not being asked for. We can reach the decision simply by judging the sign of PW corresponding to i = MARR = 15%. If PW is +ve, then the machine's ROR will be greater than 15% therefore it should be purchased.

The given cash inflows form a decreasing arithmetic series, as shown in Fig. 7.4(a). The $1,000 vector at period 10 represents salvage value at the end of useful life.

For determining PW we need to evaluate $PW_{benefits}$ and PW_{costs} from the diagram. The value of PW_{costs} is $25,000 since first cost is the only cost. Evaluation of $PW_{benefits}$ requires

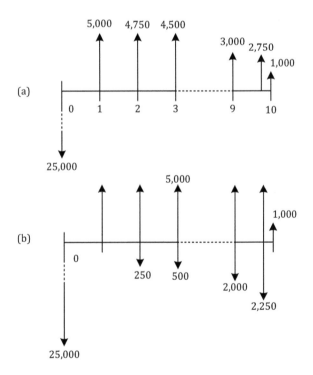

Figure 7.4 Diagram for Example 7.3.

---

[13] A +ve PW corresponding to i = MARR indicates that the project ROR is greater than MARR; hence the project is investment worthy.

[14] Salvage value is usually positive and hence a cash inflow (benefit). It happens when the market value of the resource at the end of useful life is more than the cost of its removal and disposal. At times, this may not be true and the salvage value may be negative, in which case treat it as a cash outflow (cost).

us to transfer all the benefits to period 0. To exploit the approximate pattern in the benefits, we tailor it to the standard pattern for which arithmetic series equation and function apply (Chapter 4).

The tailoring is illustrated in Fig. 7.4(b). It involves adding $5,000 upward vectors (uniform annual benefit) throughout, i.e., at each period. From the added vectors we subtract at each period the arithmetic-series downward vectors of Fig. 7.4(b). The subtraction results in cash inflows as in Fig. 7.4(a). Therefore, the original cash inflows of Fig. 7.4(a) are equivalent to the tailored cashflows of Fig. 7.4(b). The $1,000 salvage vector remains unchanged in the two figures at period 10. We can easily determine $PW_{benefits}$ from Fig. 7.4(b), as

$$P_{benefits} = \left\{ 5,000(P/A,\ 15\%,10)\ -250(P/G,15\%,10) \right\} + 1,000(P/F,15\%,10)$$

The last term in this expression accounts for the salvage value.

Thus, the machine's present worth is:

$$PW = P_{benefits} - P_{costs}$$

$$= 5,000(P/A,\ 15\%,10) - 250(P/G,15\%,10) + 1,000$$

$$(P/F,15\%,10) - 25,000$$

$$= 5,000 \times 5.019 - 250 \times 16.979 + 1,000 \times 0.2472 - 25,000$$

$$= 25,095 - 4,245 + 247 - 25,000$$

$$= -\$3,903$$

Since PW corresponding to 15% MARR is –ve, the machine's ROR is less than MARR, and hence it should not be approved[15] for purchase.

7.3 FORMULAS

As explained in Section 7.2, ROR can be evaluated from the first-principles. In this section, we extend this approach to formula-based analysis.

Based on its meaning explained earlier in Section 7.1, ROR is the interest rate that satisfies $PW_{benefits} = PW_{costs}$. In other words, $PW_{benefits} - PW_{costs} = 0$ is the condition for evaluating ROR. Since this difference is the PW of the project, the condition can be expressed as

$$PW = 0$$

Since PW is directly related[16] to AW, we can say that ROR will equal i also when AW is zero. Thus, the above condition is also satisfied by

$$AW = 0$$

and, for the same reason, by

$$FW = 0$$

In most cases, condition PW = 0 is used to evaluate ROR or apply the ROR-criterion. Where calculations for AW are simpler than for PW, condition AW = 0 is used. For the same reason use FW = 0 where appropriate.

---

[15] The alternative approach of evaluating the ROR exactly, and comparing it with the given MARR, is lengthier than the one illustrated in this example.

[16] In this text we treat PW, AW, or FW as *net*, i.e., as the excess of benefits over costs.

Example 7.4: ROR Evaluated with PW

A milling machine's first cost is $30,000. The likely benefits from the machine are $5,000 in the first year, $4,750 in the second year, and so on, decreasing annually by $250. If the machine has a useful life of 10 years and a salvage value of $1,000, what is its ROR?

SOLUTION:

We are asked to determine the value of ROR. To do this, express the PW of the given cashflows and equate it to zero. That means evaluating $PW_{benefits}$ and PW_{costs} from the machine's cashflow diagram. The diagram is in Fig. 7.4(a) except for the first cost which is $30,000.

From the tailored cash inflows in Fig. 7.4(b), as explained in the previous example,

$$P_{benefits} = \left\{ 5,000\ (P/A,i,10) - 250\ (P/G,i,10) \right\} + 1,000\ (P/F,i,10)$$

Thus,

$$PW = P_{benefits} - P_{costs}$$

$$= \left\{ 5,000(P/A,i,10) - 250(P/G,i,10) \right\} + 1,000\ (P/F,i,10) - 30,000$$

By imposing the condition PW = 0, i is evaluated from

$$5,000\ (P/A,i,10) - 250\ (P/G,i,10) + 1,000\ (P/F,i,10) - 30,000 = 0$$

Dividing throughout by 5,000, we get

$$(P/A,i,10) - 0.05\ (P/G,i,10) + 0.2\ (P/F,i,10) - 6 = 0$$

This PW-function can be solved[17] for i the way explained in Example 7.2, yielding ROR of approximately 6%.

7.4 ONE PROJECT—TWO RORs!

As mentioned earlier, ROR analysis is both lengthy and complex. It is lengthy because of the repetitive nature of trial and error calculations and interpolation, as seen in Examples 7.2–7.4. Fortunately, programmable calculators and PCs are helpful in coping with the repetitive calculations.

In this section, we discuss the complexity aspect of ROR analysis. The first complexity arises from the fact that there may at times be two or more rates of return for the same project. If so, which ROR is admissible, or are both or all of them inadmissible?

Two or more different RORs for the same project is mind-boggling, at least initially. Before we discuss how to resolve this more-than-one-answer-to-the-same-problem situation, it is helpful to know when such a situation can arise.

---

[17] The trial and error results are tabulated below:

i(%) (P/A,i,10) – 0.05(P/G,i,10) + 0.2(P/F,i,10) – 6

4 8.111 – 0.05 × 33.881 + 0.2 × 0.6756 – 6 = 0.552

5 7.722 – 0.05 × 31.652 + 0.2 × 0.6139 – 6 = 0.262

6 7.360 – 0.05 × 29.602 + 0.2 × 0.5584 – 6 = –0.008

Since the sign changes in between 5% and 6%, interpolate within this range to get

ROR = i = 5 + 0.262/(0.008 + 0.262)

    = 5 + 0.97

    = 5.97%

7.4.1 Pure Cashflows Yield One ROR

So far, we intentionally discussed only those projects that yielded only one ROR. In this subsection we discuss the characteristics of cashflows of such projects. Cashflows that yield more than one ROR are discussed later in the next section.

Cashflows encountered in engineering economics can be divided into two groups: *pure* and *hybrid*. For projects whose cashflows are *pure*, there is only one ROR. For those with *hybrid* cashflows, there *may exist* more than one ROR. Note the word "may," since hybrid cashflows can sometimes yield only one ROR.

In *pure* cashflows, one or more costs are followed by several benefits; or one or more benefits are followed by several costs. The former[18] represents *pure investment* while the latter represents *pure borrowing*.

The following cashflows are an example[19] of *pure investment*.

Year	0	1	2	3	4	5
Cashflow	−$5,000	$1,500	$1,500	$1,300	$900	$3,000

The following cashflows are an example[20] of *pure borrowing*.

Year	0	1	2	3	4
Cashflow	$10,000	−$2,500	−$2,200	−$2,000	−$4,000

The basic characteristic of *pure* cashflows is that there is *only one* sign change in the data. In the *pure* investment above, this occurs[21] at year 1 when the cashflow changes from −$5,000

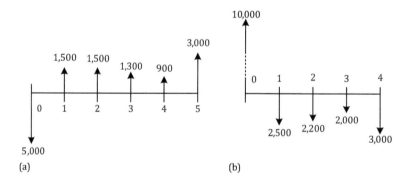

(a)   (b)

Figure 7.5 PW profiles of pure cashflows.

---

[18] In engineering economics, we come across pure investments more often than pure borrowing. Metaphorically, an engineering project can be considered a borrower and the company a lender (or investor). We usually analyze a project from company viewpoint, and hence as an investment problem.

[19] Another example is:

Year	0	1	2	3	4
Cashflow	−$100	−$400	0	$300	$350

[20] Another example is:

Year	0	1	2	3	4
Cashflow	$500	$100	−$200	−$250	−$350

[21] To determine whether the cashflow has changed sign, compare its sign with that of the previous one. In this case, at period 1 the sign is +ve, which was −ve at the previous period (at 0).

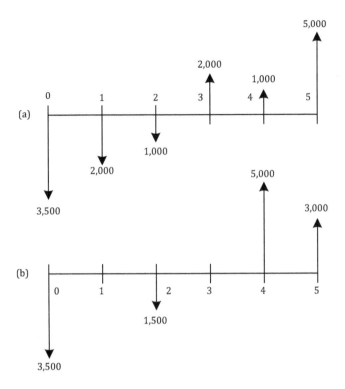

Figure 7.6 Multiple cashflows preceding the sign change.

to $1,500 (a change from −ve sign to +ve). In the *pure* borrowing above, this occurs when the cashflow changes from $10,000 to −$2,500 (a change from +ve sign to −ve). In pure cashflows, once the sign has changed it remains that way.

The sign change is visible in the cashflow diagram too. For the two pure cashflows just discussed, it is obvious in their diagrams in Fig. 7.5. Note that once the vector has changed direction, as at period 1 with respect to period 0, it continues to stay that way.

In the illustrations, the sign changed immediately following the first cashflow. This is not essential for the cashflows to be pure. There may be more than one cashflow, as long as they are all of the same sign, before the sign changes. Thus, there may be several costs in *pure* investment problems, as shown in Fig. 7.6(a), as long as they are incurred before the benefits begin to accrue. The only characteristic essential for the cashflows to be *pure* to yield one ROR is that once they change sign, they should continue to be that way. In Fig. 7.6(a) the sign changes at period 3 (from downward vector at period 2 to upward vector at 3), beyond which all vectors remain upward.

A similar situation can occur in the case of pure borrowing, i.e., there may be several benefits preceding one or more costs as long as the benefits are realized prior to the costs.

Another point to note is that for being pure the cashflows do not need to exist at each period, as seen in Fig. 7.6(b). For this diagram the cashflow table is:

Year	0	1	2	3	4	5
Cashflow	−$3,500	$0	−$1,500	$0	$5,000	$3,000

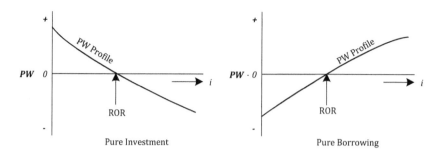

Figure 7.7 (a) Pure investment (b) Pure borrowing.

The existence of zero(s) in a cashflow table, as in the above, can be confusing in judging the number of sign changes. The zeros are simply ignored[22] since they represent the absence of vectors in the diagram. In the table therefore, there is only one sign change, at period 4, and thus only one ROR for the project the cashflows pertain to.

Why do *pure* cashflows yield only one ROR? This can be explained through the PW-profile, a plot between PW-values versus interest rates. For *pure* investments, PW-profiles are as in Fig. 7.7(a), and for *pure* borrowing as in Fig. 7.7(b). As obvious in these figures, the profile[23] curve for *pure* cashflows crosses the x-axis (*i*-axis) only once. The cross-over point represents the ROR since PW = 0 there.

7.5 HYBRID CASHFLOWS

Several engineering projects generate cashflows that are not *pure*, with costs and benefits so occurring that there is more than one sign change. We call such cashflows *hybrid*; the following is an example.

Year	0	1	2	3	4	5
Cashflow	−$3,000	$2,000	$500	−$1,000	$1,500	$2,500

Note in this table, as well as in the corresponding diagram in Fig. 7.8, that there are three sign changes: one at period 1, another at period 3, and the third at period 4. For this cashflow pattern, the number of RORs may be as many as three, i.e., none, one, two, or three. In general, *the number of RORs "may be" as many as the number of sign changes*. This is called the *cashflow-sign-change rule*.

Hybrid cashflows typically arise in contractual projects where contractors are paid on an "as-you-go" basis, since they can't invest huge sums of money, or won't like to do so for business reasons. Examples are construction of interstate highways, or manufacture of fighter planes or space shuttles. In such projects, the contractor is paid (benefit) a portion of the cost in the beginning, usually on signing the contract. Other payments follow as the

---

[22] The zero can be considered to carry the same sign as the previous non-zero cashflow. Thus, the zero at period 1 is −ve.

[23] In a PW-profile, the value of PW corresponding to $i = 0$ (where the profile meets the y-axis) is simply cash inflows minus outflows, with no regard to the time value of money. For example, for the cashflows in Fig. 7.6(b), this will be $(5,000 + 3,000) − (3,500 + 1,500) = 3,000$. This characteristic of the PW-profile can be used to check whether the profile seems correct. Also note that at $i = 0$ all functional factors become unity since $i = 0$ represents interest-free transactions.

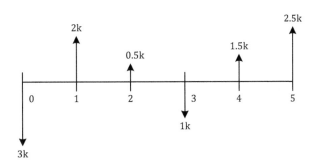

Figure 7.8 Hybrid cashflows.

work progresses. Hybrid cashflows also occur in venture capital investment if future invest-ments depend on returns for the past ones.

Prior to any ROR analysis, first check[24] the cashflows to determine whether the project is of the *pure* type. If so, then make a decision by determining[25] its ROR, or PW correspond-ing to the given MARR. If the cashflows are of the hybrid type, then the project may have more than one ROR. The determination of the correct ROR in the case of hybrid cashflows is presented in Subsection 7.5.2.

7.5.1 Internal ROR

A company rarely undertakes only one project at a time; several projects run concurrently. In general, projects are allowed to interact financially by pooling together capital funds allocated to various projects. In the rare case where the project is financially shielded, its cashflows are isolated from those of the others. In the pooled projects, on the other hand, any temporary surplus cash in one project is used for temporary needs of the others. Thus, a project with surplus cash conceptually becomes a short-term lender to another project.

Consider project A with hybrid cashflows as:

Year	0	I	2	3
Cashflow	−$3,000	$10,000	−$2,00	−$3,000

There occurs a $10,000 cash inflow at period 1. A portion of this money can be loaned to other projects as long as project A will get back $2,000 at period 2 and $3,000 at period 3 to pay for its own costs.

---

[24] In fact even before checking the cashflows for their purity, first determine whether they will yield an admis-sible (+ve) ROR at all. This is done by totaling the benefits *as they are*, i.e., ignoring the time-value of money. Similarly total the costs as they are. If the total of the benefits is less than the total of costs, obviously no +ve ROR can exist since the costs outweigh the benefits, as in the following cashflows.

Year	0	1	2	3	4
Cashflow	−$500	−$50	$100	$150	$175

In here, benefits' total ($425) is less than costs' total ($550). In the special case where their totals are equal, ROR is obviously zero. To summarize, check the algebraic sum of the given cashflows. If this sum is −ve (−$75 in the above example), no admissible ROR exists; if it is zero, ROR is zero. Only if the algebraic sum is +ve, one or more ROR is likely.

[25] As explained earlier, you can avoid the evaluation of ROR if MARR is given; simply determine the MARR-based PW and make the decision based on its sign.

The interproject shuffling of funds gives rise to what is called *external* ROR. The *external* ROR of a project is the rate of return from lending to another project within the company, or to financial institutions outside. It simply is external to the project. If the surplus cash of a project is loaned at external ROR, then the resulting ROR of the project is called *internal* ROR (IROR), to differentiate it from the ROR discussed so far in this chapter. Wherever external ROR is meant, the adjective "external" is explicitly used. But this is not true with internal ROR; sometimes, we denote IROR simply by ROR and mean it that way.

As discussed earlier, companies set up MARR to reflect the financial climate of the marketplace as well as its own financial health. MARR is the "bottom line" rate of return below which projects are considered unworthy of investment. The external ROR may at times be MARR to ensure that the surplus funds are wisely shuffled among the projects. The external ROR may differ from MARR; for example, if the financial marketplace is cash-hungry the company may lend the surplus to outside borrowers rather than to the other projects within.

7.5.2 ROR Evaluation[26]

As mentioned earlier, hybrid cashflows in which more than one sign change takes place can give rise to more than one ROR. The number of likely RORs is obvious in a PW-profile, for example two in Fig. 7.9, provided the profile is complete, i.e., covers a wide range of interest rates.

When more than one ROR is likely for a project, two analysis choices exist:

1. Abandon the ROR analysis since it is getting complex, and use one of the other five methods, such as PW or AW, to make a decision.
2. Carry out a comprehensive ROR analysis.

A comprehensive ROR analysis requires the analyst to conduct altogether four tests on the data. These tests throw more light on the characteristics of the cashflows and lead to the determination of correct ROR. By *correct ROR* we mean the one that assures profitability. The four comprehensive tests are:

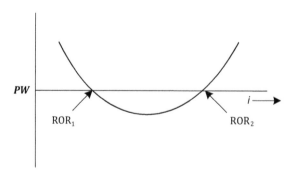

Figure 7.9 PW Profiling showing two RORs.

---

[26] The discussions in this subsection are of advanced nature. Skip this subsection if you are unlikely to encounter projects with hybrid cashflows.

Test 1 Algebraic Sum of the Cashflows
Test 2 Cashflow Sign Change
Test 3 Accumulated Cashflow Sign Change
Test 4 Net Investment

According to the first test, if the algebraic sum of the cashflows is +ve, then only one or more RORs are likely. Note that this test does not predict the likelihood of *only one* +ve ROR. A –ve algebraic sum usually[27] means no admissible (+ve) ROR. If the algebraic sum is zero, ROR will be zero.

The second test based on the cashflow-sign-change rule has already been discussed earlier in Sections 7.4 and 7.5. According to this test there can be as many RORs as the number of sign changes in the cashflows.

To conduct the third test, accumulated cashflows corresponding to each time period are posted under a new column in the cashflow table. They are then subjected to two subtests:

Subtest 3.1 *Is the accumulated cashflow at the last time period +ve?*
Subtest 3.2 *Is there only one sign change in the accumulated cashflows?*

If the answers to these two questions are both yes, then the hybrid cashflows yield only one ROR. These two conditions are thus more discriminating than the first test which can predict the likelihood of *one or more ROR*. Note that the third test combines the first two tests but on the *accumulated cashflows*, rather than on the *original (given) cashflows* to which Tests 1 and 2 are applied.

The fourth (net investment) test is applied after the ROR has been evaluated. It confirms the uniqueness of the evaluated ROR. According to this test, the evaluated ROR is the correct rate of return for the given cashflows if there is:

Subtest 4.1 *Zero net investment at the last time period*
Subtest 4.2 *Net investment throughout the project life*

These two subtests ensure that the project is always an investment[28], and at the end of its life there is no fund left, i.e., the benefits have exactly paid for all the costs.

As you may have realized by now, a comprehensive ROR analysis is quite complex compared to the other five methods. Example 7.5 illustrates such an analysis by considering hybrid cashflows.

Example 7.5: IRR Using Cashflows

Determine the internal rate of return for the project whose cashflows are given as follows. If required, use an external investment interest rate of 5%.

Year	0	1	2	3	4
Cashflow	–$3,000	$2,300	$900	–$200	–$322

SOLUTION:

Since internal ROR is to be evaluated, a comprehensive analysis is called for. We determine IROR and also confirm that it is the correct one by checking its uniqueness through the fourth test, as discussed previously. For a comprehensive analysis we conduct all the four tests; the first three in the beginning on the given cashflows, and the fourth one at the end using the evaluated ROR.

---

[27] Merritt, A. J. and Allen Sykes. *The Finance and Analysis of Capital Projects.* Longman. 1973. p. 135.
[28] Being a net investment means that at each period the current and preceding cashflows yield a –ve balance. In other words, the project never has surplus cash.

TEST 1 ALGEBRAIC SUM OF THE CASHFLOWS

In here, the algebraic sum of the given cashflows is $-3,000 + 2,300 + 900 - 200 + 322 = 322$. Since it is +ve, the benefits outweigh the costs; hence, one or more +ve RORs are likely. Note that this test ignores the time value of money.

TEST 2 CASHFLOW SIGN CHANGE

Examine the cashflows to determine whether they are pure or hybrid. The examination reveals that the given cashflows are hybrid since there is more than one sign change. Had the cashflows been pure we would have stopped here and determined the sole ROR for the project[29]. The second test predicts the maximum number of RORs possible for the project. In this case, the cashflows undergo three sign changes, which means that the project can have up to three RORs. The second test is more discriminating than the first since it can predict the likely number of RORs.

TEST 3 ACCUMULATED CASHFLOW SIGN CHANGE

The third test involves accumulated cashflows. To generate and examine them we widen the original table by creating an additional column for the accumulated cashflows, as follows.

Year	Cashflow	Acc. Cashflows
0	−$3,000	−$3,000
1	$2,300	−$ 700
2	$ 900	$ 200
3	−$ 200	$ 0
4	$ 322	$ 322

The entries in the last column have been computed as:

	A	B	C	D
1	Year	Cashflow	Acc. Cashflows	Formulas
2	0	-$3,000	-$3,000	
3	1	$2,300	-$700	=C3+D2
4	2	$900	$200	=C4+D3
5	3	-$200	$0	=C5+D4
6	4	$322	$322	=C6+D5

Once the accumulated cashflows have been posted we apply on them the two subtests.

Subtest 3.1 We note that the accumulated cashflow for the last period[30], being 322, is +ve.

Subtest 3.2 We also note that there is only one sign change in the accumulated cashflows.

On the basis of these two sufficient conditions we can say that only one (+ve) ROR is likely for the project.

---

[29] The first two tests are conducted in all ROR analyses. In fact, the first test should be conducted prior to all economic analyses, by any of the methods discussed in the text, since it reveals any inadmissibility of the given data. A −ve algebraic sum of the given cashflows means that the project is not investment worthy at all, since the costs outweigh the benefits.

[30] Note that it will equal the algebraic sum of Test 1.

TEST 4 NET INVESTMENT

The fourth test is applied later after the ROR has been evaluated. We now proceed to evaluate the ROR. Following the procedure discussed earlier and illustrated in Example 7.2, the PW-function in this case is:

$$2,300(P/F,i,1)+900(P/F,i,2)+322(P/F,i,4)-200(P/F,i,3)-3,000 = 0$$

Dividing[31] throughout by 200 to dwarf the multipliers, we get

$$11.5(P/F,i,1) + 4.5(P/F,i,2) + 1.61(P/F,i,4) - (P/F,i,3) - 15 = 0$$

We solve the above by trial and error, as explained earlier. The solution begins with a guessed beginning value of i. We can mentally work out such an i by relating the total interest with investment.

The total interest from the project, ignoring the time value of money, is $322 (obtained in Test 1 or 3) over 4 years. This is approximately $80 per year. For an initial investment of $3,000, this yields an approximate interest rate of $80/3,000 \approx 3\%$. We therefore begin the trial with $i = 3\%$.

With $i = 3\%$, the left – side of the function

$$= 11.5(P/F,3\%,1)+4.5(P/F,3\%,2)+1.61(P/F,3\%,4)-(P/F,3\%,3)-15$$

$$= 11.5\times0.9709+4.5\times0.9426+1.61\times0.8885-0.9151-15$$

$$= 11.1654+4.2417+1.4305-15.9151$$

$$= 0.9225$$

Since it is positive, we need to increase i to reduce it to zero (to equal the right side of the PW-function). So, we try $i = 4\%$, which yields the left side as 0.7051. The decrease from 0.9225 to 0.7051 for a 1% increase in i seems slow, so we increase i for the next trial by 2%. With i as 6%, the left side becomes 0.2898. Next, we try 8%, and so on. The results of the trial and error are summarized as follows.

i	Calculations Function's Left Side
3%	$11.5 \times 0.9709 + 4.5 \times 0.9426 + 1.61 \times 0.8885 - 15.9151 = 0.9225$
4%	$11.5 \times 0.9615 + 4.5 \times 0.9246 + 1.61 \times 0.8548 - 15.8890 = 0.7051$
6%	$11.5 \times 0.9434 + 4.5 \times 0.8900 + 1.61 \times 0.7921 - 15.8396 = 0.2898$
8%	$11.5 \times 0.9259 + 4.5 \times 0.8573 + 1.61 \times 0.7350 - 15.7938 = -0.1048$

As soon as the function's value changes sign, we stop the calculations, as above, and interpolate the value of i that will render the left side of function zero. In the present case, this happens between 6% and 8%. The interpolation[32] yields the value of i, which is ROR, as:

$$ROR = i = 6+2(0.2898)/(0.2898 + 0.1048)$$

$$= 6+1.47$$

$$= 7.47\%$$

---

[31] After such a division, the left side of this function is no longer the present worth.

[32] The accuracy of the results could have been improved by trying $i = 7\%$, and interpolating between 7% and 8%, as discussed earlier in this chapter.

TEST 4 NET INVESTMENT (continued)

Finally, we conduct the fourth test to confirm the correctness of the calculated 7.47% ROR. We do this by evaluating the net cashflows and examining whether or not they are net investments. These evaluations account for the time value of money at an interest rate equal to the computed ROR—7.47% in this case. The calculations and their results are summarized as follows.

Year	Cashflow	Calculations Net Cashflow
0	−$3,000	= −3,000
1	$2,300 − 3,000 (1 + 0.0747) + 2,300	= −924
2	$ 900 − 924 (1 + 0.0747) + 900	= −93
3	$ −200 − 93 (1 + 0.0747) − 200	= −300
4	$ 322 − 300 (1 + 0.0747) + 322	= 0

For period 1, net cashflow is the cashflow for year 1 ($2,300) plus the time-valued period 0 cashflow (= −3,000(1+0.0747)). Other net cashflows have been obtained the same way. For example, for period 3, it is the cashflow for period 3 (−$200) plus the time-valued net cashflow up to period 2 (= −93 (1 + 0.0747)).

After posting the net cashflows (last column), we examine them to see whether the two conditions of the fourth test are met. First, the net investment at project end should be zero, which is met in this case since the net cashflow at period 4 is zero. Second, the net cashflow at each time period should be negative, so that the project continues to be a net investment. This is also true in the last column data.

Thus, at 7.47%, the project has continued to be an investment throughout its life, never generating any surplus that could have been invested externally. With the two conditions of Test 4 satisfied, the computed 7.47% is the correct ROR for the project. In other words, this value of ROR ensures profitability as long as it is greater[33] than or equal to MARR. Thus, 7.47% is the *internal rate of return (IROR)*.

Example 7.5 was intentionally framed to keep the comprehensive analysis simple. Real-world ROR analysis can be more difficult. The following comments pertain to some of the difficulties that may arise in such analyses. The comments should help you appreciate the extent of complexities and in coping with them.

1. Net investment test (Test 4) fails, i.e., one or more net cashflows are +ve. This means that there is surplus cash in the project at certain periods. Due to the pooling of investment capital, this surplus is taken out of the project and invested elsewhere at the given external ROR (or MARR).

The ROR analysis in such a case should invest externally *only enough*[34] to render the net cashflows −ve throughout the project, still achieving zero net cashflow at project end. This manipulation allows for external investment of surpluses without "starving" the project of funds, leading to the *true* ROR (or internal ROR). For instance, if in Example 7.5 (previous table) the net cashflow for year 2 is $150 rather than −$93, the $150 surplus would be invested externally at 5%. The net investment for year 3 would then be $150 (1 + 0.05) − 200 = −$43.

External investment of the project surplus adds further realism to the decision. If not done and allowed to "sit idle" in the project, the surplus would have been assumed to earn

---

[33] In the absence of any given value we can assume the MARR to equal the external rate of return of 5%, since had there been a surplus in the project it would have earned interest at 5%.

[34] Invest externally as little as necessary to render the project a net investment throughout its life.

interest at the internal rate of return. This is not true in the real-world where funds are pooled together.

2. The aim of manipulating the hybrid cashflows is not to achieve one sign change, but to satisfy Test 4, namely zero net cashflow at project end and net investment throughout the project life.
3. In some problems, Test 2 based on the accumulated-cashflow-sign-change rule may predict fewer RORs than Test 1 does based on the cashflow-sign-change rule. This is due to sharper discriminating power of Test 2.
4. Don't stop at Test 3 just because one +ve ROR could be computed. Only when Test 4 has been gone through and its two conditions met, can we be sure of the computed ROR as a measure of profitability.
5. A project may yield ROR that is lower than the external ROR (or MARR), signifying its investment unworthiness.
6. If in Test 2 the accumulated cashflow at the last period is zero, along with more than one sign change in the accumulated cashflows, it simply means that one of the RORs is zero.
7. If in Test 2 the accumulated cashflow at the last period is –ve, along with more than one sign change in the accumulated cashflows, it means that one of the RORs is –ve.

7.6 MULTI-ALTERNATIVES

Engineering economics projects can be categorized into one of the two groups: no-alternative or multi-alternative. In no-alternative projects, there is only one candidate, and hence no choice. We have extensively discussed ROR analysis of such a project in the previous sections. It involves determining the project ROR and comparing it with the MARR set by the management. If the project ROR is financially attractive, the project is funded. As an alternative to exactly evaluating the project ROR, we may evaluate its PW using MARR as the interest rate and decide about the funding on the basis of PW-criterion (Example 7.3).

Let us begin the discussion of multi-alternative projects by first considering two-alternative[35] projects.

7.6.1 Two Alternatives

In Chapters 6 we have discussed PW-, FW-, and AW-methods of analysis. There the logical criterion in selecting the better of the two alternatives has been higher PW, FW, or AW. Unfortunately, this logic is unsound in the case of ROR analysis. The alternative with the higher ROR may not be better. The ROR-analysis of two alternatives (or more) is based on a technique called *incremental analysis*.

7.6.1.1 Incremental Analysis

Under incremental analysis the two alternatives are compared for the differences in their cashflows. The incremental data are obtained by subtracting the data of the smaller-investment alternative (lower first cost) from the corresponding data of the other alternative (higher first cost). The difference in the first cost so obtained is +ve, representing an *increment*, and

---

[35] A two-alternative project has two candidates, each being an alternative to the other. Purists may call it one-alternative project.

hence the name *incremental analysis*. Using the *incremental data (or cashflows)* we determine *incremental rate of return*, denoted by $\triangle$ROR. The computed $\triangle$ROR is then compared with the given MARR for making decisions the usual way, i.e., if $\triangle$ROR $\geq$ MARR, then select the larger-investment alternative, otherwise select the smaller-investment alternative. The concept underlying incremental analysis is explained as follows.

Suppose machine A costs $4,000 and yields certain benefits. Another machine B can also do the job, but it costs $6,000 and yields benefits that *look superior*. The question is: Is B better than A from an ROR viewpoint? In other words, we want to know whether the additional cost of $2,000 on B is worth the additional benefits expected from it. So, our goal is to determine the *incremental rate of return* $\triangle$ROR by analyzing the additional cost against the additional benefits. If $\triangle$ROR is greater than MARR, i.e., the additional $2,000 cost brings in return in excess of the MARR, then machine B is selected. Otherwise, alternative B is discarded; and A is selected provided ROR$_A$ $\geq$ MARR. Conceptually, the alternative with the higher first cost is like a *challenger*. The incremental technique judges the challenger's investment worthiness not on its own but in comparison to that of the low-investment alternative which would have been selected, in the absence of the challenger, due to its low first-cost. In the illustration under discussion, A is the defender because in the absence of its challenger B we would have selected A for investment. Example 7.6 apples incremental analysis to a two-alternative project.

Example 7.6: Two Alternatives

Two mutually exclusive alternatives A and B, whose cashflows are given as follows, exist for a project. Which alternative should be selected on the basis of ROR if MARR = 10%?

N	A	B
0	-$4,500	-$18,000
1	$2,000	$ 6,500
2	$2,700	$ 9,250
3	$2,250	$ 9,500

SOLUTION:

Since there are two alternatives and the decision criterion is ROR, the solution must be based on incremental analysis.

Which of the two alternatives costs less? It is A. So, we would prefer to invest in A. But since alternative B exists, we would like to analyze it for its additional cost and benefits to see whether investing in B makes sense. The analysis is carried out therefore on the incremental data B-A. We begin by creating another column to post the B-A data, as follows.

	A	B	C	D	E
1	N	A	B	B-A	Formulas
2	0	-$4,500	-$18,000	-$13,500	=C2-B2
3	1	$2,000	$6,500	$4,500	=C3-B3
4	2	$2,700	$9,250	$6,550	=C4-B4
5	3	$2,250	$9,500	$7,250	=C5-B5
6					
7					

From the B-A data for period 0, B costs $13,500 more than A. Other B-A data are the additional benefits from B over that from A.

At this point of the solution procedure there are two choices:

1. Determine $\triangle$ROR for the B-A incremental cashflows, or
2. Since MARR is given, check whether PW for the B-A incremental cashflows is +ve.

The second choice is usually easier in terms of calculations, and is preferred if MARR is given, as here. So, we opt for this choice. PW for the B-A cashflows corresponding to 10% MARR is given by:

$$PW_{B-A} = -13,500 + 4,500(P/F,10\%,1) + 6,550(P/F,10\%,2) + 7,250(P/F,i,3)$$
$$= -13,500 + 4,500 \times 0.9091 + 6,550 \times 0.8264 + 7,250 \times 0.7513$$
$$= -13,500 + 4,091 + 5,413 + 5,447$$
$$= 1,451$$

Since PW_{B-A} corresponding to the MARR is +ve, the actual value of $\triangle ROR$ will be greater than MARR. Hence, B is selected[36].

Were we to evaluate $\triangle ROR$ in this example, we would have used the trial and error method discussed earlier. This would have involved evaluating i that would render the PW-function zero. The result would have yielded[37] $\triangle ROR_{B-A}$ greater than MARR, selecting B over A.

7.6.2 More-Than-Two Alternatives

ROR analysis of more-than-two-alternative projects is basically an extension of that for two alternatives discussed in the previous subsection. Here too, the selection based on the highest *individual* ROR can lead to erroneous decisions. Hence incremental analysis is carried out, as illustrated in Example 7.7.

Example 7.7: Multiple Alternatives

If each of the following four mutually exclusive alternatives has 8 years of useful life, which one should be selected based on ROR if MARR = 8%?

	A	B	C	D
First Cost	$600	$500	$965	$800
Annual Benefit	$100	$120	$130	$110
Salvage Value	$375	$40	$800	$747

---

[36] One must however check that PWB is +ve, as follows.
$$PWB = -18,000 + 6,500(P/F,10\%,1) + 9,250(P/F,10\%,2) + 9,500(P/F,i,3)$$

$$= -18,000 + 6,500 \times 0.9091 + 9,250 \times 0.8264 + 9,500 \times 0.7513$$

$$= -18,500 + 5,909 + 7,644 + 7,137$$

$$\approx 2,690$$

[37] Applying Tests 1 and 2 on the incremental B-A cashflows, we can see that only one $\triangle ROR$ is likely. The relationship PW = 0 is
$$-13,500 + 4,500(P/F,i,1) + 6,550(P/F,i,2) + 7,250(P/F,i,3) = 0$$
Dividing throughout by 500,
$$-27 + 9(P/F,i,1) + 13.1(P/F,i,2) + 14.5(P/F,i,3) = 0$$
The beginning i for the trial can be approximated in a way explained in earlier examples. The approximate total incremental interest is $(4,500 + 6,550 + 7,250) - 13,500 = \$4,800$ over 3 years. So yearly interest is $1,600, giving an approximate incremental interest rate of $1,600/13,500 \approx 12\%$. Thus, the above PW-function's trial and error analysis is begun with $i = 12\%$. With a few trials, the exact value of $i (= \triangle RORB-A)$ is computed as 15.6%.

SOLUTION:

Incremental analysis is essential since ROR is the criterion in a multi-alternative project. The procedure comprises two phases.

PHASE I:

The first phase is a screening process. In this phase, we determine each alternative's ROR on its own to check whether it is greater than the given MARR. Alternatives with ROR less than the MARR are discarded, and hence do not undergo incremental analysis in phase II.

Since MARR is given, we can opt for the indirect process of checking the sign of alternatives' PW at MARR. This does away with the need to evaluate the ROR. Alternatives whose PW at MARR are +ve will yield ROR greater than MARR, and hence participate in the incremental analysis, while other don't. So,

$$PW_A = 100(P/A, 8\%, 8) + 375(P/F, 8\%, 8) - 600$$

$$= 100 \times 5.747 + 375 \times 0.5403 - 600$$

$$= 575 + 203 - 600$$

$$= 178$$

$$PW_B = 120(P/A, 8\%, 8) + 40(P/F, 8\%, 8) - 500$$

$$= 120 \times 5.747 + 40 \times 0.5403 - 500$$

$$= 690 + 22 - 500$$

$$= 212$$

$$PW_C = 130(P/A, 8\%, 8) + 800(P/F, 8\%, 8) - 965$$

$$= 130 \times 5.747 + 800 \times 0.5403 - 965$$

$$= 747 + 432 - 965$$

$$= 214$$

$$PW_D = 110(P/A, 8\%, 8) + 747(P/F, 8\%, 8) - 800$$

$$= 110 \times 5.747 + 747 \times 0.5403 - 800$$

$$= 632 + 404 - 800$$

$$= 236$$

Since the present worths of the four alternatives are +ve, they all participate in the incremental analysis.

PHASE II:

The procedure in this phase is similar to that in Example 7.6. We compare two alternatives at a time. We should prefer to select alternative B since its $500 first cost is the lowest. The next higher-first-cost alternative is A with its $600 first cost. So, we compare A with B to determine whether the additional cost of $100 on A brings in enough additional benefits to justify this cost, i.e., whether ΔROR_{A-B} is greater than MARR. The winner between A and B is compared with the next higher-first-cost alternative, and the process continues until all the alternatives have been examined incrementally for their worthiness.

The A-B incremental analysis is begun by tabulating the data under a new column, as follows. We also include a column for the time period.

Year	B	A	A-B
0	-$500	-$600	-$100
1-8	$120	$100	-$ 20
8	$ 40	$375	$335

Again, since MARR is given, we can use it as i and evaluate the incremental PW_{A-B}, based on whose sign either A or B is selected. As explained in Example 7.6, this approach is usually shorter than evaluating $\triangle ROR$ exactly[38].

Thus, we determine PW_{A-B} corresponding to 8%, as

$$PW_{A-B} = 335(P/F, 8\%, 8) - 20(P/A, 8\%, 8) - 100$$

$$= 335 \times 0.5403 - 20 \times 5.747 - 100$$

$$= 181 - 115 - 100$$

$$= -44$$

Since PW_{A-B} is -ve, $\triangle ROR_{A-B}$ will be lower than MARR. So additional investment in A is not worthwhile; B is the winner.

We now compare B with the next higher-first-cost alternative, which is D. Generate the D-B data the way we did for A-B, and determine PW_{D-B} as:

$$PW_{D-B} = 707(P/F, 8\%, 8) - 10(P/A, 8\%, 8) - 300$$

$$= 707 \times 0.5403 - 10 \times 5.747 - 300$$

$$= 382 - 57 - 300$$

$$= 25$$

Since PW_{D-B} is +ve, $\triangle ROR_{D-B}$ will be greater than MARR. So additional investment in D is worthwhile; hence D is selected over B.

Next, we compare D with C, the next higher-first-cost alternative. Generating the C-D data, we determine PW_{D-B} as:

---

[38] The analysis of the incremental A-B cashflows to evaluate $\triangle ROR$A-B is more involved. By applying Tests 1 and 2 (Example 7.5) on the cashflows we predict that, with only one sign change, one +ve $\triangle ROR$A-B is likely. This is determined from the PW-function

$-20 (P/A,i,8) + 335 (P/F,i,8) - 100 = 0$

A trial and error solution of the above yields iA-B = 4.8% (= $\triangle ROR$A-B). Since $\triangle ROR$A-B is less than the 8% MARR, the additional cost on A is not justified. Hence we decide to discard A in favor of B.

Next, B is compared with the next higher-first-cost alternative, i.e., D. The procedure is very similar. The PW-function for D-B incremental cashflows is

$-10 (P/A,i,8) + 707 (P/F,i,8) - 300 = 0$

From trial and error, $\triangle ROR$D-B is found to be 9%. Since this is greater than MARR of 8%, alternative D wins. Alternative D is finally compared with C the same way. The PW-function for C-D incremental cashflows is

$20 (P/A,i,8) + 53 (P/F,i,8) - 165 = 0$

From trial and error, $\triangle ROR$C-D is found to be 5%. Since this is less than the 8% MARR, alternative C loses to D.

Thus alternative D is selected.

$$PW_{C-D} = 53(P/F,8\%,8) + 20(P/A,8\%,8) - 165$$

$$= 53 \times 0.5403 + 20 \times 5.747 - 165$$

$$= 29 + 115 - 165$$

$$= -21$$

Since PW_{C-D} is –ve, additional investment on C can't be justified. Hence D is discarded. Therefore, alternative D should be selected.

Note that all the complexities of analyzing hybrid cashflows, some discussed in Section 7.5, can exist in the case of multi-alterative projects. If at all, they may be more pronounced because of several alternatives.

7.6.3 Analysis Period

Similar[39] to the PW analysis, ROR analysis must take care of differential useful lives of the alternatives. This is done by selecting an analysis period so that all the alternatives terminate at the same time. This may require replacing the alternatives. The cashflow data of the alternatives replaced during the analysis period are usually assumed to remain unchanged. As discussed in Chapter 5, the analysis period is normally the LCM of the alternatives' useful lives. If the LCM-based analysis period is too long, a shorter period can be used provided the data are carefully prorated. To avoid the analysis complexity, AW method may be preferred since unequal lives do not create such a difficulty in this method.

7.7 COMPUTER USE

The discussions presented in this chapter and the accompanying illustrations must have convinced you by now of the complexities of ROR-analysis. The repetitive nature of the calculations in the functional-notation method, arising from the trial and error approach, further accentuates the complexity. Programmable calculators and PCs ease the calculation efforts. Software are appropriate tools for solving problems under ROR-criterion. However, the objective of this chapter, indeed of the entire text, is the *education* focusing on the principles of engineering economics, rather than the *training* that can get emphasized in a software-centered learning environment. That is why I chose to take you through the "dirt road."

7.8 SUMMARY

The rate of return (ROR) method has been discussed extensively in this chapter. ROR can be defined as interest rate earned on the remaining balance of an investment. For a project to be investment worthy, its ROR must be greater than or equal to the minimum acceptable rate of return (MARR). MARR is set by the management based on several economic factors, both within and without the company. If MARR is known as given, a short-cut approach to ROR analysis is through project PW corresponding to MARR. If such a PW is +ve, project ROR will exceed the MARR.

---

[39] See Chapter 5 for an extensive discussion.

When the cashflows are *pure*, there exists only one ROR for a project. For hybrid cashflows, more than one ROR is possible; such cashflows involve sophisticated analysis demanding four different tests.

A project has an internal rate of return (IROR) if its surplus cash is invested elsewhere at an external rate of return. Incremental technique is essential in ROR analysis of projects with two or more alternatives. For a project with alternatives of unequal lives an analysis period equal to the LCM of the lives is normally used. Alternatively, AW method could be used to keep the decision-making simpler.

That ROR is easier to comprehend is its strength. However, a comprehensive ROR analysis is both complex and lengthy. The advent of computer technology has simplified ROR analysis to some extent, with the result that this analysis is most widely used in industry.

DISCUSSION QUESTIONS

7.1 Explain the difference between interest rate i and ROR.

7.2 Is MARR the same as interest rate i? Explain.

7.3 Why is the evaluation of ROR usually lengthy as well as complex?

7.4 Why is ROR widely used in industry in spite of its analysis complexities?

7.5 Discuss the four tests pertinent to a comprehensive ROR analysis.

7.6 Given a project's cashflows, how can you be sure of only one admissible ROR?

7.7 Explain the concept underlying incremental analysis.

7.8 Should one evaluate the exact value of ROR in ROR-analysis if MARR is given? If yes, why? If no, why not?

7.9 Explain the concept behind the net investment test.

7.10 Discuss the different between ROR and IROR.

7.11 Why is it that while computing the IROR, we must consider the sign changes in cashflows?

7.12 Discuss the best way to evaluate between multiple alternatives to a business decision.

7.13 What is the difference between rate of return and the annual interest rate?

7.14 Discuss the significance of PW, AW, and ROR.

MULTIPLE-CHOICE QUESTIONS

7.15 ROR-method of analyzing engineering economics problems is
  a. complex
  b. popular
  c. both a and b
  d. neither a nor b

7.16 For the following cashflows, the project may have up to:

Year	0	1	2	3	4	5
Cashflow	−$500	$0	$0	$400	−$50	$300

  a. one ROR
  b. two RORs
  c. three RORs
  d. four RORs

7.17 For the cashflows in Problem 7.11, the function for evaluating ROR is
  a. $500 + 50\ (P/F,i,4) - 400(P/F,i,3) + 300\ (P/F,i,5) = 0$
  b. $-500 - 50\ (P/F,i,2) + 400(P/F,i,1) + 300\ (P/F,i,3) = 0$
  c. $500 - 50\ (P/F,i,4) + 400(P/F,i,3) + 300\ (P/F,i,5) = 0$
  d. none of the above

7.18 Incremental method is *not* used in ROR-analysis of projects with:
  a. no alternative
  b. two alternatives
  c. three alternatives
  d. more than three alternatives

7.19 In multi-alternative projects, decision-making based on ROR of individual alternatives is:
  a. relatively difficult
  b. erroneous
  c. impossible without a computer
  d. widely practiced in industry

7.20 An investment of $7,500 is repaid 8 years later in a lump sum of $9,463. What is the ROR assuming annual compounding?
  a. 3%
  b. 4%
  c. 5%
  d. 6%

7.21 An investment of $5,000 is repaid 10 years later in a lump sum of $11,570. What is the ROR assuming annual compounding?
  a. 6%
  b. 7%
  c. 8%
  d. 9%

7.22 An initial investment of $205,000 is repaid 8 years later in a lump sum of $300,000. What is the ROR assuming semiannual compounding?
  a. 2.5%
  b. 3%
  c. 3.5%
  d. 4%

7.23 An initial investment of $350,000 is repaid 10 years later in a lump sum of $500,000. What is the ROR assuming semiannual compounding?
  a. 1.75%
  b. 2%
  c. 2.5%
  d. 3%

NUMERICAL PROBLEMS

7.24 The first cost of a water-jet machine to be used for slicing cheese is $60,000. The machine will generate net annual income of $12,000 during its useful life of 10 years. Determine the rate of return on investment in this machine.

7.25 Mary borrows $80,000 to buy her home. Beginning next month, she will pay the lender $850 per month for the next 20 years. What ROR will the lender be enjoying?

7.26 Cutting-Edge R&D Company is considering acquiring a new computing system for $300,000. The estimated economic life of the system is 5 years with no salvage value. If the benefit for the first year is estimated to be $100,000, and for the subsequent years $150,000 annually, should the system be acquired? Assume MARR = 20%.

7.27 American Indians sold an island in 1630 to a Dutch developer for some glass beads and trinkets worth $24. The island was worth $15 billion in 1999. What rate of return have the developer's beneficiaries enjoyed on the deal?

7.28 For the given cashflows, which of the two alternatives should be selected on the basis of rate of return if MARR = 25%?

Year	A	B
0	−$1,600	−$3,000
1–5	$ 800	$1,400

7.29 Which of the following three alternatives should be selected if MARR = 25%? Note that C is a "do-nothing" alternative (maintain the status quo).

Year	A	B	C
0	−$1,600	−$3,000	$0
1–5	$ 800	$1,400	$1,400

Year A B C

7.30 Twenty-First Century R&D Company has won a 2-year contract from NASA for a feasibility study to develop a township in space. The company will receive an advance of 30 million dollars on signing the contract and two more payments of $31.5 million and $4 million respectively on the first and second anniversaries of the contract. The costs of the project are estimated to be $35 million now, $20 million during year 1, and another $10 million during year 2. Determine the ROR for the project.

7.31 For the following cashflows what is the internal ROR, if external investment can earn 10%?

Period (n)	0	1	2	3	4	5
Cashflow	−$1,600	$1,000	$1,000	−$600	$800	$550

7.32 Three types of used robots can do a loading and unloading job. Considering these alternatives to be mutually exclusive (the selection of one precludes the others), which one should be selected on the basis of incremental rate of return if MARR = 15%?

Year	A	B	C
0	−$4,000	−$2,000	−$6,000
1	$3,000	$1,600	$3,000
2	$2,000	$1,000	$4,000
3	$1,600	$1,000	$2,000

7.33 For the following cashflows,

Year	0	1	2–9	10
Cashflow	−$1,275	$900	$300	−$2,700

(a) Compute PW at 10% MARR.
(b) Plot the PW-profile and determine the two RORs.
(c) Compute the internal ROR by considering an external interest rate of 6%, if necessary.

7.34 For the following cashflows,

Year	0	1	2	3
Cashflow	−$2,000	$1,280	$1,550	$1,300

(a) Compute the PW if $i = 8\%$.
(b) Compute the internal ROR by considering an external interest rate of 4%, if necessary.

7.35 A 1980 Dodge Lil Red Express truck was purchased off the showroom floor for $11,500. It was rarely driven and well taken care of. Today, the truck is worth $70,000. What is the ROR on the investment?

Use the following table to answer Questions 7.36–7.38.

	A	B	C
First Cost	$15,000	$5,000	$19,430
Annual Benefit	$ 8,500	$1,200	$ 9,000
Salvage Value	$ 4,500	$ 800	$ 3,000

7.36 A sandblasting company is considering buying a new blower for their operation. Which model of blower should be selected based on ROR if the MARR is 40%? Assume that each of the models have a useful life of 10 years.

7.37 Perform an incremental for the table. Based on this analysis which would be the best investment where MARR = 8%?

7.38 For Model A, the maintenance expense is $1,200 per year. Draw the cashflow diagram for the 10-year life of the model.

7.39 Suppose that you purchased ten shares of Microsoft stock in 1990 for $5 per share. Today those stocks are worth $94 per share. What is the ROR on the investment?

7.40 WorldWide Aquatic system is considering purchasing a used skimmer for oil clean-up operations. The new system would cost an estimated $375,000. The estimated useful life of the skimmer is 7 years. The first year of use would net $75,000 and the subsequent years would net $145,000 annually. The system would have a salvage value of $50,000 at the end of its useful life. Should the system be acquired if the MARR is 12%?

7.41 The Gorilla Rig Company would like to expand their business by adding an additional processing plant for their top-selling product. The new processing plant would cost $1.2 million and would have a useful life of 20 years. If the new plant would generate an additional $300,000 of revenue each year, should the company make the investment? MARR = 6%.

7.42 Now suppose that the Gorilla Rig Company could upgrade their current processing plant for $650,000, which would have a useful life of 14 years. This upgrade would result in $200,000 per year of revenue. Perform an incremental analysis of the two options for the Gorilla Rig Company; that is building a new processing plant (Problem 7.41) and upgrading the existing plant.

FE EXAM PREP QUESTIONS

7.43 If an investment of $80,000 were repaid in year 7 in a lump sum of $113,000, the rate
of return on the investment is closest to which of the following?
a. 5%
b. 10%
c. 8%
d. none of the above

7.44 Consider an investment with net cashflows of −$15,000, $7,500, $3,500, $2,000,
$3,500. The present worth indicates that the project should be accepted on interest rates:
a. less than 8%
b. between 7% and 15%
c. less than 12%
d. none of the above

7.45 If cashflow analysis produces multiple IRR, an analyst should:
a. compute present worth
b. compute modified IRR
c. use any of the compute interest rates to make a decision
d. both b and c

7.46 A sawmill procures a new pulp machine for $45,000. The machine is used for 7 years,
with increasing annual net revenue of $7,500 per year. To achieve an IROR of 12%,
what must the salvage value be?
a. less or equal to $1,000
b. greater than or equal to $1,000
c. $500
d. all of the above

7.47 A sawmill procures a new pulp machine for $45,000. The machine is used for 7 years,
with increasing annual net revenue of $7,500 per year. The PW of the pulp machine
would be what (6%/year) if the salvage value was $4,000?
a. −$ 474
b. $3,500
c. −$2,000
d. $5,000

7.48 MARR means what?
a. minimum accepted return of rate
b. multiple analysis revenue returns
c. minimum accepted rate of return
d. maximum annual rate of return

7.49 If a MARR is given by management, which means that decision makers would like a
ROR:
a. above the MARR
b. below the MARR
c. greater than or equal to MARR
d. both b and c

7.50 Consider an investment with a net cashflow of −$5,000, $1,000, $500, −$300, −$200,
and $1,200. Should the investment be made if MARR = 9%?
a. yes
b. no
c. maybe
d. none of the above

7.51 Consider an investment with a net cashflow of –$3,500, $900, $50, $700, $950, and $1,300. Should the investment be made considering a 9% MARR.

a. yes

b. no

c. maybe

d. none of the above

7.52 An oil company purchases a piece of equipment for $23,000. The loan is repaid in a lump sum of $30,000. The ROR for the lender is closet to which of the following?

a. 10%

b. 4%

c. 6%

d. 8%

7.53 Consider an investment with net cashflows of –$7,500, $500, $2,900, $1,300, $400. The present worth indicates that the project should be accepted on interest rates:

a. less than 5%

b. between 6% and 10%

c. less than 15%

d. none of the above

Chapter 8

Benefit-Cost Ratio

LEARNING OUTCOMES

- Meaning of benefit-cost ratio (BCR)
- Evaluation of project BCR
- Application of BCR-criterion
- Concept of disbenefit
- Incremental BCR analysis
- Use of input-output concept

Benefit-cost ratio (BCR) is the sixth, and the last, method of analysis. It is popular with governments which make decisions on how to spend taxpayers' money on public projects to maximize the common good. Profit-focused industries seldom use this method.

BCR is easy to comprehend. It is simply the ratio of benefits to costs. A benefit-cost ratio greater than one implies that the benefits outweigh the costs which makes the project investment worthy. As in the other criteria discussed in the last four chapters, BCR accounts for the time value of money.

8.1 MEANING

Consider a project that costs $100 today and generates a benefit of $120 next year. We might be tempted to say that BCR = $120/$100 = 1.2. But this is not correct since this way we have not accounted for the time value of money. We know that a sum of $120 next year is less today due to the interest for the year. Its equivalent value today is only 120(P/F,i,1). Let us say i is 8% per year, for which (P/F,8%,1) = 0.9259 from the interest table. The $120 benefit of the next year is therefore equivalent to $120 × 0.9259 ≈ $111 of today. Therefore, the correct BCR is $111/$100 ≈ 1.11, since the time value of money has been taken into account.

While evaluating BCR both the benefits and costs of the project must refer to the same time. In the preceding paragraph, the reference time for determining BCR was the present, i.e., t = 0. We could instead have considered t = 1 as the reference time. In that case, BCR = $120 /{$100 (F/P,8%,1)} = 120/(100 × 1.080) = 1.11. This is the same as with t = 0. Thus, a project's BCR is unique if the ratio is obtained by referring the costs and benefits to the same time period.

Consider a project whose cashflow diagram is Fig. 8.1. The investment P brings in benefits Q and R. Referring to which time period should the BCR be evaluated—period 0, period 4, or period 6 where there are vectors, or for another period where there is no vector? As explained in the previous paragraph, the BCR for the project will be the same irrespective of the time period at which it is evaluated. In general, BCR is evaluated at t = 0,

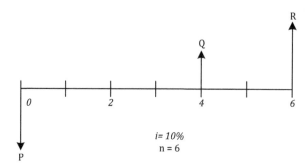

Figure 8.1 Concept of benefit-cost ratio.

referring to the present when an investment decision is being made. Doing so for the diagram in **Fig. 8.1** for $i = 10\%$,

$$BCR = \frac{Q\left(\dfrac{P}{F},10\%,4\right) + R\left(\dfrac{P}{F},10\%,6\right)}{P}$$

For projects to be investment worthy, their time-valued benefits must exceed the costs. The BCR of such a project therefore must be greater than one. For no-alternative projects (one candidate only), the higher the BCR, the more favorable the project.

8.2 EVALUATION

Generally, a project's BCR is determined by evaluating its $PW_{benefits}$ and PW_{costs} and dividing the former by the latter. Thus,

$$BCR = \frac{PW_{benefits}}{PW_{costs}} \tag{8.1}$$

Where it is easier to evaluate $FW_{benefits}$ and FW_{costs}, or where they are already known, BCR is determined from:

$$BCR = \frac{FW_{benefits}}{FW_{costs}} \tag{8.2}$$

Where it is easier to evaluate $AW_{benefits}$ and AW_{costs}, or where they are already known, BCR is determined from:

$$BCR = \frac{AW_{benefits}}{AW_{costs}} \tag{8.3}$$

We have already learned in Chapter 6 how to evaluate the numerators and denominators of the previous three equations. Thus, the BCR-analysis is a reformatted[1] PW-, FW-, or AW-analysis. Examples 8.1 and 8.2 illustrate the evaluation of BCR.

---

[1] Referring to Chapter 6 where PW-criterion was discussed, we had Equation (6.1) as:
$$PW = PW_{benefits} - PW_{costs}.$$
Since investment-worthy projects must have $PW \geq 0$, we can say
$$PW_{benefits} - PW_{costs} \geq 0$$

Example 8.1: Benefit-Cost Ratio

Should the project represented by the cashflow diagram in **Fig. 8.1** be funded if P = $350, Q = $200, and R = $400? Assume i = 10% per year.

SOLUTION:

The $350 cost is incurred now, while the two benefits are realized in future. We need to determine the present value of these benefits, which is:

$$PW_{benefits} = Q(P/F,10\%,4) + R(P/F,10\%,6)$$

$$= 200 \times 0.6830 + 400 \times 0.5645$$

$$= 136.60 + 225.80$$

$$= \$362.40$$

With the $350 cost at period 0 being the only cost, we have

$$PW_{costs} = \$350$$

Therefore,

$$BCR = \frac{PW_{benefits}}{PW_{costs}}$$

$$= \frac{362.40}{350}$$

$$= 1.04$$

Since BCR is greater than one, the project should be funded.

Example 8.2: Benefit-to-Cost Ratio

As part of on-going industrialization based on external capital, Bangladesh asks the World Bank for a loan to construct a canal to divert the waters of the Ganges. The estimated construction cost is $5 million, and the annual maintenance cost is $100,000. The canal will carry water for irrigation yielding $400,000 as annual tax from farmers. If i = 8% per year, and the canal is expected to last for 50 years, should the World Bank sanction the loan based on BCR-criterion?

SOLUTION:

The one-time construction cost is $5 million. The other two cashflows are annual. Let us assume that the maintenance cost will be paid out of the tax collected from the farmers. Thus,

$$Net\ annual\ benefit = Tax\ collected - Maintenance\ cost$$

$$= \$400,000 - \$100,000$$

$$= \$300,000$$

Dividing by PW_{costs} the above becomes

$$PW_{benefits} / PW_{costs} - 1 \geq 0$$

$$PW_{benefits} / PW_{costs} \geq 1$$

Since we define BCR as $PW_{benefits}/PW_{costs}$, it can be said that for a project to be investment-worthy BCR $\geq$ 1.

We can base the BCR-evaluation on PW, FW, or AW. Let us follow the PW-approach since the construction cost already given at period zero will not require transferring. We determine the equivalent present value of the 50-year net annual benefits as:

$$PW_{benefits} = 300,000\ (P/A, 8\%, 50)$$

$$= 300,000 \times 12.233$$

$$= \$3,669,900$$

From the construction cost, being the only cost, we have

$$PW_{costs} = \$5,000,000$$

Therefore,

$$BCR = \frac{PW_{benefits}}{PW_{costs}}$$

$$= \frac{3,669,900}{5,000,000}$$

$$= 0.734$$

Since BCR is less than one, the loan should be denied. There may however be reasons other than economic, such as social, political, or humanitarian[2], that may justify sanctioning the loan.

In Example 8.2 we might have considered the annual maintenance cost as part of the project cost, if so planned. In that case,

$$PW_{costs} = \text{First cost} + \text{Present value of maintenance costs}$$

$$= 5,000,000 + 100,000(P/A, 8\%, 50)$$

$$= 5,000,000 + 100,000 \times 12.233$$

$$= 5,000,000 + 1,223,300$$

$$= \$6,223,300$$

The \$400,000 tax collected annually is realized as benefits. Hence,

$$PW_{benefits} = 400,000(P/A, 8\%, 50)$$

$$= 400,000 \times 12.233$$

$$= \$4,893,200$$

$$BCR = \frac{PW_{benefits}}{PW_{costs}}$$

Therefore,

$$= \frac{4,893,200}{6,223,300}$$

$$= 0.786$$

---

[2] The Ganges floods the downstream villages each year during the monsoon season, resulting in thousands of deaths. The canal is expected to reduce the death toll.

Since BCR is under one, the decision is still not to sanction the loan. However, consideration of the maintenance cost as part of the project cost has changed the BCR from 0.734 (Example 8.2) to 0.786. This illustrates one of the major weaknesses of the BCR analysis. BCR is prone to misrepresentation arising from whether the operational costs are recovered from the benefits or treated as part of the first cost.

To minimize such confusion, we distinguish between the initial (or first) cost and future (running or operational) costs of a project. *In general, only the initial investment is considered project cost.* The operational expenses are paid out of the gross benefits, resulting in net benefits as in Example 8.2. It is for this reason—to avoid confusion—that operational and other future costs are called *disbenefits*. The use of this terminology distinguishes the first cost from other costs. The disbenefits[3] are subtracted from gross benefits, yielding data for the numerator of the BCR-equation. We can thus redefine BCR as:

$$BCR = \frac{Net\,Benefit}{Initial\,Cost} \qquad (8.4)$$

From Equation (8.4), BCR is simply *net benefit per unit initial investment.* The BCR-method is used mostly in analyzing public projects, as illustrated in Examples 8.3 and 8.4.

Example 8.3: Benefit-to-Cost Ratio

The twin-cities of Laurel and Hattiesburg are planning to build a municipal waste recycling plant at an initial cost of $16 million. During its life of 25 years the plant will require major overhauls every 5 years at a cost of $300,000 each. Compared to other alternatives of garbage disposal this plant will annually save the 20,000 city-households $75 each. Determine the project BCR if the interest rate is 6% per year.

SOLUTION:

Refer to the project cashflow diagram in Fig. 8.2. The one-time construction cost is $16 million. Every 5 years another $300,000 will be spent on overhaul. Thus a total of four overhauls will take place at the fifth, tenth, fifteenth, and twentieth year. The annual benefits are 20,000 × $75 = $1,500,000.

We can base the BCR-evaluation either on PW or AW, since the given cashflows are lump as well as annual sums. In the former we convert the 25-year annual benefits to their present value and divide by the total of initial cost and the present value of overhaul

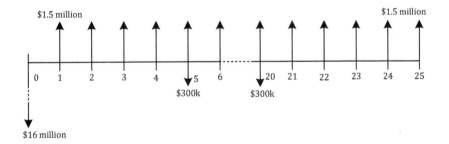

Figure 8.2 Diagram for Example 8.3.

---

[3] Where it is difficult to judge whether a cash outflow is cost or disbenefit, we use a modified version of the BCR-method. In this, rather than their ratio, the criterion is the difference between benefits and costs, called *benefit-cost difference* (BCD). This difference is unaffected by whether cash outflows are treated as costs or disbenefits. Note however that BCD is basically PW.

costs. In the AW-approach we convert the initial and overhaul costs to their equivalent annual cost, and divide the annual benefit by this equivalent cost. Based on what we have learned so far, the AW-approach will be lengthier since the overhaul costs will have to be converted first to their P-value and then to the A-value. Hence we follow the PW-approach, wherefrom

$$PW_{benefits} = 1,500,000(P/A,6\%,25)$$

$$= 1,500,000 \times 12.783$$

$$= \$19,174,500$$

The present worth of the costs is:

$$PW_{costs} = 16,000,000 + 300,000(P/F,6\%,5) + 300,000(P/F,6\%,10)$$

$$+ 300,000(P/F,6\%,15) + 300,000(P/F,6\%,20)$$

$$= 16,000,000 + 300,000\{(P/F,6\%,5) + (P/F,6\%,10)$$

$$+ (P/F,6\%,15) + (P/F,6\%,20)\}$$

$$= 16,000,000 + 300,000(0.7473 + 0.5584 + 0.4173 + 0.3118)$$

$$= 16,000,000 + 300,000 \times 2.0348$$

$$= \$16,610,440$$

Therefore[4],

$$BCR = \frac{PW_{benefits}}{PW_{costs}}$$

$$= \frac{19,174,500}{16,610,440}$$

$$= 1.15$$

Example 8.4: Benefit-to-Cost Ratio

To boost the city economy, the mayor and the council are considering building a four-lane road that will provide faster access to the nearby interstate highway. The construction cost of the road that includes a tunnel is $300 million, while its annual maintenance is estimated to cost $10 million. The annual benefits have been projected to be: additional commerce valued at $50 million, billboard advertising fees of $5 million, and reduction in accidents saving the city $500,000. The disbenefits are: $2 million yearly compensation to farmers for their land and businesses, and $300,000 annual loss in tax from the businesses along the existing dirt road. Assuming a 50-year life for the proposed road, should the project be funded if decision criterion is BCR? The city uses 6% as the annual interest rate in its funding decisions.

---

[4] Whether the overhaul costs should be treated as part of the initial investment cost or disbenefit depends on how the cities budget for their projects. We are assuming here that the benefits are paid to the city-dwellers as relief on garbage collection fee and are not available to the cities to pay for overhaul costs. If the cities plan that the overhaul costs be paid out of the savings, then these costs are disbenefits. In that case, BCR = (19,174,500 − 300,000 × 2.0348) / 16,000,000 = 1.16.

SOLUTION:

Since most of the given data are annual we are better off using AW in BCR evaluation. The net annual benefit is

$$AW_{benefits} = Total\,annual\,benefit - Total\,annual\,disbenefit$$

$$= (50,000,000 + 5,000,000 + 500,000) - (2,000,000 + 300,000)$$

$$= 53,200,000$$

The annual worth of the costs is:

$$AW_{costs} = First\,cost\,s\,annual\,worth + Annual\,maintenance\,cost$$

$$= 300,000,000(A/P,6\%,50) + 10,000,000$$

$$= 300,000,000 \times 0.0634 + 10,000,000$$

$$= 29,020,000$$

Therefore,

$$BCR = \frac{PW_{benefits}}{PW_{costs}}$$

$$= \frac{53,200,000}{29,020,000}$$

$$= 1.83$$

Since BCR is greater than one, the project should be funded.

8.3 MULTI-ALTERNATIVE PROJECTS

In the previous sections we discussed no-alternative projects. For such projects, the application of BCR-criterion involves checking whether the project BCR is more than one. In multi-alternative projects, BCR analysis becomes relatively complex, as obvious in this section.

As discussed earlier, engineering economics projects may be one of the three types:

1. Fixed input
2. Fixed output
3. Variable-input-variable-output

In *fixed input* projects, the alternatives' costs are the same, while in *fixed output* projects, their benefits are the same. Multi-alternative projects of these two types are analyzed the same way. In both, the alternatives are compared for their individual BCR value[5], and the one with the higher (if two alternatives) or highest (if more than two alternatives) BCR is selected. Example 8.5 offers an illustration.

---

[5] In fixed-input projects, since cost is the same, one can select the alternative on the basis of benefit only. Likewise, in fixed-output projects, one can do the selection on the basis of cost only.

For variable-input-variable-output projects *incremental analysis* is carried out. We have discussed such an analysis in Chapter 7 under the ROR-method. Its application to the BCR-method is presented in Section 8.4.

Example 8.5: Multiple Alternatives BCR Analysis

Either robot X or Y can carry out the loading-unloading tasks for a conveyor system, saving $20,000 annually. X costs $50,000, has a useful life of 7 years, and a salvage value of $5,000. The corresponding data for Y are: $40,000, 5 years, and $4,000. With an interest rate at 8% per year, compounded annually, which one should be selected on the basis of BCR-criterion?

SOLUTION:

This is a *fixed output* problem, because whether we select X or Y the output of the investment is the same, namely, automation of loading-unloading tasks resulting in annual savings of $20,000. Since the output is fixed the decision-making can be based on comparison of the individual BCR of X and Y.

Which of the two approaches, PW or AW, should be used for BCR evaluation in this case? There is an important reason that we adopt the AW-approach, that way we overcome the analysis difficulty due to different useful lives of X and Y. In the AW-method, analysis periods are simply the alternative's life. The PW-approach would have required LCM-based analysis period of 35 years, which is too long.

AW-based evaluation means converting for each (X and Y) the first cost and salvage value to their annual equivalents over the useful life, and subtracting the latter from the former. Thus, the robots' BCRs are:

$$BCR_X = \frac{AW_{benefits}}{AW_{costs}} = \frac{20,000}{50,000\left(\frac{A}{P},8\%,7\right) - 5,000\left(\frac{A}{F},8\%,7\right)}$$

$$= \frac{20,000}{50,000(0.1921) - 5,000(0.1121)}$$

$$= 2.21$$

$$BCR_Y = \frac{AW_{benefits}}{AW_{costs}} = \frac{20,000}{40,000\left(\frac{A}{P},8\%,5\right) - 4,000\left(\frac{A}{F},8\%,5\right)}$$

$$= \frac{20,000}{40,000(0.2505) - 4,000(0.1705)}$$

$$= 2.14$$

Since X's BCR is higher than that of Y, select[6] X.

In this example, the salvage value has been treated as cost recovery, and hence is subtracted from the first cost. If it is considered a benefit, then it becomes a part of the numerator. In that case,

---

[6] Due to the fixed output, the numerator in the BCR-equation is the same for both X and Y. Since the numerator is the same, one can decide on the basis of denominator alone, selecting the one whose denominator, i.e., PWcosts, is lower.

$$BCR_X = \frac{20,000 + 5,000\left(\frac{A}{F},8\%,7\right)}{50,000\left(\frac{A}{P},8\%,7\right)}$$

$$BCR_X = \frac{20,000 + 5,000(0.1121)}{50,000(0.1921)}$$

$$= 2.14$$

$$BCR_Y = \frac{20,000 + 4,000\left(\frac{A}{F},8\%,5\right)}{40,000\left(\frac{A}{P},8\%,5\right)}$$

$$BCR_Y = \frac{20,000 + 4,000(0.1705)}{40,000(0.2525)}$$

$$= 2.06$$

As can be seen, these BCR-values differ from those in Example 8.5, though the final decision in favor of X remains unchanged. In situations where there are doubts about whether a cash inflow, such as salvage value, be treated as benefit (part of the numerator) or as cost recovery (part of the denominator reducing the first cost), it is better to use the modified version of the BCR-method. In the modified version we look at the *benefit-cost difference* (BCD). For the case discussed, BCD for X and Y are:

$$BCD_X = 20,000 + 5,000 \times 0.1121 - 50,000 \times 0.1921$$

$$= \$10,956$$

$$BCD_Y = 20,000 + 4,000 \times 0.1705 - 40,000 \times 0.2505$$

$$= \$10,662$$

Robot X with higher BCD is selected. Thus, the decision based on BCD is the same as on BCR. BCD's merit is that BCD-values for X and Y are not affected by how we treat the salvage value—benefit or cost recovery. The BCD-approach thus avoids the confusion BCR-approach can create. Note however that the BCD-criterion is the same[7] as PW- or AW-criterion, depending on which one has been used to evaluate BCD. In here, $BCD_X = AW_X$ and $BCD_Y = AW_Y$ since BCD has been based on annual worth.

8.4 INCREMENTAL ANALYSIS

Engineering economics projects can be categorized into one of two groups: no-alternative or multi-alternative. In no-alternative projects, there is only one candidate, and hence no

---

[7] Except that the analysis period in BCD evaluation is not LCM-based as in PW-analysis. This deficiency does not exist if BCD is AW-based.

choice. We have discussed BCR-analysis of such projects in earlier sections, which involves determining the project BCR and approving it if BCR ≥ 1.

As mentioned earlier, multi-alternative projects can be of three types: fixed input, fixed output, and variable-input-variable-output. We have discussed the first two types in Section 8.3. This section presents variable-input-variable-output problems that *must be* solved by the *incremental analysis* method.

8.4.1 Two Alternatives

In the incremental BCR-analysis, the two alternatives are compared for the differences in their costs and benefits. Since differential data are generated by subtracting the data of the smaller investment from the other, the difference in investment (first cost) is +ve, i.e., incremental—hence the name *incremental* analysis. Using the *incremental costs and benefits* data we determine *incremental BCR*, denoted by $\triangle BCR$. If $\triangle BCR \geq 1$, then the additional cost on the larger-investment alternative is justifiable. The concept behind the incremental analysis can be explained as follows.

Suppose machine A costs $4,000 and yields certain benefits. Consider that another machine B can also do the job, but it costs $6,000 and will yield benefits that look superior. Is B really better than A from the BCR-viewpoint? We want to know whether the additional cost of $2,000 on B is worth the benefits expected of it. Our goal is to determine the $\triangle BCR$—the *incremental benefit-cost-ratio* (additional benefits divided by additional costs). If $\triangle BCR$ is greater than one, then machine B is selected; otherwise A is selected. If $\triangle BCR$ is exactly one, then either can be selected.

Conceptually, the alternative needing higher investment is a *challenger*. The incremental technique judges a challenger's investment worthiness not on its own but relative to that of the smaller-investment alternative. In the illustrative discussion just mentioned, machine A due to its lower first cost is the obvious choice, and B is the challenger. Example 8.6 illustrates the application of incremental analysis to a two-alterative project.

Example 8.6: Two Alternatives

Two mutually-exclusive alternatives A and B exist for a project. Given the following cash-flows which one should be selected on the basis of BCR if i = 10% per year?

n	A	B
0	−$4,500	−$18,000
1	$2,000	$ 6,500
2	$2,700	$ 9,250
3	$2,250	$ 9,500

SOLUTION:

With different costs and benefits, alternatives A and B represent a variable-input-variable-output problem[8], and hence the solution *must be* based on incremental analysis.

Which of the two alternatives has the lower first cost? With $4,500, it is A, which becomes the obvious choice. We prefer to select A unless the $13,500 additional cost on

---

[8] The input of A, being $4,500, differs from that of B which is $18,000. Their outputs (benefits) during their lives also differ.

B (= 18,000 – 4,500) proves to be beneficial. We therefore analyze the B-A increment data by creating an additional column to post them, as follows.

n	A	B	B-A
0	–$4,500	–$18,000	–$13,500
1	$2,000	$ 6,500	$ 4,500
2	$2,700	$ 9,250	$ 6,550
3	$2,250	$ 9,500	$ 7,250

Next, we evaluate $\triangle BCR_{B-A}$ based on the incremental data. Between PW or AW as the basis of evaluation, the AW-approach is not simple since the cash inflows are not uniform. We thus follow the PW-approach, wherefrom:

$$\triangle BCR_{B-A} = \frac{\triangle PW_{benefits}}{\triangle PW_{costs}}$$

$$= \frac{4,500(P/F,10\%,1) + 6,550(P/F,10\%,2) + 7,250(P/F,i,3)}{13,500}$$

$$= \frac{4,500(0.9091) + 6,550(0.8264) + 7,250(0.7513)}{13,500}$$

$$= \frac{4,091 + 5,413 + 5,447}{13,500}$$

$$= 1.11$$

Since $\triangle BCR_{B-A}$ is greater than one, the additional benefits from B are worth the additional cost on it. Hence, B is selected[9].

---

[9] The analysis based on individual BCR of A and B would have led to a wrong decision, as illustrated as follows.

$$BCR_A = \frac{PW_{benefits}}{PW_{costs}}$$

$$= \frac{2,000(P/F,10\%,1) + 2,700(P/F,10\%,2) + 2,250(P/F,i,3)}{4,500}$$

$$= \frac{2,000(0.9091) + 2,700(0.8264) + 2,250(0.7513)}{4,500}$$

$$= \frac{1,818 + 2,231 + 1,690}{4,500}$$

$$= 1.28$$

$$BCR_B = \frac{PW_{benefits}}{PW_{costs}}$$

$$= \frac{6,500(P/F,10\%,1) + 9,250(P/F,10\%,2) + 9,500(P/F,i,3)}{18,000}$$

$$= \frac{2,000(0.9091) + 2,700(0.8264) + 2,250(0.7513)}{18,000}$$

$$= \frac{5,909 + 7,644 + 7,137}{18,000}$$

$$= 1.15$$

8.4.2 More-than-Two Alternatives

BCR analysis of more-than-two-alternative projects is basically an extension of that for the two-alternative projects just discussed. Here too, selection based on highest individual BCR can lead to an erroneous decision. Hence, the incremental technique *must be* applied, as illustrated in Example 8.7.

Example 8.7: Multiple Alternatives

If each of the following five mutually exclusive alternatives has 8 years of useful life, which one should be selected based on BCR-criterion? Assume $i = 8\%$ per year.

	A	B	C	D	E
First Cost	$600	$500	$965	$ 80	$750
Annual Benefit	$100	$120	$130	$110	$105
Salvage Value	$375	$ 40	$800	$747	$230

SOLUTION:

For this problem incremental analysis is essential since BCR-criterion is the basis of decision in a multi-alternative project of the variable-input-variable-output type.

However, prior to the incremental analysis of such problems, it is always desirable to check the alternatives' individual BCR. This helps eliminate alternatives with BCR under one. The elimination may reduce the number of alternatives to be incrementally analyzed. Let us therefore determine the individual BCRs. For project A we have:

$$BCR_A = \frac{100\left(P/A,8\%,8\right) + 375\left(P/F,8\%,8\right)}{600}$$

$$BCR_A = \frac{100\left(5.747\right) + 375\left(0.5403\right)}{600}$$

$$BCR_A = \frac{574.7 + 202.6}{600}$$

$$BCR_A = \frac{777.3}{600}$$

$$= 1.30$$

In a similar way, we determine the individual BCR of project B, C, D, and E to be 1.42, 1.22, 1.29, and 0.97 respectively. Since alternative E's BCR is less than one, it is eliminated. The remaining four participate in the incremental analysis, as follows.

The incremental analysis procedure for this example is similar to that under the ROR-method (Example 7.7,). We compare two alternatives at a time. Alternative B is the obvious first choice since its investment of $500 is the lowest. The next higher-first-cost alternative is A with its $600 investment. So we compare B with A to find whether the additional cost on A brings in enough additional benefits to yield $\triangle BCR_{A-B} \geq 1$. The winner between A and B will be compared with the next alternative, and so on. This process of comparing two alternatives at a time continues until all the alternatives have been examined for their worthiness.

On the basis of individual BCR, we would have selected A since its BCR is greater. Note that this decision contradicts the one based on incremental analysis. Selecting A would have been a wrong decision, depriving the company of the opportunity to invest in B. The $13,500 additional investment in B, as shown in Example 8.6, is attractive.

For A-B analysis, we tabulate the incremental data, as in the last column as follows:

Year	B	A	B-A
0	-$500	-$600	-$100
1–8	$120	$100	-$ 20
8	$ 40	$375	$335

We next analyze the incremental A-B cashflows to determine $\triangle BCR$. Basing the analysis on PW, we have[10]:

$$\triangle BCR_{A-B} = \frac{335\,(P\,/\,F,8\%,8) - 20\,(P\,/\,A,8\%,8)}{100}$$

$$= \frac{335\,(0.5403) - 20\,(5.747)}{100}$$

$$= \frac{147 - 115}{100}$$

$$= 0.66$$

Since $\triangle BCR_{A-B}$ is less than one, the additional cost on A is not justifiable. Hence we discard A; B being the winner continues to remain the choice.

B is now compared with the next higher-first-cost alternative, i.e., D. The procedure is very similar to before. Create a table to post D-B data, if necessary. The incremental BCR for D-B is found to be:

$$\triangle BCR_{D-B} = \frac{707\,(P\,/\,F,8\%,8) - 10\,(P\,/\,A,8\%,8)}{300}$$

$$= \frac{707\,(0.5403) - 10\,(5.747)}{300}$$

$$= \frac{382 - 57.5}{300}$$

$$= 1.08$$

Since $\triangle BCR_{D-B}$ is greater than one, alternative D wins (B is discarded).

Alternative D is next compared with C, the last alternative. Following the same procedure,

$$\triangle BCR_{C-D} = \frac{53\,(P\,/\,F,8\%,8) - 20\,(P\,/\,A,8\%,8)}{165}$$

$$= \frac{53\,(0.5403) - 20\,(5.747)}{165}$$

$$= \frac{28.64 - 114.94}{165}$$

$$= 0.87$$

Since $\triangle BCR_{C-D}$ is less than one, alternative C loses to D.

Thus, alternative D should be selected.

---

[10] The salvage value is being considered a benefit since only the first cost is usually treated as investment. Also note that the incremental annual benefit can be negative, as here.

8.5 SUMMARY

The benefit-cost ratio (BCR) of a project represents its time-valued benefit per unit investment (first cost). BCR is commonly used in economic analysis of government projects which aim at benefitting as many people as possible. PW, FW, or AW, whichever is easier to set up depending on the given cashflow pattern, is used as the basis of BCR-evaluation. However, in projects with alternatives of unequal lives, AW-based BCR evaluation is preferred since it does away with the need for a LCM-based analysis period. BCR-analysis can at times be confusing, especially if it is difficult to decide whether the cash outflow(s) other than the first cost should be treated as cost or disbenefit. Under the concept of disbenefit, all costs other than the first cost are paid out of project benefits. Similar confusion arises with salvage value, i.e., whether to treat it as benefit or recovery of the first cost (investment). If such confusion can't be resolved, then BCD-analysis should be carried out instead. However, note that BCD-analysis is similar to PW-analysis. For fixed-output or fixed-input multi-alternative projects, alternatives' individual BCRs are ranked, and the alternative of the highest BCR is selected. For variable-input-variable-output multi-alternative projects, incremental analysis must be carried out. Prior to the incremental analysis, individual BCRs are evaluated and examined, and alternatives with BCR less than one are discarded.

DISCUSSION QUESTIONS

8.1　Why is BCR most popular in analyzing government projects for funding?

8.2　If there are controversies about whether a cash outflow is cost or disbenefit, what can be done in BCR-analysis?

8.3　Explain the concept underlying *disbenefit*.

8.4　What role, if any, does the input-output concept play in BCR-analysis?

8.5　How does the BCR-method account for time value of money?

8.6　How do you decide whether PW or AW should be used in determining BCR?

8.7　Does the incremental BCR analysis conceptually differ from the incremental ROR analysis? If yes, how? If no, why not?

8.8　What criteria must a challenger meet in order to justify investment?

8.9　Why can decisions based on BCR of more than two alternatives lead to erroneous decisions?

MULTIPLE-CHOICE QUESTIONS

8.10　If alternatives' useful lives are different, BCR-analysis should preferably be based on
　　a.　PW
　　b.　FW
　　c.　AW
　　d.　ROR

8.11　For which multi-alternative project incremental analysis is essential?
　　a.　fixed input
　　b.　fixed output
　　c.　variable-input-variable-output
　　d.　fixed-input-fixed-output

8.12　In the BCR-equation, the disbenefit data belong to the
　　a.　numerator
　　b.　denominator
　　c.　both a and b, depending on interest rate
　　d.　either a or b, depending on salvage value

8.13 When there is doubt about a data item—whether it is disbenefit or cost—we generally use a modified version of the BCR-method which is basically
    a. PW method
    b. AW method
    c. ROR method
    d. FW method

8.14 In BCR-analysis, if the salvage value is used to recover the first cost, it is being considered as:
    a. cost
    b. disbenefit
    c. benefit
    d. −ve cost

8.15 While analyzing a multi-alternative project for decision-making under BCR criterion, the alternatives' individual BCRs are evaluated _____ the incremental analysis.
    a. prior to
    b. after
    c. a or b, depending on the type of project
    d. a or b, depending on the cashflow pattern

8.16 If $\triangle BCR_{C-D}$ is 1.25, alternative
    a. C is selected
    b. D is selected
    c. a or b, depending on whose first cost is lower
    d. a or b, depending on whose annual benefit is higher

8.17 A solution *must* be based on incremental analysis if it is what type of problem:
    a. variable input, variable output
    b. fixed input
    c. fixed output
    d. all of the above

8.18 What is the PW of a project if for the next 25 years the net annual benefit is $15,000 at 10% interest?
    a. $200,000
    b. $123,941
    c. $136,155
    d. none of the above

8.19 What is the BCR of a project if the $PW_{benefits}$ is equal to $100,000 and the PW_{costs} is equal to $250,000?
    a. .895
    b. .4
    c. .34
    d. .2

8.20 What is the BCR of a project if the $PW_{benefits}$ is equal to $680,000 and the PW_{costs} is equal to $1,800,000?
    a. .377
    b. .298
    c. .41
    d. .5

8.21 In general, which cost(s) are considered the total project cost(s)?
    a. operating expenses
    b. yearly insurance expenses
    c. initial project cost
    d. all of the above

8.22 What is the AW_{costs} of a project if the first cost is $7,000 at 7% interest and the annual maintenance cost is $100? Assume the project has a 25-year expected life.
 a. 80
 b. 50
 c. 68.9
 d. 70.1

8.23 What is the $AW_{benefits}$ of a project is the total annual benefits of a project are $2,000 and the total annual disbenefit is $800?
 a. $2,800
 b. $0
 c. $400
 d. $1,200

8.24 What is the BCR of a project assuming the inputs from question 8.20 and 8.21?
 a. 17.11
 b. 35
 c. 8
 d. 24

8.25 If $\Delta BCR_{A-B} = .94$ which, if any, of the alternatives should be chosen?
 a. alternative A should be chosen
 b. alternative B should be chosen
 c. neither A nor B should be chosen
 d. alternative B, depending on AW_{costs}

8.26 If $\Delta BCR_{B-D} = 1.01$ which, if any, of the alternatives should be chosen?
 a. alternative B should be chosen
 b. alternative D should be chosen
 c. neither D nor B should be chosen
 d. alternative B, depending on $AW_{benefits}$

NUMERICAL PROBLEMS

8.27 If in Example 8.4 a court order decrees the city to pay the farmers an additional one-time compensation of $35 million when the project begins, should the city still construct the road?

8.28 Mauritius asks the World Bank for a loan to finance the construction of a canal. The estimated construction cost is $6.5 million. The estimated maintenance cost is $200,000 per year. The annual benefit from irrigation taxes is projected to be $450,000. Assuming the canal to last for 50 years, should the World Bank fund the project if the decision criterion is BCR? The bank uses a subsidized interest rate of 2% for third-world development.

8.29 What is the BCR for the project whose cashflow diagram is shown in Fig. 8.3? The first two cash outflows are investments, while the last two are overhaul costs. Assume $i = 8\%$ per year, and project life = 5 years.

8.30 The new landfill is estimated to cost the city of Joy $3 million and to last for 60 years. It will annually save the city $375,000 in fees presently paid to a private contractor for garbage disposal outside the city limits. The environmental disbenefit of having a landfill within the city limits is estimated to be $135,000 per year. If the city can borrow money at a federally subsidized annual rate of 3%, should the city fund the new landfill based on BCR-criterion?

8.31 An EPA-imposed water cleaning system is to be installed in a plant. There are two options: basic and deluxe. The basic model costs $2,500, is to be maintained at an annual cost of $200, and has no salvage value. The deluxe model costs $4,000, is maintenance-free, and has a salvage value of $500. Assuming that both are expected to last for 10 years, and $i = 12\%$ per year, which one should be selected based on BCR-criterion?

8.32 Two models (P and Q) of a machine are under consideration for procurement. Their cashflows are given in Fig. 8.4 which does not show the salvage values. If the salvage values are $5,000 for P and $15,000 for Q, which model is better based on BCR-criterion? Assume $i = 8\%$ per year.

8.33 Under a court decree, Universal Forgers has to erect a noise barrier to shield the nearby residential community from airport noise. The barrier could be made of wood or plastic. Their costs are:

	Plastic	Wood
First Cost	$765,000	$550,000
Annual Maint. Cost	$ 50,000	$120,000

As part of the court settlement the community will share the cost by annually paying Universal Forgers $50,000 for the plastic barrier or $40,000 for the wood barrier. On the

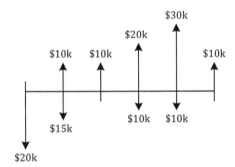

Figure 8.3 Diagram for Problem 8.29.

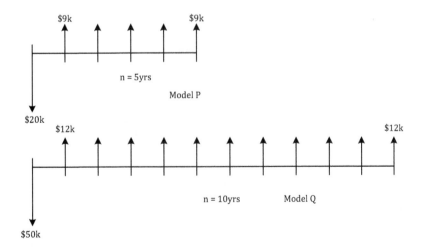

Figure 8.4 BCR analysis of two-alternative projects.

basis of BCR, which material is better? Assume barriers' lives to be 20 years, and i to be 10% per year.

8.34 Which one of the following five mutually exclusive alternatives should be selected on the basis of BCR-criterion? The figures are in dollars.

	A	B	C	D	E
First Cost	$4,000	$3,500	$6,500	$5,000	$4,800
PW_{benefits}	$7,500	$6,000	$5,750	$6,500	$8,000

8.35 A small city by the Mississippi River has occasionally experienced flooding. As a solution to the problem, the city is considering the following four options. Which one of these should be selected if the decision criterion is BCR?

	AW_{costs}	Annual Damage
No Flood Control	$0	$400,000
Construct Levees	$ 50,000	$250,000
Small Reservoir	$100,000	$120,000
Large Reservoir	$200,000	$ 20,000

(Hint: Annual benefit from an option is the saving from reduced damage.)

8.36 The National Parks Service is considering building a visitors center at one of their wilderness areas in the Midwest. With an initial investment of $750,000 the park is projected to have 20,000 visitors in the first year. On average visitors stay 3 hours in a park before leaving. What is the BCR of the project if each hour represents $10 profit/hour per visitor. Assume that the wilderness area would have an expected useful life of 22 years, a salvage value of $75,000, $32,000 of maintenance costs, and a 5% annual interest rate.

8.37 County Officials are analyzing whether or not they should improve 15 miles of county roads in order to reduce the number of annual traffic accidents and fatalities. Repaving the roads and emplacing guardrails would cost $1,000,000 on the 15 miles of road. The improvements would last 12 years and have no salvage value. This investment would result in 17 fewer traffic accidents (valued at $15,000 per accident) and three fewer fatalities (valued at $650,000 per loss). What is the BCR of the project if the interest rate is 7%? Should the project be undertaken?

8.38 WLTV News 7 is in the process of revamping news and weather coverage in order to better compete against larger national stations. One option that the channel has is to buy into a network of Doplar Radar in their region for $675,000 that would make weather predictions more accurate. The buy-in would last 5 years and have no salvage value. Assume that 70,000 people would use the enhanced service each day for 250 days each year. Of those, 12% are able to change their plans, valued at $3 per transaction (for value of information), and 55% would receive satisfaction from the service valued at $2 per transaction. Should the investment be undertaken? What is the BCR of the Doplar Radar buy-in if the interest rate is 7%?

8.39 WLTV News 7 could also improve their current radar network within their state. In order to bring the system up to date the news channel would have to invest $200,000 and the improvements would have a useful life of 7 years. The annual maintenance costs of the system would be $40,000 annually. Assume that 25,000 people would use the upgraded service each day for 250 days each year. Of those, 7% are able to change their plans, valued at $3 per transaction (for value of information), and 35% would receive satisfaction from the service valued at $2 per transaction. Compare the results to Problem 8.38. Which project should be undertaken, considering the BCR analysis?

8.40 System B and C can carry out the necessary sorting functions of a warehouse, saving $15,000 annually. B costs $35,000 and has a useful life of 8 years and a salvage value of $3,000. System C costs $40,000, has a useful life of 9 years, and a $3,200 salvage value. With an interest rate of 9% per year, compounded annually, which system should be purchased based on the BCR criterion?

8.41 Three mutually exclusive alternatives D, C, and E exist for a project. Given the following cashflows, which one should be selected based on the basis of BCR if $i = 12$?

Year	D	C	E
0	−$10,000	−$18,000	−$25,000
1	$ 4,000	$ 2,500	$ 8,452
2	$ 3,500	$ 8,342	$10,270
3	$ 2,700	$ 8,765	$ 6,347

8.42 Two mutually exclusive alternatives Z and X exist for a project. Given the following cashflows, which should be selected on the basis of BCR if $i = 8\%$ per year?

Year	Z	X
0	−$35,000	−$50,000
1	$ 8,000	$ 5,000
2	$12,000	$26,000
3	$15,000	$20,000

8.43 A regional power company is planning on building a new substation in a rural area in which they provide service. The initial investment would total $500,000. During the substation's 30-year life it will require major overhauls every 6 years at a cost of $75,000 each. Compared to other alternatives the new substation would save 68,000 households $30 annually in utility costs. Determine the BCR if the interest rate is 6%.

FE EXAM PREP QUESTIONS

8.44 BCR analysis should:
   a. take the ratio of $PW_{benefits}$ to the user over AW_{costs} of facilitator
   b. take the ratio of PW_{costs} to the user over $AW_{benefits}$ of facilitator
   c. analyze ratio of equivalent benefits over equivalent costs to the user
   d. none of the above

8.45 Consider an investment with net cashflows of −$20,000, $8,000, $4,500, and $10,000. The PW suggests that the project should be accepted for interest rates:
   a. less than 10%
   b. between 8% and 12%
   c. between less than 15%
   d. all of the above

8.46 What is the AW_{costs} of a project if the first cost is $6,547 at 15% interest and the annual maintenance cost is $500? Assume the project has a 15-year expected life.
   a. $1,620
   b. $1,349
   c. $2,134
   d. $ 800

8.47 What is the BCR of a project with AW$_{benefits}$ of $13,000 and AW$_{costs}$ of $5,467?
   a. .97
   b. 3
   c. .4205
   d. 2.377

8.48 What is the PW of a project if for the next 20 years the net annual benefit is $34,765 at 6% interest?
   a. $300,000
   b. $398,755
   c. $405,678
   d. none of the above

8.49 If $\triangle BCR_{A-D}$ is 1.25, alternative
   a. a is selected
   b. d is selected
   c. a or b, depending on whose first cost is lower
   d. a or b, depending on whose annual benefit is higher

8.50 If $\triangle BCR_{B-C}$= .94 which, if any, of the alternatives should be chosen?
   a. alternative A should be chosen
   b. alternative C should be chosen
   c. neither A nor C should be chosen
   d. alternative C, depending on AW$_{costs}$

8.51 What is the BCR of a project if the PW$_{benefits}$ is equal to $83,000 and the PW$_{costs}$ is equal to $765,313?
   a. .234
   b. b. .108
   c. .43
   d. .63

8.52 What is the BCR of a project in the AW$_{benefits}$ is equal to $114,000 and the AW$_{costs}$ is equal to $675,000?
   a. .37
   b. .94
   c. .168
   d. .43

8.53 What is the PW of a project if for the next 30 years the net annual benefit is $153,786 at 9% interest?
   a. $1,200,349
   b. $1,579,997
   c. $1,634,270
   d. none of the above

Chapter 9

Comparison

LEARNING OBJECTIVES

- Comparison among the six methods
- Why and when to prefer a method
- Relative popularity of the methods
- Computer impact on method popularity
- Computer-aided analyses

In Chapters 5 through 8, we have extensively discussed the six methods commonly used in economic analysis of engineering projects. An obvious question is: Why so many methods? Others are: Which one to prefer for analyzing a given problem? And why? How to ensure that the decision made on the basis of a particular criterion is reliable? Which method is commonly used in industry and why? In this last chapter of Part II we seek answers to such questions.

The discussions in this chapter are comparative in nature. Though the major merits and limitations of the six methods may have been obvious or briefly pointed out in the respective chapters, a detailed comparative study prepares engineers and technologists to apply the methods properly. It enables them to understand the methods' intricacies.

Engineering economics is basically decision-making, as pointed out in the very beginning of the text (Chapter 1). Decision-making is much more than solving a numerical problem. An early and crucial step in the decision-making process is to select the analysis method to use. In Chapters 5 through 8 we learned how to apply the methods in solving numerical problems; in here we discuss the why and when of these methods.

9.1 COMPARATIVE OVERVIEW

An important point to keep in mind, while analyzing engineering economics problems, is that the same problem may sometimes yield different answers depending on the method used for analysis. In a multi-alternative project, for example, alternative A might be the best using one method while alternative C using the other. How can engineering economists be sure of the decision in such a situation?

Different answers to the same problem beg the obvious question: Which of the answers is correct, or are they all incorrect? We discuss this issue throughout Chapter 7 by comparatively reviewing what we have learned in Chapters 5 and 6.

Of the six, the payback period method is the simplest, straightforward, and the least confusing. This is also the only method that normally does not account for the time value of money, though it can be enhanced to do so as illustrated in the last section of Chapter 5.

The PW-, FW-, and AW-method discussed in Chapters 6 are all based on the worth of the project. They differ only in the reference time to which the worth corresponds. In the PW-method, reference time is the present[1], while in the FW-method it is the future. The future could be located anywhere beyond the present, though it is normally at the end of useful life. In the AW-method, rather than a reference time marker, the entire duration of the useful life or analysis period becomes the time window. These three analysis methods usually lead to the same decision, and hence can be grouped under what may be called the worth-based method.

The other two analysis methods, namely ROR and BCR, discussed in Chapter 6, are complex[2] and may create confusion in decision-making. They can yield different answers for the same problem, which may differ from those using the payback period or worth-based methods.

Thus, the six analysis methods form four different groups. After discussing each group in the following four sections (9.2 through 9.5), we study them comparatively in Section 9.6.

9.2 PAYBACK PERIOD METHOD

The payback period method sometimes leads to decisions that are incompatible with those by the other methods, as illustrated in Example 9.1. What can analysts do in such a situation? They should discuss the discrepancy with project manager and apply one or more of the other methods so as to achieve more confidence in the final decision.

Example 9.1: Payback Period Method

(a) Which of the following two projects should be selected on the basis of payback period?

Year	A	B
0	−$400	−$425
1	$250	$150
2	$250	$150
3	$0	$150
4	$0	$150

(b) What's the decision under PW-criterion for $i = 12\%$ per year?
(c) Explain any discrepancy in the results of (a) and (b).

SOLUTION:

(a) The payback period for A is easily determined as 400/250 = 1.6 years, the time duration in which the $400 investment is recovered. For project B, it is 425/150 = 2.83 years. With shorter payback period as decision criterion, alternative A is selected.

---

[1] In engineering economics, the sense of the present is relative. The present is prior to the future and can be located at any time marker, though it is usually at t = 0.
[2] Although BCR is based on PW, and therefore on FW or AW, BCR evaluation may at times be confusing as discussed in Chapter 8.

(b) Next we analyze the problem under PW-criterion. The PWs for the two alternatives are

$$PWA = 250(P/F,12\%,1) + 250(P/F,12\%,2) - 400$$

$$= 250 \times 0.8929 + 250 \times 0.7972 - 400$$

$$= 223 + 199 - 400$$

$$= \$22$$

$$PWB = 150(P/A,12\%,4) - 425$$

$$= 150 \times 3.037 - 425$$

$$= 456 - 425$$

$$= \$31$$

With higher (+ve) PW as decision criterion, alternative B is selected.

(c) The PW-based decision in (b) is opposite of that under payback period in (a).

Payback period and PW can yield contradictory results for a project if its benefits are skewed (concentrated) to the beginning or end of the analysis period. In this example, the $250 benefit for alternative A is skewed to the beginning, being zero during the latter part of project life. For alternative B, the $150 benefit is uniformly spread over its 4-year life.

In general, if the cashflows are fairly distributed throughout the project life, the payback and PW methods are likely to yield the same result, as illustrated as follows.

In the same example, let us consider for alternative A a cashflow pattern similar to that of B, such as:

Year	A	B
0	−$400	−$425
1	$125	$150
2	$125	$150
3	$125	$150
4	$125	$150

Following the calculation procedure in Example 9.1, this data yield:

$$\text{Payback period for A} = \frac{400}{125} = 3.2\,\text{years}$$

$$\text{Payback period for B} = 2.83\,\text{years}\,(\text{as in Example 9.1})$$

$$PWA = 125(P/A,12\%,4) - 400$$

$$= 125 \times 3.037 - 400$$

$$= 380 - 400$$

$$= -\$20$$

$$PWB = \$31\,(\text{as in Example 9.1})$$

We thus see that both methods lead to the same result, i.e., selection of alternative B, if cashflow patterns are similar.

9.3 WORTH-BASED METHODS

The three worth-based methods are very similar, and that is why they always lead to the same decision. Their similarities arise from the duality of functional equations:

$$FW = PW(F/P,i,n) \quad \text{or} \quad PW = FW(P/F,i,n)$$

$$FW = AW(F/A,i,n) \quad \text{or} \quad AW = FW(A/F,i,n)$$

$$AW = PW(A/P,i,n) \quad \text{or} \quad PW = AW(P/A,i,n)$$

From these equations it is clear that any of the three worths is simply scaled versions of the other two. The scaling is due to multiplication with the (translating) functional factor which is constant for given i and n. It is this scaling that leads to the same conclusion by any of the three worth methods. This is true for any cashflow pattern. Example 9.2 illustrates the compatibility among the three worth-based methods.

Example 9.2: Worth-Based Methods Comparisons

Given the following cashflows, which alternative should be selected if decision criterion is: (a) PW, (b) FW, (c) AW? Assume $i = 10\%$ per year.

	A	B	C
First Cost	$200	$600	$450
Annual Benefit	$110	$150	$140
Useful Life (yrs)	2	6	4

SOLUTION:

Since the alternatives' useful lives are different, a common analysis period is decided upon on the basis of LCM. The LCM of their 2-, 6-, and 4-year lives is 12. We thus analyze the alternatives over 12 years[3].

(a) PW-analysis:

During the analysis period, alternative A is replaced five times, at periods 2, 4, 6, 8, and 10, as seen in the cashflow diagram Fig. 9.1(a). For B and C, replacement occurs once and twice, as seen in Figs. 9.1(b) and (c). Assuming that the costs and benefits for all replacements remain unchanged during the analysis period, the alternatives' PWs are:

$$PWA = PW_{benefits} - PW_{costs}$$

$$= 110(P/A,10\%,12) - \{200 + 200(P/F,10\%,2) + 200(P/F,10\%,4)$$

$$+ 200(P/F,10\%,6) + 200(P/F,10\%,8) + 200(P/F,10\%,10)\}$$

$$= 110 \times 6.814 - 200(1 + 0.8264 + 0.6830 + 0.5645 + 0.4665 + 0.3855)$$

$$= 750 - 200 \times 3.9259$$

$$= -\$35$$

---

[3] See the discussion later, following Example 9.3. Under the AW-method the analysis period of the alternatives, being their useful lives, can be different.

$$PWB = PW_{benefits} - PW_{costs}$$

$$= 150(P/A, 10\%, 12) - \{600 + 600(P/F, 10\%, 6)\}$$

$$= 150 \times 6.814 - 600(1 + 0.5645)$$

$$= 1,022 - 600 \times 1.5645$$

$$= \$83$$

$$PWC = PW_{benefits} - PW_{costs}$$

$$= 140(P/A, 10\%, 12) - \{450 + 450(P/F, 10\%, 4) + 450(P/F, 10\%, 8)\}$$

$$= 140 \times 6.814 - 450(1 + 0.6830 + 0.4665) = 954 - 450 \times 2.1495$$

$$= -\$13$$

With highest +ve PW as decision criterion, alternative B is selected.

(b) FW-analysis:

Based on the three cashflow diagrams in Fig. 9.1, and on the assumption that the costs and benefits for all replacements remain unchanged, the alternatives' FWs at the twelfth period are:

$$FWA = FW_{benefits} - FW_{costs}$$

$$= 110(F/A, 10\%, 12) - \{200(F/P, 10\%, 12) + 200(F/P, 10\%, 10)$$

$$+ 200(F/P, 10\%, 8) + 200(F/P, 10\%, 6) + 200(F/P, 10\%, 4)$$

$$+ 200(F/P, 10\%, 2)\}$$

$$= 110 \times 21.384 - 200(3.138 + 2.594 + 2.144 + 1.772 + 1.464 + 1.210)$$

$$= 2,352 - 200 \times 12.322$$

$$= -\$112$$

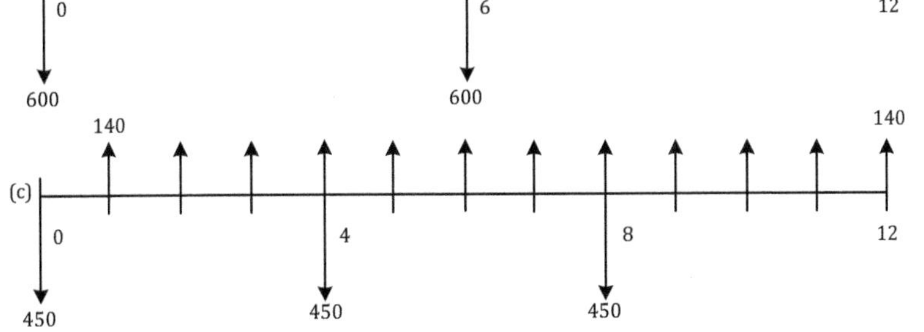

Figure 9.1   Diagram of Example 9.2.

$$FWB = FW_{benefits} - FW_{costs}$$

$$= 150(F/A, 10\%, 12) - \{600(F/P, 10\%, 12) + 600(F/P, 10\%, 6)\}$$

$$= 150 \times 21.384 - 600(3.138 + 1.772)$$

$$= 3,208 - 600 \times 4.910$$

$$= \$262$$

$$FWC = FW_{benefits} - FW_{costs}$$

$$= 140(F/A, 10\%, 12) - \{450(F/P, 10\%, 12)$$

$$+450(F/P, 10\%, 8) + 450(F/P, 10\%, 4)\}$$

$$= 140 \times 21.384 - 450(3.138 + 2.144 + 1.464)$$

$$= 2,994 - 450 \times 6.746$$

$$= -\$42$$

With highest +ve FW as criterion, alternative B is selected. Thus, the decision under FW-criterion is the same as under PW-criterion.

(c) AW-analysis:

Based on the cashflow diagrams in Fig. 9.1, and on the assumption of unchanged costs and benefits for all replacements, the alternatives' AWs are[4]:

$$AWA = AW_{benefits} - AW_{costs}$$

$$= 110 - \{200(A/P, 10\%, 12) + 200(P/F, 10\%, 2)(A/P, 10\%, 12)$$

$$+200(P/F, 10\%, 4)(A/P, 10\%, 12) + 200(P/F, 10\%, 6)(A/P, 10\%, 12)$$

$$+200(P/F, 10\%, 8)(A/P, 10\%, 12) + 200(P/F, 10\%, 10)(A/P, 10\%, 12)\}$$

$$= 110 - 200(A/P, 10\%, 12)\{1 + (P/F, 10\%, 2) + (P/F, 10\%, 4)$$

$$+(P/F, 10\%, 6) + (P/F, 10\%, 8) + (P/F, 10\%, 10)\}$$

$$= 110 - 200 \times 0.1468 \ (1 + 0.8264 + 0.6830 + 0.5645 + 0.4665$$

$$+0.3855)$$

$$= 110 - 200 \times 0.1468 \times 3.9259$$

$$= 110 - 115$$

$$= -\$5$$

---

[4] In the evaluation of AWA, the factor (P/F,10%,2) in the third term on the right-hand side of the = sign yields the P-value at period 0 of the 200-cashflow at period 2, while factor (A/P,10%,12) converts this P-value into its equivalent annual worth over the 12-year analysis period.

$$\text{AWB} = \text{AW}_{\text{benefits}} - \text{AW}_{\text{costs}}$$

$$= 150 - 600(A/P, 10\%, 12) - 600(P/F, 10\%, 6)(A/P, 10\%, 12)$$

$$= 150 - 600 \times 0.1468 - 600 \times 0.5645 \times 0.1468$$

$$= 150 - 88 - 50$$

$$= \$12$$

$$\text{AWC} = \text{AW}_{\text{benefits}} - \text{AW}_{\text{costs}}$$

$$= 140 - 450(A/P, 10\%, 12) - 450(P/F, 10\%, 4)(A/P, 10\%, 12)$$

$$-450(P/F, 10\%, 8)(A/P, 10\%, 12)$$

$$= 140 - 450(A/P, 10\%, 12)\{1 + (P/F, 10\%, 4) + P/F, 10\%, 8)\}$$

$$= 140 - 450 \times 0.1468(1 + 0.6830 + 0.4665)$$

$$= 140 - 142$$

$$= -\$2$$

With highest +ve AW as the criterion, alternative B is selected. Thus, the decision under AW-criterion is the same as under PW- or FW-criterion.

As seen in this example, the decision remains the same irrespective of which of the three worth-based methods is used for analysis.

In Example 9.2, FWs and AWs were evaluated directly from the given cashflow diagrams, with no attempt to utilize the relationships that exist among the three worths. This resulted in lengthy calculations for parts (b) and (c). Had the relationships been exploited, the solutions would have been briefer and simpler as illustrated in Example 9.3.

Example 9.3: Worth-Based Methods Comparison

How can the solution of Example 9.2, especially of parts (b) and (c), be made concise?

SOLUTION:

To achieve brevity in the solution of Example 9.2, we should attempt to benefit from the relationship among the three worths.

(a) PW-analysis:

Since FW or AW is not known at this stage, evaluation of PW can't be made any simpler than that in Example 9.2. However, parts (b) and (c) can be made concise, as illustrated as follows, since the values of PWs, as determined in part (a), are by then known.

(b) FW-analysis:

There are basically two approaches to solve this part. The first one is, based on the given cashflow diagram, as done in Example 9.2. The second approach exploits the functional relationship between PW and FW, keeping the solution brief. Since PWs are already known from part (a) we simply convert them to their FWs, as follows.

$$FWA = PWA(F/P,10\%,12)$$

$$= -\$35 \times 3.138$$

$$= -\$110$$

$$FWB = PWB(F/P,10\%,12)$$

$$= \$83 \times 3.138$$

$$= \$260$$

$$FWC = PWC(F/P,10\%,12)$$

$$= -\$13 \times 3.138$$

$$= -\$41$$

The conciseness offered by the second approach is obvious when these FW-evaluations are compared with those in part (b) of Example 9.2.

(c) AW-analysis:

Again, the first approach involves direct analysis of the given cashflow diagram, as in Example 9.2. The second approach, based on the functional relationship between PW[5] and AW, yields a brief solution. Since PWs are already known, from part (a), simply convert them into their AWs, as follows.

$$AWA = PWA(A/P,10\%,12)$$

$$= -\$35 \times 0.1468$$

$$= -\$5$$

$$AWB = PWB(A/P,10\%,12)$$

$$= \$83 \times 0.1468$$

$$= \$12$$

$$AWC = PWC(A/P,10\%,12)$$

$$= -\$13 \times 0.1468$$

$$= -\$2$$

A comparison of these AW-evaluations with those in part (c) of Example 9.2 illustrates that exploiting the functional relationship between the worths leads to briefer and simpler solutions.

There is still another way in the case of the AW-method to solve Example 9.2. We can apply the basic concept of annual worth learned in Chapter 6. Since AW represents the net annual cashflow that is equivalent to all the costs and benefits, the analysis period in the AW-method does not have to be the same for all the alternatives. That means we can

---

[5] Known FWs determined in part (b) of Example 9.2 or 9.3 can be used instead.

consider just one complete cycle[6] of the cashflows for each alternative. In other words, the analysis period of an alternative is simply its useful life. This is a unique advantage of the AW-method, non-existent with PW- and FW-methods.

Based on this explanation, we can solve part (c) of Example 9.2 by considering just the first cycle of cashflows for each alternative. Referring to **Fig. 9.1**, from the first complete cycle of the alternative we get

$$AWA = 110 - 200(A / P, 10\%, 2)$$

$$= 110 - 200 \times 0.5762$$

$$= 110 - 115$$

$$= -\$5$$

$$AWB = 150 - 600(A / P, 10\%, 6)$$

$$= 150 - 600 \times 0.2296$$

$$= 150 - 138$$

$$= \$12$$

$$AWC = 140 - 450(A / P, 10\%, 4)$$

$$= 140 - 450 \times 0.3155$$

$$= 140 - 142$$

$$= -\$2$$

These AW-values are the same[7] as in part (c) of Examples 9.2 and 9.3.

Based on the previous discussions along with those in Examples 9.2 and 9.3, we can say that the three worth-based methods lead to the same decision. However, remember that:

1. The AW-method frees you from the inconvenience alternatives' unequal lives create in PW- and FW-analysis. AW is evaluated by considering just one complete cycle of cashflows for each alternative. This is quite a relief in multi-alternative projects, especially where the LCM-based analysis period becomes long.
2. A major attraction of worth-based methods is that they do not require incremental analysis technique. Even in multi-alternative projects, the selection is based simply on alternatives' individual worths.

Of the three worth-based methods, FW is the least used in engineering economic analysis. This is primarily due to the nature of most engineering projects, requiring decision based on investment considerations as of now, or on annual operational advantage[8]. The

---

[6] By complete cycle we mean the span of one useful life, which avoids any consideration of replacement. Thus, for alternative A, the cycle comprises the first $200 cost followed by the two $100 benefits. For B, it comprises the first $600 cost followed by the six $150 benefits, while for C the first $450 cost followed by the four $140 benefits.

[7] If a difference occurs in spite of correct calculations, it is due to the rounding-off of functional factors' values in the tables.

[8] The PW-method suits projects where decision is based on economic consideration as of now. The AW-method is suitable for projects where recurring cashflows are of paramount importance, as in maintaining equipment or operating a facility.

FW-method, on the other hand, focusing on the future, is applicable mostly in the financial sector. FW-based decisions are made mostly by financial institutions; it is therefore basically non-engineering in nature. Engineers and technologists normally do not deal with financial investments which fall within the scope of business economics. Thus, PW and AW are directly applicable to engineering economics, while FW is "tangentially" so.

In the following comparative discussions we use only PW, as representative of the worth-based methods.

9.4 ROR METHOD

The attraction of the ROR-method lies in the fact that ROR is measured in terms of percentage, which is understood by most professional people. The other attraction is that in its analysis it does not need the value of applicable interest rate, or MARR, which is usually difficult to assess. In fact, interest rate i is the unknown to be evaluated.

The ROR-method's recent popularity in industry is attributed to the ubiquitous growth in computer technology. Computer systems have rendered the ROR-based decisions manageable. As seen in Chapter 7, a non-computer approach to ROR analysis is lengthy and error-prone in comparison to the other five types of analysis. Enhanced user-friendliness of modern engineering economics software aids ROR analysis the most.

ROR-based decisions may at times contradict those by the other methods. This happens when alternatives' cashflow patterns differ significantly from each other. The skewness in cashflows, for example heavy in the early years of the project, can give rise to discrepancy in results, as illustrated in Example 9.4. For projects whose cashflows are fairly distributed, the ROR method usually leads to the same decision as the other methods, as illustrated in Example 9.5.

Example 9.4: Payback Period and ROR

(a) Which of the following two projects should be selected on the basis of the payback period?

Year	A	B
0	−$400	−$400
1	$250	$150
2	$250	$150
3	$0	$150
4	$0	$150

(b) Does the decision in (a) change if ROR is the criterion?

(d) Explain any contradiction in the results of (a) and (b).

SOLUTION:

(a) The payback period for A is easily determined to be 400/250 = 1.6 years, the duration in which the $400 investment is recovered. For project B, it is 400/150 = 2.67 years. With shorter payback period as decision criterion, alternative A is selected.

(b) Since a two-alternative project is being analyzed under ROR-criterion, we first need to establish whether incremental analysis is necessary. Since the two alternatives have the same input—an investment of $400—the project is of the fixed-input type, and hence an individual-ROR-ranking-based decision will suffice. In other words, there is no need for incremental analysis.

The alternatives' RORs are obtained by solving their PW-functions for i. For A, the PW-function is $250\,(P/A,i,2) - 400 = 0$, whose solution[9] yields i to be greater than 15%, but less than 18%. Thus, $15\% < ROR_A < 18\%$. The PW-function for B is $150\,(P/A,i,4) - 400 = 0$, whose solution[10] yields $ROR_B > 18\%$.

Thus, based on higher ROR criterion, alternative B is selected. This decision contradicts the one in part (a) based on payback period.

(c) This inconsistency in the ROR and payback results can be explained by the difference in cashflow distribution for A and B. Alternative A is better under payback criterion since its $250 annual benefit is large in the beginning, returning the investment faster. Alternative B's $150 annual benefit being relatively smaller and spread widely over four periods takes longer to recover the investment. The inconsistency in results usually arises when alternatives' cashflow distributions are significantly different, as in this case.

Since A returns the investment faster while B earns higher ROR, each has its own attraction. Companies with long-term investment strategies will prefer B to A. Capital-tight companies, on the other hand, may select A since a shorter payback replenishes the limited capital for other investments.

Example 9.5: Multiple Alternatives

Assuming 8-year useful life for each of the following four mutually exclusive alternatives and 8% MARR, which one should be selected under (a) PW-criterion, (b) ROR-criterion?

	A	B	C	D
First Cost	$600	$500	$450	$800
Annual Benefit	$100	$120	$130	$110
Salvage Value	$375	$ 40	$800	$747

SOLUTION:

(a) The selection is based on the largest +ve PW corresponding to an interest rate equal to MARR, i.e., 8% per year. The PWs of the four alternatives are:

$$PW_A = 100(P/A,8\%,8) + 375(P/F,8\%,8) - 600$$
$$= 100 \times 5.747 + 375 \times 0.5403 - 600$$
$$= 574.70 + 202.61 - 600$$
$$= \$177$$

[9] From the PW-function, $(P/A,i,2) = 400/250 = 1.6$. If the PW-function comprises just one factor, as here, the range in which ROR lies can be found just by thumbing through the interest tables. Turn to any page of the table and look at the value of $(P/A,i,2)$. Is it 1.6? If not, thumb through the pages until it is 1.6 or near about. The i corresponding to the page becomes ROR_A. In this case, $(P/A,15\%,2) = 1.626$, while $(P/A,18\%,2) = 1.566$. So $i_A > 15\%$ since $(P/A,i,2) = 1.6$, but less than 18%. Thus, $ROR_A < 18\%$ but greater than 15%. Until we check where ROR_B lies, there is no need to determine the exact value of ROR_A (which by the way is 16.3% through interpolation). As seen in the next footnote, ROR_B is greater than 18%. With ROR_A less than 18%, we therefore select alternative B based on its higher rate of return. Note that we did not have to determine the exact values of RORs in this problem, where our interest is simply to know whose ROR is higher. Such judicious actions reduce the time and effort in reaching a decision, and should be taken wherever appropriate.

[10] Here we have $(P/A,i,4) = 400/150 = 2.67$. Thumb through the pages looking at the P/A column for n = 4. You will find that for $i = 18\%$, the entry is 2.69, which is greater than the 2.67 we are interested in. Only a higher i will reduce this P/A-value to 2.67. So we can say that $ROR_B > 18\%$. Sometimes, if you are lucky, the exact value being searched for is found on a page, yielding an exact ROR. The exact value of ROR_B, though not essential here, is through interpolation 18.46%.

$$PW_B = 120(P/A,8\%,8) + 40(P/F,8\%,8) - 500$$

$$= 120 \times 5.747 + 40 \times 0.5403 - 500$$

$$= 689.64 + 21.61 - 500$$

$$= \$211$$

$$PW_C = 130(P/A,8\%,8) + 800(P/F,8\%,8) - 965$$

$$= 130 \times 5.747 + 800 \times 0.5403 - 965$$

$$= \$214$$

$$PW_D = 110(P/A,8\%,8) + 747(P/F,8\%,8) - 800$$

$$= 110 \times 5.747 + 747 \times 0.5403 - 800$$

$$= \$236$$

With largest +ve PW as the criterion, D is selected.

Since it is a multi-alternative problem, let us first check whether incremental analysis is essential. As explained in Chapter 6, incremental analysis is necessary in variable-input-variable-output problems. Since the alternatives' first costs (input) and benefits (output) are different, this problem is the variable-input-variable-output type. So we need to carry out incremental analysis.

Under the incremental analysis we first check the alternatives' individual RORs to ensure that they are above MARR. Any alternative with individual ROR less than MARR is discarded from further (incremental) analysis. In this case, all the alternatives' individual RORs are greater than MARR since their PWs corresponding to the given MARR have been determined in part (a) to be +ve. So, all the four alternatives participate in incremental analysis, which begins by comparing two alternatives at a time. Since alternative B's first cost is the lowest, we prefer to select it. The next higher-first-cost alternative is A. So, we compare B with A to determine whether the additional cost on A and additional benefits from A yield a rate of return greater than MARR. The winner between A and B is compared with the next higher-first-cost alternative. This process continues until all the alternatives have been examined, one at a time.

For the first comparison generate the A-B incremental data, as follows in the last column:

Year	B	A	A-B
0	−$500	−$600	−$100
1–8	$120	$100	−$ 20
8	$ 40	$375	$335

By applying[11] Tests 1 and 2 on the incremented data we can predict that only one +ve ROR is likely, since there is one sign change in the A-B cashflows. This incremental rate of return $\Delta ROR_{A\text{-}B}$ is determined[12] from the PW-function:

$$-20(P/A,i,8) + 335(P/F,i,8) - 100 = 0$$

---

[11] See Chapter 7, if necessary.

[12] Since MARR is known, one can use the MARR-based short approach over the lengthier procedure of determining the exact ROR. One simply has to check the sign of the project PW corresponding to i = MARR. If it is +ve, then ROR>MARR; otherwise ROR<MARR. In this case, we have
    PWA-B   = 335(P/F,8%,8) − 20(P/A,8%,8) − 100
             = 335 × 0.5403 − 20 × 5.747 − 100
             = −$33.9
Since PWA-B is negative, RORA-B will be less than MARR (8%). Thus, alternative A is not better than B. It is therefore discarded, and B continues to remain selected.

A trial and error solution of the above yields $i \approx 5\%$, which equals $\triangle ROR_{A-B}$. Since $\triangle ROR_{A-B}$ is less than the 8% MARR, the additional cost on A is not justified. Hence we decide to continue to prefer B.

Next, B is compared with the ensuing higher-first-cost alternative, i.e., D. The procedure is very similar. The PW-function for D-B incremental cashflows is:

$$-10(P/A, i, 8) + 707(P/F, i, 8) - 300 = 0$$

From trial and error, $\triangle ROR_{D-B}$ is evaluated to be almost 9%. Since this is greater than MARR of 8%, alternative D wins.

Alternative D is finally compared with C, the last alternative. The PW-function for C-D incremental cashflows is:

$$20(P/A, i, 8) + 53(P/F, i, 8) - 165 = 0$$

From trial and error, $\triangle ROR_{C-D}$ is found to be 5%. Since this is less than the 8% MARR, alternative C loses to D. Thus, alternative D is finally selected.

Note that it was also the choice on the basis of PW. Thus, both PW- and ROR-criterion led in this case to the same decision, primarily because the alternatives' cashflow distributions are alike.

9.4.1 Common Error

A common error in the ROR-method is *not to apply* incremental analysis in multi-alternative projects of the variable-input-variable-output type. Let us reconsider Example (9.5) and solve it without incremental analysis. It involves evaluation of alternatives' individual RORs by solving their PW-function for the unknown i. For alterative A, the PW-function is

$$100(P/A, i, 8) + 375(P/F, i, 8) - 600 = 0$$

Its solution yields ROR_A just under 14%. Solving the PW- functions of other alternatives, their rates[13] of return are $ROR_B = 18\%$, $ROR_C \approx 12\%$, and $12\% < ROR_D < 15\%$. On the basis of largest +ve ROR we would have selected alternative B—obviously a wrong decision compared to that in Example 9.5. Therefore, *under ROR-criterion one must carry out incremental analysis in selecting the best of two or more variable-input-variable-output alternatives.* This is also true for BCR-criterion.

Why does an individual ROR-based decision sometimes differ from that based on incremental analysis? Let us review the previous discussion along with the results of Example 9.5. Based on individual ROR alternative B is the best, but on the basis of incremental analysis D is the

---

One can adopt this approach to other pairs as well, for example,

PWD-B = 707(P/F,8%,8) – 10(P/A,8%,8) – 300
= 707 × 0.5403 – 10 × 5.747 – 300
= $24.5

Since PWD-B is positive, RORD-B will be greater than 8%. So, alternative D is better than B. Thus, discard B and select D.

Next, compare D with C. For the incremental C-D data, we have

PWC-D = 53(P/F,8%,8) + 20(P/A,8%,8) – 165
= 53 × 0.5403 + 20 × 5.747 – 165
= –$21.4

Since PWC-D is negative, RORC-D will be less than 8%. Alternative C is not better than D. Thus, D is finally selected.

[13] We did not bother to determine the exact values of RORs, where it involved interpolation, to save time and effort. Since we are interested only in knowing which ROR is the largest—rather than in their absolute values—so that the corresponding alternative can be selected, we may save time by determining the ROR values only as approximately as essential. For example, we saved time by avoiding interpolation for alternative D whose ROR was under 15%; less than the 18% already found for B. Carry out the interpolation(s) only if necessary.

best. In general, engineers will prefer to choose the alternative that costs the least, which in this case is alternative B. But, the company may be looking for opportunities to invest capital as long as it can earn return in excess of MARR. Assuming this to be the case, the engineer has to check whether any of the other higher-first-cost alternatives (A, C, or D) offers such an opportunity. This is where incremental analysis becomes indispensable—in helping determine whether any of the other alternatives offers an investment opportunity. If so, the company seizes the opportunity. To keep the additional investment low, obviously, the engineer first tries the next higher-first-cost alternative; A in this case. Since the additional $100 cost on A over B did not yield a rate of return greater than MARR, alternative A was discarded. Such a process of elimination continued until the best alternative "filtered out" to be D.

Had the incremental analysis been not applied, and therefore D not chosen, an opportunity to invest to earn a return in excess of the set MARR would have been lost. Obviously, no company would like that.

9.5 BCR METHOD

The benefit-cost-ratio method is used primarily in evaluating government projects. As seen in Chapter 6, it is closely related[14] to the worth-methods. Due to its "ratio" measure, the BCR method may at times yield misleading results. This arises from the ambiguity whether cash outflows other than the investment be treated as costs or disbenefits, or some as costs and others as benefits. As explained in Chapter 6, any such doubt affecting BCR-based results can be avoided by adopting the BCD-method which is based on the difference between benefits and costs, much like the PW-method.

A point of contention in the BCR-method is the interest rate to use in the analysis. Assuming benevolence as a hallmark of civilian governments, interest rates for public investments are usually subsidized, i.e., below the cost of capital in the free (financial) market.

The BCR-method also can lead to decisions that contradict those by other methods, as illustrated in Example 9.6. The incompatibility in decisions is likely to arise when alternatives' cashflow distributions are not alike.

As in ROR-analysis, BCR also requires incremental analysis for variable-input-variable-output multi-alternative problems. The discussions in Subsection 9.4.1 apply to the BCR method as well. The other four methods (payback and the three worth-based) do not require incremental analysis at all.

Example 9.6: Benefit-To-Cost Ratio

Assuming $i = 12\%$ per year, a 6-year useful life, and no salvage value, which of the following four alternatives should be selected based on payback period? Does the decision change if the criterion is BCR?

---

[14] The BCR is a ratio while the worths are difference. Each worth is related to BCR as follows:
$$PW = PW_{benefits} - PW_{costs}$$
Dividing by
$$PW_{costs}, PW/PW_{costs} = PW_{benefits}/PW_{costs} - 1$$
$$= BCR - 1$$
Therefore,
$$PW = PW_{costs} (BCR - 1)$$
Similarly,
$$FW = FW_{costs} (BCR - 1)$$
$$AW = AW_{costs} (BCR - 1)$$
Note that in the above with BCR = 1, PW = 0, FW = 0, and AW = 0, as expected.

	A	B	C	D
First Cost	$80	$50	$15	$90
Annual Benefit	$18	$15	$ 5	$25

SOLUTION:

For the first part of the problem, we evaluate the payback period for each alternative and select the one of the shortest period.

Payback period for A = 80/18 = 4.44 years
Payback period for B = 50/15 = 3.34 years
Payback period for C = 15/5 = 3 years
Payback period for D = 90/25 = 3.6 years

On the basis of shortest payback period, alternative C should be selected.

For the second part of the problem, first note that the alternatives have different inputs (costs) and outputs (benefits). Thus, the solution of this variable-input-variable-output problem requires incremental analysis. But before we carry out this analysis, we should check whether any of the alternatives has BCR less than one. Such alternatives are eliminated from further analysis. Noting that BCR = $PW_{benefits}/PW_{costs}$,

$$BCR_A = \frac{18(P/A, 12\%, 6)}{80} = \frac{18 \times 4.111}{80} = 0.92$$

$$BCR_B = \frac{15(P/A, 12\%, 6)}{80} = \frac{15 \times 4.111}{50} = 1.23$$

$$BCR_C = \frac{5(P/A, 12\%, 6)}{80} = \frac{5 \times 4.111}{15} = 1.37$$

$$BCR_D = \frac{25(P/A, 12\%, 6)}{80} = \frac{25 \times 4.111}{90} = 1.14$$

Based on the results, alternative A with BCR less than one is discarded. The incremental analysis is carried out with the remaining three alternatives B, C, and D.

We begin the incremental analysis by choosing alternative C as our favorite since its first cost is the lowest. Since the next higher-first-cost alternative is B, analyze the incremental B-C data as:

$$\Delta BCR_{B-C} = \frac{\Delta PW_{Benefits}}{\Delta PW_{Costs}} = \frac{(15-5)\left(\frac{P}{A}, 12\%, 6\right)}{50-15} = \frac{10 \times 4.111}{35} = 1.17$$

Since ΔBCR_{B-C} is higher than one, alternative B wins, i.e., B is selected and C discarded. We now carry out the next incremental analysis, by comparing alternative B with D.

$$\Delta BCR_{D-B} = \frac{\Delta PW_{Benefits}}{\Delta PW_{Costs}} = \frac{(25-15)\left(\frac{P}{A}, 12\%, 6\right)}{(90-50)} = \frac{10 \times 4.111}{40} = 1.03$$

Since ΔBCR_{D-B} is higher than one, alternative D is preferred to B.

Thus, on the basis of BCR, alternative D should be selected. Note that the BCR-based decision of selecting D is different from that based on the payback criterion.

9.6 COMPREHENSIVE COMPARISON

An overall comparison of the six analysis methods is provided in this section through a table and two examples. Table 9.1 presents a comprehensive summary of these methods in four groups. Examples 9.7 and 9.8 offer a comprehensive illustration of their application. Example 9.7 shows that for similar cashflow distributions the decision is the same irrespective of the method used. However, where the distributions are not reasonably alike, the decisions may be different.

The BCR-method is almost exclusively used for analyzing government and non-profit projects. The other five are practiced mostly in for-profit industries.

Example 9.7: Multiple Alternatives, Multiple Methods

Which of the following three alternatives should be selected as per the six analysis methods discussed in this text? Explain any contradiction(s) in the selection. Assume MARR = 10% per year, and zero salvage value for each.

	A	B	C
First Cost	$50	$100	$61
Annual Benefit	$30	$ 40	$37
Useful Life (yrs)	2	3	2

SOLUTION:

We consider the six methods in the order presented in the text.

(a) Payback Period

Under this method we evaluate the payback period for each alternative and select the one with the shortest period.

Payback period for A = 50/30 = 1.67 years
Payback period for B = 100/40 = 2.5 years
Payback period for C = 61/37 = 1.65 years

On the basis of shortest payback period, alternative C should be selected.

Table 9.1 Summaries of the Analysis Methods

Comparison Basis	Payback	Worth[a]-Based	ROR	BCR
Analysis Complexity	Low	Moderate	High	Moderate
Need for Computers	Low	Moderate	High	Low
Incremental Analysis	No	No	Yes[b]	Yes[c]
Industrial[d] Popularity	High	Moderate	High	Little
Usage by Governments[e]	Little	Little	Little	High
Multi-Answers[f]	No	No	Yes[g]	Yes[h]
Values of _i_ as Input	Yes	Yes	No	Yes
External Investment	No	No	Yes[i]	No

[a] Includes PW-, FW-, and AW-methods.
[b] Only in case of variable-input-variable-output multi-alternative projects.
[c] Only in case of variable-input-variable-output multi-alternative projects.
[d] Profit-making industries.
[e] As well as non-profit organizations.
[f] Same problem has more than one answer.
[g] Apply the four tests to determine IROR, see Chapter 9.
[h] Depends on whether future cash outflows are treated as disbenefits or costs.
[i] The available surplus cash is invested external to the project, yielding IROR.

(b) PW

Since the alternatives' useful lives are different, a common analysis period based on their LCM should be used. The LCM of their lives (2, 3, and 2 years) is 6. So we analyze the alternatives over a 6-year time window.

During the 6-year analysis period, alternatives A and C will be replaced twice, at periods 2 and 4, as shown in Figs. 9.2(a) and (c). Alternative B will be replaced only once, at period 3, as seen in Fig. 9.2(b). Assuming the replacement cost(s) and the resulting benefits remain unchanged during the analysis period, the alternatives' PWs are

$$PW_A = 30(P/A,10\%,6) - \{50 + 50(P/F,10\%,2) + 50(P/F,10\%,4)\}$$

$$= 30(P/A,10\%,6) - 50\{1 + (P/F,10\%,2) + (P/F,10\%,4)\}$$

$$= 30 \times 4.355 - 50(1 + 0.8264 + 0.6830)$$

$$= 130.7 - 125.5$$

$$= \$5.2$$

$$PW_B = 40(P/A,10\%,6) - \{100 + 100(P/F,10\%,3)\}$$

$$= 40(P/A,10\%,6) - 100\{1 + (P/F,10\%,3)\}$$

$$= 40 \times 4.355 - 100(1 + 0.7513)$$

$$= 174.2 - 175.1$$

$$= -\$0.9$$

$$PW_C = 37(P/A,10\%,6) - \{61 + 61(P/F,10\%,2) + 61(P/F,10\%,4)\}$$

$$= 37(P/A,10\%,6) - 61\{1 + (P/F,10\%,2) + (P/F,10\%,4)\}$$

$$= 37 \times 4.355 - 61(1 + 0.8264 + 0.6830)$$

$$= 161.1 - 153.1$$

$$= \$8$$

With largest +ve PW as decision criterion, alternative C is selected. This decision is the same as under payback criterion in (a). The compatibility in these two results is attributable to the similarity in alternatives' cashflow distributions.

(c) FW

The future is selected to be at the end of the analysis period. Based on the three cashflow diagrams in Fig. 9.2, and on the assumption that the costs and benefits for all replacements remain unchanged, the alternatives' FWs at the sixth period are[15]:

---

[15] We could have followed the shorter approach illustrated in Example 9.3. By multiplying the PWs determined in part (b) with (F/P,10%,6), we could have determined FWs as

$\quad$ FWA = PWA(F/P,10%,6) = 5.3 × 1.772 = 9.4

$\quad$ FWB = PWB(F/P,10%,6) = −0.9 × 1.772 = −1.6

$\quad$ FWC = PWC(F/P,10%,6) = 8 × 1.772 = 14.2

Note that these FWs are approximately the same as those based on the diagrams.

$$FW_A = 30(F/A, 10\%, 6) - \{50(F/P, 10\%, 6) + 50(F/P, 10\%, 4)$$

$$+50(F/P, 10\%, 2)\}$$

$$= 30(F/A, 10\%, 6) - 50\{(F/P, 10\%, 6) + (F/P, 10\%, 4) + (F/P, 10\%, 2)\}$$

$$= 30 \times 7.716 - 50(1.772 + 1.464 + 1.210)$$

$$= 231.5 - 222.3$$

$$= \$9.2$$

$$FW_B = 40(F/A, 10\%, 6) - \{100(F/P, 10\%, 6) + 100(F/P, 10\%, 3)\}$$

$$= 40 \times 7.716 - 100(1.772 + 1.331)$$

$$= 308.6 - 310.3$$

$$= -\$1.7$$

$$FW_C = 37(F/A, 10\%, 6) - \{61(F/P, 10\%, 6) + 60(F/P, 10\%, 4)$$

$$+60(F/P, 10\%, 2)\}$$

$$= 37(F/A, 10\%, 6) - 61\{(F/P, 10\%, 6) + (F/P, 10\%, 4)$$

$$+(F/P, 10\%, 2)\}$$

$$= 37 \times 7.716 - 61(1.772 + 1.464 + 1.210)$$

$$= 285.5 - 271.2$$

$$= \$14.3$$

On the basis of largest +ve FW, alternative C is selected. As expected, this decision is compatible with that based on PW.

(d) AW

As we learned in Chapter 6 and elsewhere in the text, there is no need for the analysis period to be the same for each alternative when annual worth is the decision criteria. One can simply use the alternatives' individual life as its analysis period, resulting in concise calculations as shown as follows. This is a strong point in favor of AW-analysis when alternatives' useful lives are different. Referring to the first complete cycle of each cash-flow diagram in Fig. 9.2, the alternatives' AWs are evaluated[16] as follows.

For A, the relevant diagram comprises the $50 first cost, followed by the two $30 benefits. Thus,

$$AW_A = 30 - 50(A/P, 10\%, 2)$$

$$= 30 - 50 \times 0.5762$$

$$= 30 - 28.8$$

$$= \$1.2$$

---

[16] We could have used the shorter approach illustrated in Example 9.3. By multiplying the PWs determined in part (b) with (A/P,10%,6), we could have determined AWs as:

AWA = PWA(A/P,10%,6) = 5.3 × 0.2296 = 1.2
AWB = PWB(A/P,10%,6) = −0.9 × 0.2296 = −0.2
AWC = PWC(A/P,10%,6) = 8 × 0.2296 = 1.8

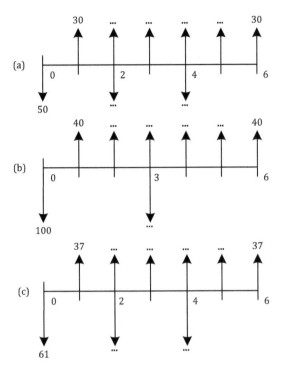

Figure 9.2 Diagram for Example 9.7.

For B, the relevant diagram comprises the $100 first cost, followed by the three $40 benefits. Thus,

$$AW_B = 40 - 100(A/P, 10\%, 3)$$

$$= 40 - 100 \times 0.4021$$

$$= 40 - 40.2$$

$$= -\$0.2$$

For C, the relevant diagram comprises the $61 first cost, followed by the two $37 benefits. Thus,

$$AW_C = 37 - 61(A/P, 10\%, 2)$$

$$= 37 - 61 \times 0.5762$$

$$= 37 - 35.1$$

$$= \$1.9$$

With largest +ve AW as criterion, alternative C is selected. As expected, this decision is compatible with that by the PW- or FW-method in parts (b) and (c). It is also compatible with the payback period method for the reason explained in part (b).

(e) ROR

With the three alternatives inputs (costs) and outputs (benefits) being different, the given problem is of the variable-input-variable-output type, and hence its solution must be based on incremental analysis. But, prior to this analysis we need to determine whether any of the alternatives have individual ROR less than MARR so that they can be eliminated from incremental analysis. There are two ways to proceed: (i) Since MARR is given, one can check the sign of the alternatives' PW (if +ve, then ROR>MARR), or (ii) Exactly evaluate each ROR by setting PW = 0. We follow the first approach which is usually simpler.

Since PWs evaluated in part (a) are based on MARR, we do not need to carry out any new calculations. Of the three present worths, PWB is –ve. Hence B is discarded. Thus, the three-alternative problem has reduced to a two-alternative problem. Under the incremental analysis we compare alternatives A and C. Since A's first cost is lower, we prefer to select it. We thus evaluate C over A to check whether the additional cost on C and the resulting additional benefits yield a rate of return $\triangle ROR_{C-A}$ greater than MARR.

For determining $\triangle ROR_{C-A}$, generate the C-A incremental data and carry out the calculations. Since the useful life of these two alternatives is the same at 2 years, the analysis period is 2 years. From the given data, we get:

Year	A	C	C-A
0	–$50	–$61	–$11
1	$30	$37	$ 7
2	$30	$37	$ 7

By applying[17] Tests 1 and 2 on these C-A cashflows we predict a +ve rate of return since there is only one sign change. Next, we evaluate PW_{C-A} to check its sign.

$$PW_{C-A} = 7(P/A,10\%,2) - 11$$

$$= 7 \times 1.736 - 11$$

$$= 12.152 - 11$$

$$= 1.152$$

Since PW_{C-A} is +ve, investment-wise C is better than A. Hence C should be selected. Note that it was also the choice based on all the previous methods. No contradiction in results has occurred so far since the cashflow distributions for the three alternatives are alike.

(f) BCR

Since this is a multi-alternative problem, we need to first check whether incremental analysis is essential. Looking at the cashflows, the alternatives are of the variable-input-variable-output type. Hence the solution must be based on incremental analysis. Prior to the analysis, however, we need to check for any alternative with BCR less than one. Such alternatives are excluded from incremental analysis. Considering one complete cycle of the cashflows, and noting that BCR = $PW_{benefits}/PW_{costs}$, we get:

$$BCR_A = \frac{30(P/A,10\%,2)}{50} = \frac{30 \times 1.736}{50} = 1.04$$

$$BCR_B = \frac{40(P/A,10\%,2)}{100} = \frac{40 \times 2.487}{100} = 0.99$$

$$BCR_C = \frac{37(P/A,10\%,2)}{61} = \frac{37 \times 1.736}{61} = 1.05$$

---

[17] See Chapter 9 if necessary.

Based on these results, alternative B with BCR less[18] than one is discarded from any further consideration. In other words, incremental analysis encompasses alternatives A and C only. With lower first cost, alternative A is our initial choice. We therefore analyze the C-A data to evaluate ΔBCR_{C-A}. The calculations should be simpler since both A and C have the same useful life. Had it been different we would have needed to analyze over LCM-based analysis period.

$$\Delta BCR_{C-A} = \frac{\Delta PW_{Benefits}}{\Delta PW_{Costs}} = \frac{(37-30)\left(\frac{P}{A},10\%,2\right)}{(61-50)} = \frac{7 \times 1.736}{11} = 1.10$$

Since ΔBCR_{C-A} is higher than one, alternative C wins. Thus, on the basis of BCR too, alternative C should be selected. This decision is compatible with those in (a) through (d).

In this example, all the six methods led to the same decision since the alternatives' cashflows were alike. Where the cashflows are significantly different, the decisions may be contradictory.

9.7 INDUSTRIAL USAGE

Companies use one or more of the six methods discussed in the text for analyzing engineering economics projects. The analysis method is decided upon by the company management in consultation with engineering economists. Engineers working in small companies, with fewer people knowledgeable in engineering economics, or in their own business may select the method by themselves.

The choice of a method suitable for economic analysis of engineering projects is influenced by several factors, such as:

1. level of investment
2. simplicity of analysis sought
3. time horizon of the project: short-term, long-term, etc.
4. financial health of the company
5. economic environment and market stability
6. government incentives for engineering investment
7. type of business (computer chips or potato chips!)

According to fragmented surveys on their popularity, the ROR method is the most widely used in industry, as shown in Fig. 9.3. It is closely followed by PW and AW; the two engineering-oriented worth-based methods. The payback period method is also common in industry in spite of its non-consideration of the time value of money. In companies with limited capital pool, it may probably be the only one in use. The BCR method is most prevalent in government and non-profit organizations.

Most companies use only one of the six methods, as recommended by the upper management. If this method gives rise to any doubt about the result of the analysis, then other methods may be used to enhance the reliability of the final decision. Several companies use more than one method, especially payback period for initial "screening" of the alternatives.

---

[18] Since B's BCR is almost equal to one, the engineer should check its data once more to ensure that they are accurate. Even a slight variation in the data can render B acceptable.

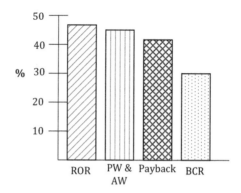

Figure 9.3 Industrial usage of the methods.

9.8 SUMMARY

A comparison among the six methods—payback period, PW, FW, AW, ROR, and BCR—commonly used in analyzing engineering economics projects has been presented in Chapter 9. We discussed why a method may yield results that contradict those by the other methods. The three worth-based methods, namely PW, AW, and FW, always lead to the same decision. This is not true when all the six methods are applied. In general, for projects with heavily skewed cash inflows (more benefits either in the beginning or at the end of the project life), the payback period-, ROR-, and BCR-methods may yield results that differ from those by the worth-based methods. If the cash inflows are fairly distributed, all the methods lead to the same decision. The payback period method favors projects with heavy cash inflows in the beginning of the project. Of the six, the ROR- and BCR-methods may involve incremental analysis. The ROR-method is widely used in industry, especially in financially healthy companies with long-term investment strategies. The advent of computer systems has rendered the otherwise-complex ROR method popular. The simpler payback period method, which does not account for the time value of money, is commonly used by companies struggling to survive, i.e., with short-term investment strategy. This method is sometimes used to "filter" out some of the alternatives, prior to detailed economic analysis of the more promising ones.

DISCUSSION QUESTIONS

9.1 Why do engineers sometimes apply more than one method to analyze an investment project?

9.2 Explain the concept of incremental analysis to a manager who has a college degree but did not study engineering economics.

9.3 Discuss why engineering economics is much more than solving a given numerical problem.

9.4 Describe an engineering project for which FW analysis might be more appropriate. (Hint: Consider expansion of an existing manufacturing facility that will take 3 years to complete.)

9.5 If the payback period method is both simple and popular in industry, why do we need the other methods, especially ROR which is complex and lengthy?

9.6 Can there be a cashflow distribution for which decision by the three worth-based methods may differ? If yes, when? If no, why not?

9.7 When is incremental analysis applied? Explain through an example.

9.8 Explain one advantage of the annual worth analysis over future worth and present worth analyses? Also, how do their cashflow cycles differ?

9.9 Briefly explain what determines the need for incremental analysis when using the ROR method.

9.10 Explain why BCR analysis is controversial.

MULTIPLE-CHOICE QUESTIONS

9.11 Of the six methods for analyzing engineering economics problems, the most widely used is
   a. PW
   b. AW
   c. ROR
   d. payback period

9.12 The PW-method always leads to the same decision as the
   a. ROR-method
   b. BCR-method
   c. payback period method
   d. AW-method

9.13 Incremental analysis may have to be carried out while analyzing a problem under the criterion of
   a. PW
   b. BCR
   c. payback period
   d. AW

9.14 The least likely method for analyzing an engineering project is
   a. FW
   b. PW
   c. ROR
   d. payback period

9.15 Computer technology has popularized the _____ in industry.
   a. PW method
   b. BCR method
   c. ROR method
   d. payback period method

9.16 If MARR is given, ROR analysis can be simplified by evaluating
   a. BCR
   b. PW
   c. ROR
   d. payback period

9.17 ROR is evaluated by setting
   a. payback period = 3 years
   b. MARR = 10%
   c. BCR = 0
   d. PW = 0

9.18 Which method is used almost exclusively to evaluate governmental projects?
   a. incremental
   b. ROR
   c. BCR
   d. payback period method

9.19 Which of the following can lead to skewed results of decision-making analysis?
   a. concentration of benefits to beginning or end of period
   b. equally distributed cashflows
   c. human error
   d. unequal lives of alternatives

9.20 Which of the following would be the best decision based on AW: $AW_A = -3$, $AW_B = -11$, $AW_C = 4$.
   a. AW_A
   b. AW_B
   c. AW_C
   d. all of the above

9.21 For multiple alternative projects, which of the following are the same?
   a. fixed output
   b. fixed input
   c. variable input, variable output
   d. all of the above

9.22 Which of the following always must result in incremental analysis?
   a. fixed output
   b. fixed input
   c. variable input, variable output
   d. none of the above

9.23 If the alternative decisions have unequal lives, the common analysis period is based on what?
   a. whatever works the best
   b. 15 years
   c. useful life
   d. LCM

9.24 Which of the following methods would yield the best result given the cashflows for a period?
   a. PW method
   b. FW method
   c. AW method
   d. all of the above

NUMERICAL PROBLEMS

9.25 Which of the two projects whose cashflow diagrams are given in Fig. 9.4 should be selected on the basis of PW criterion if the annual interest rate is 9% compounded yearly? Does the decision change under FW and AW criteria?

9.26 If the criterion in the Problem 9.25 is payback period, does the decision change?

9.27 There are three alternatives in a productivity project. Alternative A involves the use of robots, B of modular fixtures, and C of upgraded machine. The alternatives' first costs in dollars are 3,600, 5,000, and 8,400 respectively. If their net annual benefits are $1,890, $2,680, and $3,200, and useful lives are 2, 2, and 3 years, which alternative should be selected on the basis of payback period? Does the decision change if ROR-criterion is used? Explain any discrepancy in the decision. Assume annual compounding and 12% MARR.

9.28 Which of the two projects, A or B, whose cashflow diagrams are given in Fig. 9.5 should be selected on the basis of payback criterion? Does the decision change if the criterion is BCR? Assume 9% interest rate compounded annually.

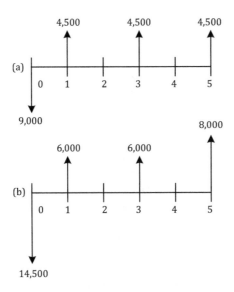

Figure 9.4 Diagram for Problem 9.27.

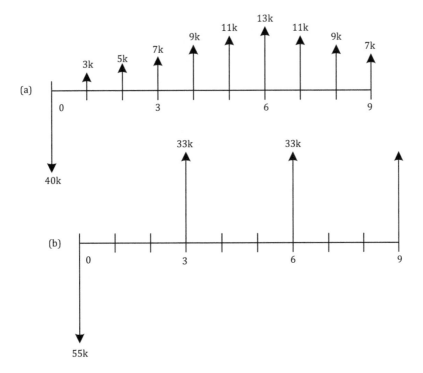

Figure 9.5 Diagram for Problem 9.30.

9.29 For the following cashflows which alternative should be selected on PW-basis if i = 20% per year? Is the decision affected if the criterion is ROR?

Year	A	B
0	−$80	−$4150
1–5	$40	$ 70

9.30 For the following cashflows which alternative should be selected on the basis of BCR? Is the decision affected if ROR is the criterion? Assume MARR = 12%.

Year	A	B
0	−$900	−$900
1	$ 50	$600
2	$350	$300
3	$700	$200
4	$400	$150

9.31 Which of the two projects, whose cashflow diagrams are given in Fig. 9.6, should be selected on the basis of PW if the annually compounded interest rate is 5% per year? Does the decision change under BCR-criterion?

9.32 Which of the two projects, whose cashflow diagrams are given in Fig. 9.7, should be selected on the basis of PW if MARR = 15%? Does the decision change under ROR or BCR criteria?

9.33 If, in Example 9.7, alternative B's cash inflows are $10, $40, and $90 respectively for the 3 years of its life, instead of $40 annual, how are the various results affected? Discuss any contradictions.

9.34 Which of the following alternatives should be selected using payback period and PW-methods discussed in the text? Explain any contradiction(s) in the results. Assume MARR = 10%, and zero salvage value. Note that the useful lives of alternatives A and C are 2 years, and that of B 3 years. If necessary, assume an external interest rate of 8% per year.

Year	A	B	C
0	−$50	−$100	−$61
1	$30	$ 90	$37
2	$30	$ 20	$37
3	$10		

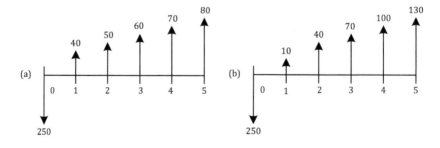

Figure 9.6 Diagram for Problem 9.33.

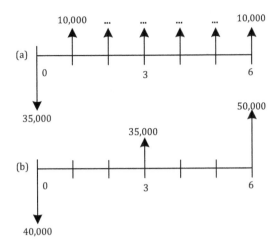

Figure 9.7 Diagram for Problem 9.34.

Use the following cashflow diagram to answer Questions 9.35–9.37.

Year	Make A	Make B	Make C
0	–$5,000	–$7,000	–$9,000
1	$2,000	$3,000	$4,560
2	$2,000	$2,500	$3,750
3	$2,000	$2,000	$3,500

9.35 Which of the following alternatives would be the best decision based on the PW of cashflows? Assume $i = 8\%$.

9.36 Perform the incremental analysis of the following cashflow diagram. Assuming $i = 8\%$.

9.37 Compute the BCR for alternatives A and B. Would Make A or B provide the greatest amount of common good? Assume that $i = 8\%$, a 3-year useful life and no salvage value.

Use the following cashflow diagram to answer Questions 9.38–9.39.

Year		0	1	2	3	4
Cashflow	Alternative A	–$5,000	$2,200	$2,420	$2,662	$2,928
	Alternative B	–$7,500	$2,444	$3,502	$3,200	$3,900

9.38 What is the BCR for Alternatives A and B? $i = 6\%$.

9.39 Using the ROR method:
a. determine the payback period
b. determine the ROR on each of the two alternatives. MARR = 5%.

Use the follow diagram to answer Questions 9.40–9.42.

Year	Make A	Make B	Make C
First Cost	$6,000	$7,500	$8,000
Annual Benefit	$2,400	$3,500	$5,000
Useful Life (yrs)	4	3	2

9.40 Which of the following alternatives should be selected based on PW analysis? Assume $i = 15\%$.

9.41 Which alternative should be selected based on the payback period analysis?

9.42 Apply the AW method to the previous table. Perform incremental analysis if necessary. Assume MARR of 15% and no salvage value.

Use the following diagram to answer Questions 9.43–9.44.

Year	System Z	System X
0	−$3,000	−$3,000
I	$2,300	−$ 700
2	$ 900	$ 200
3	−$ 200	$ 0
4	$ 322	$ 322

9.43 Which of the following alternatives should be selected based on PW analysis? Assume $i = 9\%$.

9.44 Apply the BCR method to the previous diagram. Perform incremental analysis if necessary. Assume MARR of 9% and no salvage value.

FE EXAM PREP QUESTIONS

9.45 A manufacturing company produces a wind-powered generator for $200,000. The machine is used for 10 years, increasing net annual revenues by $30,000 each year. A salvage value of $15,000 produces a present worth (6% per year) closest to:
   a. $1,400,000
   b. $ 345,671
   c. $ 239,619
   d. $1,133,893

9.46 A manufacturing company produces a wind-powered generator for $200,000. The machine is used for 10 years, increasing net annual revenues by $30,000 each year. To achieve a ROR on the investment of 10% the first year's benefits must be closest to which?
   a. $ 15,000
   b. $458,612
   c. $ 34,000
   d. $ 20,500

9.47 A conveyor system is purchased for $17,349 and is retained for 7 years and then salvaged for $0. It produces net annual revenues of $5,000 and annual costs of $900. If the system is depreciated using the straight-line method and the effective tax rate is 25%, what is the present value of salvage value?
   a. $0
   b. $15,647
   c. $ 9,399
   d. $ 3,290

9.48 Consider an investment with net cashflows of −$45,000, $10,000, $20,500, and $25,000. The PW suggests that the project should be accepted for interest rates:
   a. less than 10%
   b. between 8% and 12%
   c. between less than 15%
   d. all of the above

9.49 Consider an investment with net cashflows of −$35,000, $11,719, $16,409, and $23,916. The ROR for the investment is:
a. 10%
b. 25%
c. 20%
d. none of the above

9.50 Consider an investment with net cashflows of −$29,550, $13,460, $8,756, $7,346, $14,000, and $6,547. The payback period for the investment is:
a. 1.3
b. 2.99
c. 3.05
d. 5

9.51 Multiple alternatives to decisions are mutually exclusive if:
a. both may be chosen
b. one may be chosen by not the other
c. neither may be chosen
d. one must be chosen prior to the other

9.52 An investment of $100,000 produces net revenues of $29,000 for 9 years and has no salvage value. For a 10% interest rate, a 2% error in the estimate of the investment produces an error in present worth closest to (error indicates 2% higher than predicted):
a. $20,134
b. $45,278
c. $ 9,009
d. $12,499

9.53 Consider an investment with net cashflows of −$55,321, $24,657, $16,218, $9,836, $12,837, and $19,478. The payback period for the investment is:
a. 3.4
b. 2.75
c. 4
d. 3.9

9.54 Consider an investment with net cashflows of −$185,000, $65,160, $45,459, $32,455, and $75,900. The ROR for the investment is:
a. 7%
b. 8%
c. 12%
d. d 30%

Part III

Realism

In Part II, we discussed the application of six commonly used analysis methods. However, we did not include any consideration of two important facts of the engineering industry, namely depreciation and income tax. We cover them in Part III along with replacement analysis.

The organization of this text in three parts, as mentioned earlier, is metaphorically similar to the construction of a house. Part I was the foundation, while Part II is the structure comprising floors, walls, and ceilings. Part III represents the furnishings such as drapes, carpets, and kitchen appliances. We discuss here some essential refinements to the analyses covered in the text.

Part III therefore adds realism to what we have learned so far, by incorporating "real-world" considerations. In Chapter 10, we learn to account for depreciation of resources. One of the ways governments guide national economies is through taxation, whose impact on economic analyses is discussed in Chapter 11. An important and distinct area of engineering economics is resource replacement—the subject matter of Chapter 12. Two very important investment decision analyses are break-even analysis and risk analysis. Engineers can make effective decision-making using these tools; it is well covered in Chapters 13 and 14. Finally, in Chapter 15 we discuss some practical aspects, including the effect of inflation on analyses and also summarize advanced topics relevant to engineering economics.

Chapter 10

Depreciation

LEARNING OBJECTIVES

- Meaning of *depreciation*
- Net cashflows to account for depreciation
- An asset's book value
- How to cost depreciation
- A depreciation schedule
- Straight-line method
- SOYD method
- Declining balance method
- Double-declining balance method
- Unit-of-production method
- MACRS as government-approved method
- Composite depreciation method
- Analysis of depreciation-modified net cashflows

Machines, equipment, and other resources or assets in which companies invest capital deteriorate with use, resulting in a decrease in the value. The loss in value is described by the term depreciation and expressed as depreciation cost. Depreciation costs are also called depreciation charges because they are deducted from, or charged to, the profit realized from the invested resource. As a financial incentive, governments allow companies to deduct depreciation cost from the profit before it is taxed. In this chapter we study three interrelated aspects of depreciation:

1. Determination of depreciation costs
2. Effect of depreciation costs on cashflows
3. Economic analysis with depreciation accounted for

10.1 MEANING

Depreciation measures the *decrease in value* of an asset. It is the opposite of *appreciation* which means an increase in value. In periods of peace and tranquility, engineering assets generally depreciate since better ones are continually developed, and marketed, through technical innovations. In periods of war and uncertainty, assets may appreciate due to the eminent danger of destruction and slowdown of the economy. In the 20th century, we experienced appreciation during World War II. Since then, the world economy has operated with depreciation as a fact.

Consider that XYZ Company procured a robot for $20,000 to improve its assembly line productivity. At the end of the first year of use, the robot's value due to wear, tear, and obsolescence is $16,000. This decrease of $4,000 in the robot's value is its depreciation for the first year. Let us say that the robot saved $9,000 during the first year of its use. This saving adds to the company profit which is taxable. The investment in the robot thus generated a $9,000 gross profit, but only at a loss of $4,000 in the robot's value. The company therefore accounts this $4,000 as a cost, resulting in a *net profit* of $5,000 (= $9,000 – $4,000). The income tax is due on the net profit of $5,000, not on the gross profit of $9,000. The provision of charging depreciation to the income from an invested resource thus reduces the tax burden[1].

Capital equipments depreciate due to obsolescence or deterioration from use, or both. A machine tool for example depreciates due to wear and tear from its use. The same is true for an automobile. Some equipments such as personal computers depreciate primarily due to obsolescence created by new and better products. With the exception of land, almost every asset needed by the engineering industry is depreciable and subject to depreciation costs. With an infinite life, land is non-depreciable even if its market value for the year has decreased.

For depreciation purposes, an asset's useful life is considered finite. The asset's initial value reduces to its salvage value at the end of useful life. We can illustrate this through a graph as in Fig. 10.1. The values are plotted along the y-axis and the useful life along the x-axis. The equipment costs P at period 0 when its use begins. Over time, at the end of useful life N its value drops to the salvage value S. Thus, total depreciation over the useful life is P-S, the difference between the first cost and salvage value. If this total depreciation is charged uniformly, then the depreciation cost is (P-S)/N per year.

10.2 TERMINOLOGIES

In the course of depreciation costing we come across several terms that seem familiar. The important terms are discussed here because we need to understand their meanings precisely to avoid any error in costing.

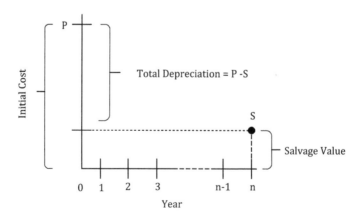

Figure 10.1 Total depreciation equal P-S.

---

[1] What is the company's gain through lower tax is the government's loss through reduced collection. Governments therefore limit the amount of depreciation that can be charged in a year.

10.2.1 Asset

An asset is any resource used in business. It may be tangible or intangible. Equipment, machines, computers, and buildings are examples of tangible assets. Goodwill, prompt after-sale service, quality reputation, and a community-caring image are examples of intangible assets. Tangible assets can easily be quantified in monetary terms, while intangibles can't.

10.2.2 Useful Life

Useful life is the duration of time equipment or resource is expected to be in use. Its value is usually known, but not with 100% certainty since it is futuristic. It is only an estimate, usually based on past data. Insurance companies normally keep track of useful lives of a variety of equipment.

For a given first cost, shorter useful lives allow larger annual depreciation charges, thus reducing the income tax more. While faster and larger depreciation charges are desirable for the company, governments lose since they collect less in income tax. To avoid any controversy on the amount of allowable depreciation, largely due to the difference in *expected* useful life as judged by the company and taxing authority, government may fix the value of allowable useful life for an asset. We discuss this further in Subsection 10.3.5.

The term *useful life* differs from similar terms such as service life, economic life, market life, and shelf life. We discuss the difference between useful life and economic life later in Chapter 12.

10.2.3 Market Value

We have earlier defined depreciation as a loss in value, without specifying the type of value. An asset has *market value* and *book value*. The market value is what the asset can be disposed of at the time of selling. It is what others, especially prospective buyers, are prepared to pay for the asset. It is therefore determined by the marketplace, irrespective of how much the owner thinks the asset is worth. It also depends on the timing of asset disposal. While the market value of a capital asset may not change as abruptly as the value of a company stock, it nevertheless fluctuates.

10.2.4 Book Value

The term *book* is derived from the age-old practice of maintaining company books to record asset costs. An asset's first cost is its initial book value. With use, the value decreases[2] each year, thus reducing the asset's book value. The book value is what the owner needs to sell the asset for to avoid a loss. However, depending on the market's need for the used asset, its market value may be above or below the book value.

The market value and book value of an asset usually differ. There are several factors that influence this difference. For products under a continual onslaught of evolving technologies, the market value of a used asset is usually lower than its book value. Personal computers and electronic products are good examples. For most engineering assets however, such as machines, concrete mixers, electric motors, market value and book value are about the same. For an antique, market value is much higher than its book value which is usually zero.

---

[2] Depreciation is a charge to compensate for this.

10.2.5 Depreciation Schedule

A depreciation schedule is simply a listing of the annual[3] depreciation costs (or charges) for an asset. A depreciation schedule looks like:

Year	1	2	3	4	5
Depr. Exp	$9,000	$9,000	$9,000	$9,000	$9,000

If the depreciation cost is constant during the asset life, the entries in the schedule are the same, as can be seen in the table. Note that the tabulated depreciation schedule looks like a cashflow table.

The book value is a function of depreciation costs. An asset's cost is usually entered in appropriate company books at the time of procurement. This entry is known as first cost, initial cost, or procurement cost. Every year certain depreciation is charged to the asset. By subtracting the depreciation charge for the year from the book value at the beginning of the year, we get the book value at the end of the year. The year-end book value becomes the book value at the beginning of the next year, as illustrated in the following table for an asset whose first cost is $10,000.

	A	B	C	D
1	Year	Beginning BV	Depr Exp	Year-End BV
2	1	$10,000.00	$2,000.00	$8,000.00
3	2	$8,000.00	$1,500.00	$6,500.00
4	3	$6,500.00	$1,000.00	$5,500.00
5	4	$5,500.00	$750.00	$4,750.00
6	⋮	⋮	⋮	⋮

As can be seen, the book value at the end of the year is equal to the beginning book value minus the depreciation cost; for example, for year 3 it is $6,500 − $1,000 = $5,500. Note that the first and third columns together form the depreciation schedule.

Oftentimes we keep track of the *cumulative depreciation* cost to know the to-date depreciation. This is done by adding another column to the table, as follows.

	A	B	C	D	E
1	Year	Beginning BV	Depr Exp	Year-End BV	Cumulative Depr.
2	1	$10,000.00	$2,000.00	$8,000.00	
3	2	$8,000.00	$1,500.00	$6,500.00	$3,500.00
4	3	$6,500.00	$1,000.00	$5,500.00	$4,500.00
5	4	$5,500.00	$750.00	$4,750.00	$5,250.00
6	⋮	⋮	⋮	⋮	⋮

The book value at any time during the asset's life is the difference between the first cost and cumulative depreciation up to that time. Thus,

Book value = First cost − Cumulative depreciation

---

[3] Could depreciation charge be other than annual? In theory yes, but in practice no. Since taxes are paid annually, depreciation costs are also charged on an annual basis.

For example, as posted in the table, the year-end book value for year 3

$$= \text{First cost} - \text{Cumulative depreciation up to year 3}$$

$$= \$10,000 - \$4,500$$

$$= \$5,500$$

Example 10.1: Depreciation Schedule

A machine costs $11,500. At the end of its 5-year useful life its salvage value is estimated to be $1,500. Prepare its depreciation schedule if annual depreciation is (a) 20% of the first cost, (b) 20% of the book value. The depreciation charge for the last (fifth) year must be adjusted so as not to exceed the maximum allowable.

SOLUTION:

Total[4] allowable depreciation for the machine, being the difference between the first cost ($11,500) and salvage value ($1,500) is $10,000 over 5 years of its useful life.

(a) The annual depreciation being 20% of the first cost is $0.2 \times \$11,500 = \$2,300$. A complete depreciation schedule is shown[5] in the following table. Note, however, that the depreciation for the fifth year is $800—remainder[6] of the total allowable. Since by the end of the fourth year cumulative depreciation is already $9,200, the fifth-year depreciation = $10,000 − $9,200 = $800.

	A	B	C	D
1	Year	Beginning BV	Depr Exp	Cumulative Depr.
2	0	$11,500.00		
3	1	$11,500.00	$2,300.00	$2,300.00
4	2	$11,500.00	$2,300.00	$4,600.00
5	3	$11,500.00	$2,300.00	$6,900.00
6	4	$11,500.00	$2,300.00	$9,200.00
7	5	$11,500.00	$800.00	$10,000.00

(b) In this case depreciation is 20% of the book value. The complete schedule is shown as follows. Note that the book values have been tabulated in the last column to facilitate the evaluation of depreciation for the subsequent year.

	A	B	C	D
1	Year	Depr Exp	Cumulative Depr.	Beginning BV
2	0			$11,500
3	1	$2,300	$2,300	$9,200
4	2	$1,840	$4,140	$7,360
5	3	$1,472	$5,612	$5,888
6	4	$1,178	$6,790	$4,710
7	5	$3,210	$10,000	$1,500

Note that the depreciation cost for each of the first 4 years is 20% of the book value at the end of the previous[7] year. For example, for year 3, it is 20% of $7,360 = $1,472. The fifth-

---

[4] The word *total* or *maximum* as an adjective to *allowable depreciation* means the same.
[5] e data in depreciation schedule tables usually correspond to period-ends, as in cashflow tables.
[6] The cumulative depreciation cannot exceed $10,000—the total allowable.
[7] The end of the previous year marks the beginning of the current year. So the book value at the end of the previous year is the book value at the beginning of the current year.

year depreciation of $3,210 is the remainder that could be charged, i.e., total allowable minus the cumulative depreciation up to the fourth year ($10,000 – $6,790 = $3,210).

If you find the entries in the table confusing, an alternative table format may be the following in which book value is posted in the second column.

	A	B	C	D
1	Year	Beginning BV	Cumulative Depr.	Depr Exp
2	0	$11,500		
3			$2,300	
4	1	$9,200		$2,300
5			$1,840	
6	2	$7,360		$4,140
7			$1,472	
8	3	$5,888		$5,612
9			$1,178	
10	4	$4,710		$6,790
11			$3,210	
12	5	$1,500		$10,000

In this format the depreciation costs have been posted in between the years since they relate to the duration rather than to a particular point in time. The entries are perhaps more comprehensible in this format. For example, the depreciation cost of $1,472 for the duration in between years 2 and 3 is 20% of $7,360—the book value at the beginning of year 2. As another example, the $4,710 book value at the beginning of year 4 is simply the book value at the beginning of year 3 ($5,888) minus the depreciation cost for the duration in between years 3 and 4 ($1,178). Note that for any year the total of the book value and cumulative depreciation equals the first cost ($11,500). For example, for year 3, it is $5,888 + $5,612 = $11,500.

10.3 METHODS

There are several methods to determine the annual depreciation charge for an asset. They differ primarily in how the total allowable depreciation, P-S, is distributed over the asset's useful life N; in other words, referring to Fig. 10.1, in how to arrive from point P to point S. Obviously, there can be several paths connecting P to S, the simplest one being a straight line.

10.3.1 Straight Line

The straight-line method (SL-method) is commonly used in depreciation costing. This method distributes the total depreciation P-S *uniformly* over the useful life N. In other words,

$$\text{Annual Depreciation} = \frac{1}{N}(P - S) \tag{10.1}$$

The factor 1/N is called *SL depreciation rate* since the annual depreciation is 1/N times the maximum allowable depreciation P-S. For example, for an asset of 5-year useful life, this rate is 1/5 = 0.20 = 20%. Thus, the annual depreciation is 20% of the total allowable. We

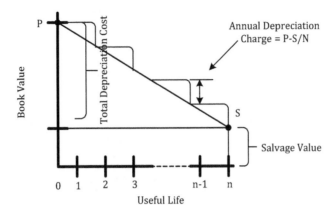

Figure 10.2 Straight-line depreciation.

used straight-line depreciation in part (a) of Example 10.1 without mentioning it, since the annual depreciation charge was constant.

Why is it called the straight-line method? Refer to Fig. 10.2 which shows annual depreciation charge according to the straight-line method. Note that the charges are constant and equal to (P-S)/N, as represented by the vertical lines of the stairs pattern. Since this pattern rests on a straight-line connecting P to S, it is called the straight-line method.

The merit of the straight-line method is its simplicity—equal charge each year. This simplicity is its disadvantage too. It does not allow for higher charges in the beginning of the useful life when an asset loses its value more. Higher charges in the beginning would have enabled faster recovery of the investment. The straight-line method thus delays the recovery of investment. The other methods, discussed in the following subsections, do not suffer from this limitation.

Example 10.2: Straight-Line Depreciation

Universal Arm (a robot) costs $25,000. At the end of its 4-year useful life its salvage value is estimated to be $5,000. Prepare its depreciation schedule under the straight-line method.

SOLUTION:

In here, P = $25,000, S = $5,000, and N = 4 years.
From Equation[8] (10.1),

$$\text{Annual Depreciation} = \frac{P - S}{N}$$

$$= \frac{\$25,000 - \$5,000}{4}$$

$$= \$5,000$$

---

[8] Straight-line depreciation problems can be solved without needing the equation, as illustrated in Example 10.1.

The depreciation schedule for Universal Arm can be prepared the way explained in Example 10.1, and is tabulated as follows.

Year[a]	Beginning Book Value	Depreciation	Ending Book Value
0–1	$25,000	$ 5,000	$20,000
1–2	$20,000	$ 5,000	$15,000
2–3	$15,000	$ 5,000	$10,000
3–4	$10,000	$ 5,000	$ 5,000
	Total:	$20,000	

[a] The entry as a range can remove confusion. For example, for year 2-3, $15,000 is the book value at the end of year 2 or beginning of year 3, while $10,000 is the value at the end of year 3.

Note that the total depreciation charged over the 4 years of useful life is $20,000. The resulting book value of $5,000 at the end of the fourth year (useful life) being equal to the salvage value confirms the accuracy of the results.

We are often interested in an asset's book value at the end of a specific year. One can read this value off a complete depreciation schedule table if it exists. The alternative is to use an equation, which is derived as follows.

We intend to determine the book value at the end of jth year. Since the annual depreciation is (P-S)/N,

$$\text{Cumulative Depreciation up to jth year} = \frac{J \times (P - S)}{N}$$

Since the asset's first cost is P, of which J(P–S)/N has been recovered through depreciation charges, its remaining value, or the book value, must be the difference between the two. Thus,

$$\text{Book Value at jth year} = P - \frac{J \times (P - S)}{N} \tag{10.2}$$

Substituting from Equation (10.1), this can also be expressed as:

$$\text{Book Value at jth year} = P - J \times \text{Annual Depreciation}$$

We can apply Equation (10.2) to Example 10.2. The book value at the end of third year (J = 3) from this equation[9] is:

$$= \$25,000 - \frac{3 \times (\$25,000 - \$5,000)}{4}$$

$$= \$25,000 - \$15,000$$

$$= \$10,000$$

---

[9] Rather than use the equation, we can follow the first-principles. The annual depreciation is (25,000 – 5,000)/4 = $5,000. The total depreciation over three years will thus be $15,000, resulting in a book value of $25,000 – $15,000 = $10,000. The first-principles approach works well only for the SL-method. For other depreciation methods, the equation-based approach is more efficient.

As expected, this book value is the same as that in the last column of Example 10.2's depreciation schedule for year 2–3.

10.3.2 Sum-of-Years-Digits (SOYD)

As mentioned earlier, the SL-method results in constant annual depreciation during the asset life. Since most assets depreciate faster in the beginning of their lives, companies charge depreciation accordingly. Methods that enable higher depreciation charges in the early years of the asset life are therefore more realistic. The sum-of-years-digits (SOYD) method is one of them. It charges higher depreciation in the beginning by factoring in the asset's remaining life.

SOYD is simply the sum of all the digits up to the useful life. For example, if an asset's useful life is 5 years, then its SOYD = 1 + 2 + 3 + 4 + 5 = 15 years. Similarly, for an 8-year asset it is 36 years (1 + 2 + ... + 8). An asset's SOYD can be determined by either of the following two approaches.

10.3.2.1 Short *Useful Life (N small)*

If the useful life N is short, then simply add[10] all the digits up to N. For example, if the useful life is 4 years, then its SOYD = 1 + 2 + 3 + 4 = 10 years. For most engineering economics problems, this approach suffices since N is usually small.

10.3.2.2 Long *Useful Life (N large)*

If the useful life is long, then adding the digits may be lengthy and prone to error. In that case, use the equation derived as follows.

For an asset of useful life N years,

$$SOYD = 1 + 2 + 3 + ... + (N - 1) + N$$

Rewriting the right-side of this in reverse order,

$$SOYD = N + (N - 1) + (N - 2) + ... + 2 + 1$$

Adding these two expressions, we get

$$SOYD + SOYD = (N + 1) + (N + 1) + (N + 1) + ... + (N + 1) + (N + 1)$$
$$2(SOYD) = N \text{ times} (N + 1)$$
$$= N(N + 1)$$

Therefore,

$$SOYD = \frac{N(N + 1)}{2} \qquad (10.3)$$

---

[10] SOYD calculation for a long-life asset is simpler if the SOYD for a shorter life is known. Let us say that we need to determine the SOYD of a 10-year asset and that the SOYD of an 8-year asset is known to be 36 years. Since $SOYD_{10 \text{ years}} = 1 + 2 + 3 + 4 + 5 + 6 + 7 + 8 + 9 + 10$, we can say that $SOYD_{10 \text{ years}} = (1 + 2 + 3 + 4 + 5 + 6 + 7 + 8) + 9 + 10 = SOYD_{8 \text{ years}} + 9 + 10 = 36 + 19 = 55$ years

For example, for a 15-year asset this equation yields,

$$SOYD_{15} = 15 \times 16/2$$

$$= 120 \text{ years}$$

Under SOYD, the asset's depreciation for the year is obtained by multiplying the total depreciation P-S by a factor that equals *(Remaining Useful Life)*/SOYD. The remaining useful life must be reckoned from the beginning of the year for which depreciation is being calculated. For the first year, the remaining life equals the useful life N. For the third year, the remaining life equals N–2 since the asset has already been used for 2 years.

Consider an asset with first cost P and salvage value S, i.e., with a total depreciation of P-S. If N is the useful life in years, then

$$\text{First year Depreciation} = \frac{\text{Remaining Useful Life}}{\text{SOYD}} \times (P-S)$$

$$= \frac{N}{\text{SOYD}} \times (P-S)$$

$$\text{Second year Depreciation} = \frac{N-1}{\text{SOYD}} \times (P-S)$$

$$\text{Third year Depreciation} = \frac{N-2}{\text{SOYD}} \times (P-S)$$

$$\vdots$$

$$\vdots$$

$$\text{Nth year Depreciation} = \frac{N-(N-1)}{\text{SOYD}} \times (P-S)$$

$$= \frac{1}{\text{SOYD}} \times (P-S)$$

Note that the numerator in these relations has decreased from N to 1. We illustrate the application of the SOYD method through Example 10.3.

Example 10.3: SOYD Depreciation Schedule

Universal Arm (a robot) costs $25,000. At the end of its 4-year useful life its salvage value is estimated to be $5,000. Prepare its depreciation schedule under the SOYD method.

SOLUTION:

In here, P = $25,000, S = $5,000, and N = 4 years. We first determine SOYD. Since the useful life is short, we can do without the equation. From first-principles,

$$SOYD = 1 + 2 + 3 + 4 = 10 \text{ years}$$

Total depreciation = P − S

$$= \$25,000 - \$5,000$$

$$= \$20,000$$

Using the relationship,

$$\text{First year Depreciation} = \frac{\text{Remaining Useful Life}}{\text{SOYD}} \times (P - S)$$

we get,

$$\text{First year Depreciation} = \frac{4}{10} \times (\$20,000)$$

$$= \$8,000$$

$$\text{Second year Depreciation} = \frac{3}{10} \times (\$20,000)$$

$$= \$6,000$$

$$\text{Third year Depreciation} = \frac{2}{10} \times (\$20,000)$$

$$= \$4,000$$

$$\text{Fourth year Depreciation} = \frac{1}{10} \times (\$20,000)$$

$$= \$2,000$$

The SOYD-based depreciation[11] schedule for Universal Arm can be tabulated the way explained in Examples 10.1–10.2, and is given as follows.

Year	Beginning Book Value	Depreciation	Ending Book Value
1	$25,000	$ 8,000	$17,000
2	$17,000	$ 6,000	$11,000
3	$11,000	$ 4,000	$ 7,000
4	$ 7,000	$ 2,000	$ 5,000
	Total:	$20,000	

The ending book value of $5,000, being equal to the salvage value, confirms the accuracy of the results. This can also be done by adding the depreciation charges to check their total which should equal P-S, as in the table ($20,000 total).

In comparison to the schedule under the SL-method (Example 10.2), note that the:

1. Yearly SOYD depreciation charge is not constant
2. Depreciation charge is higher in the beginning

---

[11] We could have shortened the calculations by first determining the value of (P-S)/SOYD as 20,000/10 = 2,000. The depreciations would then have been simply (N)2,000 for the first year, (N − 1)2,000 for the second year, and so on. This might have simplified the calculations as:1st year depreciation = 4 × 2,000 = $8,0002nd year depreciation = 3 × 2,000 = $6,0003rd year depreciation = 2 × 2,000 = $4,0004th year depreciation = 1 × 2,000 = $2,000

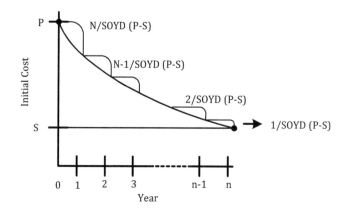

Figure 10.3 SOYD depreciation.

The characteristics of SOYD-based depreciation costing can be illustrated graphically, as in Fig. 10.3. Note how the depreciation charge reduces with time, and thus is not constant as in Fig. 10.2 (SL-method). The base of the staircase pattern in Fig. 10.3 is nonlinear, and hence the SOYD method can be called a *nonlinear* method. The nonlinearity is also obvious in the yearly depreciation equation, where the remaining life as a multiplier reduces each year.

10.3.3 Declining Balance

The declining balance method charges a fixed (constant) percentage of the asset's book value as depreciation for the year. Since the book value decreases each year, the annual depreciation charge[12] also decreases, being highest for the first year. For this reason, the declining balance method is also nonlinear, as the SOYD method.

The percentage is decided by the taxing authority, usually a department of the federal or central government. In the U.S. two rates are allowed: 150% or 200% of the straight-line rate. The 200% based approach is often called the *double declining balance (DDB) method*, the term *double* signifying the fact that the rate is twice that by the SL-method. Since the DDB enables faster depreciation than that with the 150% rate, DDB is preferred.

Declining balance depreciation may be determined by the first-principles, as illustrated in Example 10.4. For the mathematically inclined readers who prefer equations, the following derivations should be of interest. Equation-based analysis has been illustrated in Example 10.5 by redoing Example 10.4.

Consider an asset of initial cost P, salvage value S, and useful life N. The straight-line rate is 1/N since, as explained earlier in conjunction with Equation (10.1), the annual SL depreciation is obtained by multiplying the total depreciation (P-S) with 1/N. In the declining balance method, the rates are 150% and 200% of the straight-line rate, i.e., 1.5/N or 2/N. However, these rates are multiplied with the *book value*, not with the *total depreciation* (P-S) as in the straight-line method. The procedure for determining declining balance depreciation for the 200% rate[13] is as follows.

---

[12] Charge and cost are synonymous in the context of depreciation.
[13] For the 150% rate, simply replace 2/N throughout the derivation with 1.5/N.

<u>Year 1</u>
Book value at the *beginning* of the 1st year

$= \text{First(initial)cost}$

$= P$

1st year depreciation $= (2/N) \times$ 1st year book value

$= (2/N) \times P$

$= 2P/N$

Book value at the *end* of 1st year

$= \text{Book value at the beginning of 1st year} - \text{1st year depreciation}$

$= P - 2P/N$

$= P(1 - 2/N)$

<u>Year 2</u>
Book value at the *beginning* of the 2nd year

$= \text{Book value at the end of 1st year}$

$= P(1 - 2/N)$

2nd year depreciation $= (2/N) \times$ 2nd year book value

$= (2/N) \times P(1 - 2/N)$

$= (2P/N)(1 - 2/N)$

Book value at the *end* of 2nd year

$= \text{Book value at the beginning of 2nd year} - \text{2nd year depreciation}$

$= P(1 - 2/N) - (2P/N)(1 - 2/N)$

$= P\{1 - 2/N - 2/N + (2/N)^2\}$

$= P\{1 - 4/N + (2/N)^2\}$

$= P(1 - 2/N)^2$

Year 3
Book value at the *beginning* of the 3rd year

$$= \text{Book value at the end of 2nd year}$$

$$= P(1 - 2/N)^2$$

3rd year depreciation $= (2/N) \times$ 3rd year book value

$$= (2/N) \times P(1 - 2/N)^2$$

$$= (2P/N)(1 - 2/N)^2$$

Continuing the procedure for other years,

$$\vdots$$

$$j\text{th year depreciation} = (2P/N)(1 - 2/N)^{j-1}$$

$$\vdots$$

$$N\text{th year depreciation} = (2P/N)(1 - 2/N)^{N-1}$$

(10.4)

Note that Equation (10.4) has been derived from first-principles and that it is valid for the double declining balance method. For declining balance rate of 150%, 2/N should be replaced in this equation, and elsewhere, by 1.5/N.

There are two approaches to solve a declining balance problem:

1. Follow the first-principles, or
2. Use Equation (10.4), or its variation for the 150% rate.

In the first-principles approach, we basically follow the steps that led to Equation (10.4), but using the given data. This is illustrated in Example 10.4. The depreciation calculations can also be based on the equation-approach, as illustrated in Example 10.5.

Where a complete depreciation schedule is to be prepared, both the first-principles and the equation-approach are equally suitable. However, when the depreciation charge for only a specific year, or a few intermediate years, is to be determined, the equation-approach is better, as illustrated in Example 10.6, since it does away with the intervening steps essential in the first-principles approach.

Example 10.4: DDB METHOD

Universal Arm (a robot) costs $25,000. At the end of its 4-year useful life its salvage value is estimated to be $5,000.

(a) Prepare its DDB depreciation schedule.
(b) Discuss any discrepancy in the schedule, especially between book value at the end of useful life and the salvage value.

SOLUTION:

The given data are: $P = \$25,000$, $S = \$5,000$, and $N = 4$ years

(a) For DDB depreciation the rate is 2/N. Since the entire depreciation schedule is being asked for, we can either follow the first-principles or use Equation (10.4). Let us follow the first-principles, especially since N is small[14]. This involves following the steps used in deriving Equation (10.4), but with the given data. Thus,

Year 1
Book value at the *beginning* of the 1st year

$$= \text{First}\,(\text{initial})\,\text{cost}$$

$$= \$25,000$$

1st year depreciation[15] = $(2/N) \times$ 1st year book value

$$= (2/4) \times \$25,000$$

$$= \$12,500$$

Book value at the *end* of 1st year

$$= \text{Book value at the beginning of 1st year}$$

$$-\text{1st year depreciation}$$

$$= \$25,000 - \$12,500$$

$$= \$12,500$$

Year 2
Book value at the *beginning* of the 2nd year

$$= \text{Book value at the end of 1st year}$$

$$= \$12,500$$

2nd year depreciation = $(2/N) \times$ 2nd year book value

$$= (2/4) \times \$12,500$$

$$= \$6,250$$

Book value[16] at the *end* of 2nd year

$$= \text{Book value at the beginning of 2nd year} - \text{2nd year depreciation}$$

$$= \$12,500 - \$6,250$$

$$= \$6,250$$

---

[14] For large values of N, the first-principles approach becomes lengthy.

[15] One can evaluate 2/N as 2/4 = 0.5 at this stage, and use this 0.5, instead of 2/N, throughout the calculations to achieve simplicity.

[16] Since the given data, rather than symbols, are being used as the calculation proceeds, the results get simplified; we do not face any lengthy equation as while deriving Equation (10.4). This illustrates that the first-principles approach can be simpler.

Year 3
Book value at the *beginning* of the 3rd year

= Book value at the end of 2nd year

= $6,250

3rd year depreciation = (2/N) × 3rd year book value

$$= (2/4) \times \$6,250$$

$$= \$3,125$$

Book value at the *end* of 3rd year

= Book value at the beginning of 3rd year − 3rd year depreciation

= $6,250 − $3,125

= $3,125

Year 4
Book value at the *beginning* of the 4th year

= Book value at the end of 3rd year

= $3,125

4th year depreciation $= (2/N) \times$ 4th year book value

$$= (2/4) \times \$3,125$$

$$= \$1,563$$

Book value at the *end* of 4th year

= Book value at the beginning of 4th year

−4th year depreciation

= $3,125 − $1,563

= $1,562

The DDB depreciation schedule can be summarized in a table as:

◢	A	B	C	D
1	Year	Beginning BV	Depr. Exp	Ending BV
2	1	25,000	12,500	12,500
3	2	12,500	6,250	6,250
4	3	6,250	3,125	3,125
5	4	3,125	1,563	1,562
6		**Total Depr. Exp:**	**23,438**	

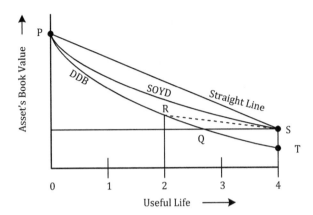

Figure 10.4 Linear and nonlinear depreciation.

(b) We notice a discrepancy between the expected salvage value $5,000 and the book value $1,562 at the end of useful life. These two should have been the same. This discrepancy is also reflected in the total depreciation cost $23,438 being greater than the maximum allowable of $20,000 (P-S = $25,000 − $5,000). Thus, there is something wrong with the results.

The discrepancy can be explained through Fig. 10.4. In any depreciation method the book value should reduce from the initial cost P to the salvage value S by the end of the useful life. Conceptually, a "depreciation journey" begins at point P and should end at S. The SL-method leads from P to S through the shortest path—a straight line. The SOYD method also leads to the destination S, but via a nonlinear path. The declining balance method also follows a nonlinear path, but may miss the destination S. In the present case (above results), the depreciation journey ends at point T, which is below point S. In other words, the book value at the end of useful life (point T) is less than the salvage value (point S). In some cases, the declining balance method may lead to a point above S (book value greater than salvage value). In rare cases (if the analyst is lucky!), this method leads exactly to the destination S (book value equal to salvage value).

What can be done about this likely discrepancy with the declining balance method? The choices are:

Point T below S—Stop charging depreciation once the salvage value has been reached. At the appropriate year (point Q in Fig. 10.4) charge just enough depreciation to render the book value to match the salvage value.

Point T above S—If the resource is used beyond the useful life, keep charging depreciation until salvage value is reached.

A third choice, mostly practiced, is to switch over to the SL-method well in time so that the destination point S is reached at the end of useful life. In Fig. 10.4 this happens at point R, wherefrom RS is the linear path. We discuss this switching over later in Section 10.5. Since it involves two methods, one before and the other after the switch-over, the procedure is called the *composite method*.

For the example under discussion, with Point T below S, the depreciation schedule of Example 10.4 is modified by reducing the total depreciation to the maximum allowed. This necessitates charging none for the fourth year and only $1,250 for the third year, as explained as follows.

We need to reduce the total charged from $23,438 to $20,000. Our attempt of reduction should begin backward from the end of useful life, i.e., from year 4. Assuming no charge for the fourth year, the total charged up to the third year is $23,438 − $1,563 = $21,875. Since this is still higher than the maximum allowed ($20,000), we next reduce the charge for the third year. This reduction must equal $21,875 - $20,000 = $1,875. So, the third-year depreciation is $3,125 - $1,875 = $1,250.

The resulting schedule is as follows, where we have added another column for the cumulative depreciation for the sake of completeness.

	A	B	C	D	E
		Beginning			Cumulative
1	Year	BV	Depr. Exp	Ending BV	Depr.
2	1	$25,000	$12,500	$12,500	
3	2	$12,500	$6,250	$6,250	$18,750
4	3	$6,250	$3,125	$5,000	$20,000
5	4	$5,000	$0	$5,000	$20,000

Example 10.5: Equation-Approach

Solve Example 10.4 using the equation-approach.

SOLUTION:

The given data are: $P = \$25,000$, $S = \$5,000$, and $N = 4$ years

(a) Since DDB depreciation is required, the rate to use is 2/N. The relevant equation is Equation (10.4),

$$jth\ year\ depreciation = (2P/N)(1 - 2/N)^{j-1}$$

By substituting the given values of P ($25,000) and N (4 years), and the appropriate value of j (= 1, 2, 3, and 4), depreciation for each year is determined as:

$$1st\ year\ depreciation = (2 \times \$25,000/4)(1 - 2/4)^{1-1}$$
$$= (\$12,500) \times (1 - 0.5)^{0}$$
$$= (\$12,500) \times 0.5^{0}$$
$$= \$12,500 \times 1$$
$$= \$12,500$$

$$2nd\ year\ depreciation = (2 \times \$25,000/4)(1 - 2/4)^{2-1}$$
$$= (\$12,500) \times (1 - 0.5)^{1}$$
$$= (\$12,500) \times 0.5^{1}$$
$$= \$12,500 \times 0.5$$
$$= \$6,250$$

$$3rd\ year\ depreciation = (2 \times \$25,000/4)(1 - 2/4)^{3-1}$$
$$= (\$12,500) \times (1 - 0.5)^{2}$$
$$= (\$12,500) \times 0.5^{2}$$
$$= \$12,500 \times 0.25$$
$$= \$3,125$$

$$4\text{th year depreciation} = \left(2 \times \$25{,}000/4\right)\left(1 - 2/4\right)^{4-1}$$

$$= \left(\$12{,}500\right) \times \left(1 - 0.5\right)^{3}$$

$$= \left(\$12{,}500\right) \times 0.5^{3}$$

$$= \$12{,}500 \times 0.125$$

$$= \$1{,}563$$

The discussions in part (b) of Example 10.4 requiring modification in the third and fourth year depreciation charges remain applicable. Hence, the final depreciation schedule is:

	A	B	C	D	E
		Beginning			Cumulative
1	Year	BV	Depr. Exp	Ending BV	Depr.
2	1	$25,000	$12,500	$12,500	$12,500
3	2	$12,500	$6,250	$6,250	$18,750
4	3	$6,250	$1,250	$5,000	$20,000
5	4	$5,000	$0	$5,000	$20,000

Note that the use of equation rendered the solution in Example 10.5 more concise in comparison to that in Example 10.4. In addition, the yearly depreciation computation was "decoupled" from each other, i.e., could be calculated without needing the previous year's results. These make the equation-approach better where depreciation charge for a specific year is to be determined, as illustrated in Example 10.6. This approach is desirable also when a few intermediate years are involved.

Example 10.6: DDB

What is the fifth-year DDB depreciation charge for a machine whose first cost is $50,000 and estimated salvage value at the end of its 8-year useful life is $2,000?

SOLUTION:

Since we need to determine the depreciation charge for a specific (fifth) year only, the use of equation will be simpler than the first-principles. The latter would have required determining depreciation for the previous years too to get the book value for year 5. The relevant formula, Equation (10.4), is

$$j\text{th year depreciation} = \left(2P/N\right)\left(1 - 2/N\right)^{j-1}$$

For P = $50,000, N = 8 years, and j = 5, we get

$$5\text{th year depreciation} = \left(2 \times \$50{,}000/8\right)\left(1 - 2/8\right)^{5-1}$$

$$= \left(\$12{,}500\right) \times \left(1 - 0.25\right)^{4}$$

$$= \$12{,}500 \times 0.75^{4}$$

$$= \$12{,}500 \times 0.3164$$

$$= \$3{,}955$$

Using Equation (10.4), two more equations are derived in the following two subsections. These equations are handy in determining specific year-end cumulative declining balance and the asset book value.

10.3.3.1 Cumulative Declining Balance

The cumulative DDB depreciation at the end of jth year is the total of all the annual depreciation charges up to the jth year.

From Equation (10.4),

$$1st\,year\,DDB\,depreciation = (2P/N)(1-2/N)^{1-1}$$

$$= (2P/N)\,(1-2/N)^{0}$$

$$= (2P/N)\,\times\,1$$

$$= 2P/N$$

$$2nd\,year\,DDB\,depreciation = (2P/N)(1-2/N)^{2-1}$$

$$= (2P/N)\,(1-2/N)^{1}$$

$$= (2P/N)\,(1-2/N)$$

$$3rd\,year\,DDB\,depreciation = (2P/N)(1-2/N)^{3-1}$$

$$= (2P/N)(1-2/N)^{2}$$

$$\ldots$$

$$\ldots$$

$$jth\,year\,DDB\,depreciation = (2P/N)(1-2/N)^{j-1}$$

By adding the above, we get

Cumulative DDB depreciation up to year j

$$= 2P/N + (2P/N)(1-2/N) + (2P/N)(1-2/N)^{2} +$$

$$\ldots + (2P/N)(1-2/N)^{j-1}$$

$$= (2P/N)\{1 + (1-2/N) + (1-2/N)^{2} + \ldots + (1-2/N)^{j-1}\}$$

Assume that the cumulative DDB depreciation up to year j is Y. Then we have

$$Y = (2P/N)\,\{1 + (1-2/N) + (1-2/N)^{2} + \ldots + (1-2/N)^{j-1}\}$$

Multiplying both sides of the above by (1–2/N)

$$Y\,(1-2/N) = (2P/N)\,\{(1-2/N) + (1-2/N)^{2} + \ldots + (1-2/N)^{j}\}$$

Subtracting the expression for Y from that for Y (1–2/N), we get[17]

---

[17] All the terms on the right-hand side of the two expressions, except (1 – 2/N)j and 1, cancel with each other.

$$Y\left(1-2/N\right) - Y = \left(2P/N\right)\left\{\left(1-2/N\right)^j - 1\right\}$$

$$Y - 2Y/N - Y = \left(2P/N\right)\left\{\left(1-2/N\right)^j - 1\right\}$$

$$-2Y/N = \left(2P/N\right)\left\{\left(1-2/N\right)^j - 1\right\}$$

$$-Y = P\left\{\left(1-2/N\right)^j - 1\right\}$$

$$Y = P\left\{1 - \left(1-2/N\right)^j\right\}$$

i.e.,

$$\text{Cumulative DDB depreciation} = P\left\{1-\left(1-2/N\right)^j\right\} \tag{10.5}$$

For a declining balance rate of 150%, in the equation replace 2 with 1.5.

10.3.3.2 Book Value

The book value of an asset at the end of a specific year j will simply be its initial cost P minus the cumulative depreciation up to the jth year (Equation (10.5)). For double declining balance, therefore

$$\text{Book value at the end of year j} = P - P\left\{1-\left(1-2/N\right)^j\right\}$$

$$= P - P + P\left(1-2/N\right)^j$$

$$= P\left(1-2/N\right)^j$$

Thus,

$$\text{Book value at jth year} - \text{end} = P\left(1-2/N\right)^j \tag{10.6}$$

Example 10.7: DDB and Book Value

Using the appropriate equations, determine the values of cumulative DDB depreciation and book value in Example 10.6 at the end of third year.

SOLUTION:

The equations derived in the previous two subsections are applicable. The cumulative DDB depreciation at the end of the third year is obtained by substituting j = 3 in Equation (10.5). For the given values of P ($25,000) and N (4 years), this yields

$$\text{Cumulative DDB depreciation at 3rd year-end} = P\left\{1 - \left(1-2/N\right)^3\right\}$$

$$= 25,000\left\{1-\left(1-2/4\right)^3\right\}$$

$$= 25,000\left\{1 - 1/8\right\}$$

$$= 25,000 \times 7/8$$

$$= \$21,875$$

The asset's book value at the end of the third year is obtained from Equation (10.6), as

$$\text{Book value at 3rd year} - \text{end} = P(1 - 2/N)^j$$

$$= 25{,}000(1 - 2/4)^3$$

$$= 25{,}000 \times 1/8$$

$$= \$3{,}125$$

The book value could have been determined easily from first-principles (without needing the equation), since the cumulative DDB depreciation at the third year-end is already known[18]. Merely subtracting the cumulative DDB depreciation $21,875 from the asset's first cost of $25,000 would have yielded $3,125.

10.3.4 Unit-of-Production

The straight-line annual depreciation is based on the fraction a year is of the asset's useful life. The SL-method, as well as the others, works well if the asset is in regular use. For assets that are used irregularly, wear and tear do not relate to the physical time duration. In such cases we prorate depreciation on the basis of actual use, i.e., on actual units of production (UOP) for the year.

Consider an automobile, for example, whose first cost is $16,000. It is expected to last 7 years during which it will be driven 100,000 miles. If its salvage value is $2,000, then the annual straight-line depreciation as one-seventh of the total depreciation $14,000 (= P-S = 16,000 – 2,000) will be $2,000. This annual charging assumes that the automobile is driven 100,000/7 = 14,286 miles per year. Consider however that the automobile belongs to a retiree who uses it for local driving of 5,000 miles a year. In that case, rather than the time-based depreciation, we should use mileage-based depreciation, i.e., use the UOP method. Based on the actual usage, the depreciation for 5,000 miles of yearly driving will be one-twentieth (5,000/100,000 = 1/20) of $14,000, i.e., $700. If during a particular year the retiree takes up a traveling salesperson's job that increases driving to, say, 20,000 miles, then for that year the depreciation charge will be one-fifth (20,000/100,000 = 1/5) of $14,000, i.e., $2,800.

The UOP method is suitable for depreciating natural resources based on annual extraction or harvesting as a proportion of the total available. It is basically similar to the straight-line method, except that the charge for the year is prorated by the ratio: *(production or use for the year)/(expected lifetime production or use)*. For an asset of initial cost P and salvage value S, the prorating of total depreciation P-S yields:

$$\text{UOP Depreciation for the Year} = \frac{\text{Production for the Year}}{\text{Expected Lifetime Production}} \times (P - S)$$

Example 10.8 illustrates the application of the UOP method.

Example 10.8: UOP Method

To support a road construction project, the management decides to install at the site a concrete mixer for an initial cost of $35,000. At the end of the construction period of 5

---

[18]   The use of Equation (10.6) is more effective if the corresponding cumulative DDB depreciation is unknown.

years the machine is likely to be sold for $3,000. The annual concrete needs during the 5 years are: 300, 400, 800, 200, and 100 tons. Prepare the mixer's depreciation schedule based on the UOP method, as well as its book values.

SOLUTION:

It is given that:

$$P - S = \$35,000 - \$3,000$$

$$= \$32,000$$

We need to know the lifetime use of the mixer to prorate P-S on the basis of annual use. Adding the needs of the 5 years,

$$\text{Total usage} = 300 + 400 + 800 + 200 + 100$$

$$= 1,800 \text{ tons}$$

Thus,

	A	B	C	D	E	F	G	H
		Initial	Salvage		Usage	Total Usage		
1	Year	Cost (P)	Value (S)	(P-S)	(tons)	(tons)	Yearly Depr.	Formulas
2	1	$35,000	$3,000	$32,000	$300	$1,800	$5,333	=E2/F2*D2
3	2			$32,000	$400	$1,800	$7,111	=E3/F3*D3
4	3			$32,000	$800	$1,800	$14,222	=E4/F4*D4
5	4			$32,000	$200	$1,800	$3,556	=E5/F5*D5
6	5			$32,000	$100	$1,800	$1,778	=E6/F6*D6

The depreciation schedule for the mixer is a summary of the above charges in a table, as follows. Check that the total depreciation equals P-S ($32,000), confirming the accuracy of the results.

Year	Depreciation
1	$ 5,333
2	$ 7,111
3	$14,222
4	$ 3,556
5	$ 1,778
Total Depr:	$32,000

The year-end book values can be determined by subtracting the yearly depreciation from the beginning book values, as shown as follows. The last year's (fifth year) book value of $3,000 being equal to the salvage value offers another check on the accuracy of results.

	A	B	C	D	E
1	Year	Beginning BV	Yearly Depr.	Ending BV	Formulas
2	1	$35,000	$5,333	$29,667	=B2-C2
3	2	$29,667	$7,111	$22,556	=B3-C3
4	3	$22,556	$14,222	$8,333	=B4-C4
5	4	$8,333	$3,556	$4,778	=B5-C5
6	5	$4,778	$1,778	$3,000	=B6-C6
7		total:	$32,000		

10.3.5 Accelerated Cost Recovery System (ACRS)

We have discussed several methods of depreciation costing in the previous sections. In general, companies are free to choose the method for their use. But for tax purposes they must follow the one recommended by the taxing authority. In the U.S., the federal government's Internal Revenue Service (IRS) collects income tax. Prior to 1954 the IRS required companies to follow the straight-line method. Beginning 1954 the accelerated methods of DDB and SOYD were required. In 1981 these methods were superseded by a simpler procedure called Accelerated Cost Recovery System (ACRS), which was modified later under the Tax Reform Act 1986 as MACRS where M stands for *Modified*. It may be noted that not all countries practice systems such as ACRS or MACRS, and where they do the systems may be different. Non-U.S. readers should refer to the cost recovery systems, if any, prevalent in their countries, and to the associated documents obtainable from the taxing arm of their governments.

10.3.5.1 MACRS

The Modified Accelerated Cost Recovery System (MACRS, pronounced "makers") simplifies depreciation costing for both the government and the companies. The simplification has been achieved by grouping all the depreciable assets into certain property classes of definite lives, called *recovery periods*. Altogether there are eight classes[19] with recovery periods of 3, 5, 7, 10, 15, 20, 27.5, and 39 years. For each property class the IRS has established fixed depreciation allowance percentages which yield, when multiplied with the first cost, the depreciation charge for each year of the asset's useful life.

Some common items and their property class are given in Table 10.1, along with the *recovery periods*—the term used in the MACRS method for useful life.

The recovery percentages for different property classes are given in Table 10.2. For percentages for the real-estate property classes (last two in Table 10.1) refer to the original IRS document.

Example 10.9 illustrates the application of the MACRS method. Its simplicity—a depreciation method that is confusion-free and easy to use—has been the goal of the IRS.

Example 10.9: MACRS

XYZ Company puts in use a computer workstation that costs $20,000. Determine the depreciation schedule as per MACRS.

Table 10.1 MACRS Property Classes

Property Type	Recovery Period (yrs)
Fabricated metal products, special tools for the manufacturing of plastic parts	3
High-tech equip, R&D equip, automobiles, light trucks	5
Office furniture, manufacturing equip, fixtures	7
Railroad cars, barges, vessels, tugs, barges	10
Telephone dist. plant and similar utility products, wastewater plants	15
Electrical power plants, municipal sewers	20
Residential rental property	27.5
Nonresidential property, elevators, escalators	39

---

[19] For complete information refer to the latest edition of: IRS Publication 534. *Depreciation*. U.S. Government Printing Office: Washington, D.C.

SOLUTION:

First refer to Table 10.1 to determine the class the computer workstation belongs to; from there it is a 5-year property. Next refer to Table 10.2 for recovery percentages for a 5-year property which are:

Recovery Year	1	2	3	4	5	6
Allowable, %	20.00	32.00	19.20	11.52	11.52	5.76

For the workstation costing $20,000, the first-year depreciation will thus be 20% of $20,000 = $4,000. For the second year, it will be 32% of $20,000 = $6,400. Similarly, for the subsequent years they will be $3,840, $2,304, $2,304, and $1,152. The resulting depreciation schedule is:

Year	1	2	3	4	5	6	Total:
Depreciation	$4,000	$6,400	$3,840	$2,304	$2,304	$1,152	**$20,000**

Note that the total depreciation charged is $20,000, exactly equal to the investment cost to be recovered. Checking the total depreciation to ensure the accuracy of results is always desirable.

Table 10.2 MACRS Recovery[1] Percentages

Class	3-yr	5-yr	7-yr	10-yr	15-yr	20-yr
			Year			
1	33.33	20.00	14.29	10.00	5.00	3.750
2	44.45	32.00	24.49	18.00	9.50	7.219
3	14.81*	19.20	17.49	14.40	8.55	6.677
4	7.41	11.52*	12.49	11.52	7.70	6.177
5		11.52	8.93*	9.22	6.93	5.713
6		5.76	8.92	7.37	6.23	5.285
7			8.93	6.55*	5.90*	4.888
8			4.46	6.55	5.90	4.522
9				6.56	5.91	4.462*
10				6.55	5.90	4.461
11				3.28	5.91	4.462
12					5.90	4.461
13					5.91	4.462
14					5.90	4.461
15					5.91	4.462
16					2.95	4.461
17						4.462
18						4.461
19						4.462
20						4.461
21						2.231

[1] The percentages begin as declining balance. The 15-year and 20-year properties are based on 150%, while the other four on 200% (DDB). The percentages change to straight-line at the asterisked point.

The recovery percentages in Table 10.2 are based[20] on DDB depreciation with timely switch-over to the SL-method. Another fact underlying MACRS is that the depreciation for the first year is half of that by the DDB-method. This is because MACRS assumes the asset to have been put in service at midyear. The effect of charging only half for the first year is a spillover of depreciation into the year following the end of useful life. For example, a 5-year property has, as illustrated in Example 10.9, a depreciation charge in the sixth year. This is true for other property classes too, as seen in Table 10.2.

10.4 COMPOSITE METHOD

A depreciation method that enables faster recovery of the invested capital is always desirable. The SL-method offers constant recovery; that is why it has been superseded by nonlinear methods that yield faster recovery during the early years of asset life. In general, it is acceptable to switch from one method to another during the asset life. Such a combination of two or more methods is termed *composite depreciation method.*

Referring to Fig. 10.5, the SL-method represented by line PS yields constant annual charge. For nonlinear methods, such as SOYD or declining balance, the depreciation follows a curve like PqrS. The depreciation charge depends on the slope of the line or the curve, depending on the method used. In the case of straight-line this slope is fixed, while it varies along the curve. In the early years, depreciation is large if the curve is followed. For example, the curve's slope at point q is greater than the fixed slope of the straight-line PS. During the latter years depreciation charges along the curve are lower (slope milder) than that with the SL-method. For example, at point r the curve's slope is milder than that of the straight-line PS.

Thus, a nonlinear method (curve) is better than the SL-method in the early years and worse in the latter years. To keep the depreciation charge high, it is thus essential to begin with the nonlinear method and switch to SL-method somewhere along the curve, for example

---

[20] To understand how the recovery percentages have been arrived at, consider a five-year property as illustration. We begin charging depreciation by DDB and switch over to straight-line when beneficial. The salvage value in MACRS is always considered zero, and hence the total depreciation equals the first cost. To calculate the DDB depreciation rate, we need to determine first the straight-line rate. The straight-line rate for a 5-year property is $1/5 = 0.20$. The DDB rate being twice of the straight-line rate will be 0.40, or 40%. MACRS allows half depreciation in the first year on the assumption that the asset is put to use in the middle of the year. In this case, therefore, first-year depreciation is half of 40%, i.e., 20% as seen in Table 12.2 for the 5-year property. For the subsequent years we determine both DDB and SL depreciations to check which one is greater, so that we can decide to switch over. The relevant calculations are illustrated as follows. Note that for SL depreciation during the second year we use a ratio of 1/4.5 because the first-year depreciation corresponded to half-year only.

Year	Calculations	MACRS %
1	Half-year DDB depreciation = 0.50 × 40%	20
2	DDB depreciation = 0.40 (100% − 20%)	32
	SL depreciation = (1/4.5) (100% − 20%)	17.78
3	DDB depreciation = 0.40 (100% − 52%)	19.20
	SL depreciation = (1/3.5) (100% − 52%)	13.71
4	DDB depreciation = 0.40 (100% − 71.20%)	11.52
	SL depreciation = (1/2.5) (100% − 71.20%)	11.52
5	SL depreciation = (1/1.5) (100% − 82.72%)	11.52
6	SL depreciation = remaining 100% − 94.24%	5.76

The underlined percentage being greater of the two for the year becomes the MACRS recovery percentage for the year, as seen in Table 10.2 for the 5-year property class. The recovery percentages for other property classes have been determined the same way. Note that for all property classes the percentage for the last year is half of that for the previous year.

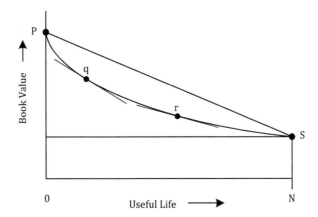

Figure 10.5 Concept of composite depreciation.

in between q and r in Fig. 10.5. Obviously, the appropriate time to switch corresponds to the point where the curve's slope equals that of the line PS. This is the concept underlying composite depreciation, as illustrated in Example 10.10.

Example 10.10: Composite Method

The first cost of a battery-operated machine is $2,500, and its salvage value at the end of 5-year useful life is estimated to be $100. Determine its depreciation schedule based on the DDB-SL[21] composite method.

SOLUTION:

From the discussions in Sections 10.4 and 10.5 as well as in the earlier sections, and also from Fig. 10.5, depreciation will be most rapid under the DDB-SL composite method.

It is given that the total depreciation = P–S = $2,500 – $100 = $2,400. With the SL-method, the depreciation rate for a 5-year asset is 1/5, and therefore the yearly SL-charge[22] will be (1/5) × $2,400 = $480.

The DDB depreciation rate is twice that of the SL-rate, i.e., 2 × (1/5) = 2/5, and therefore the charges for the 5 years will be:

$$1\text{st year depreciation} = (2\,/\,N) \times \text{Book Value}$$

$$= (2\,/\,5)(2,500)$$

$$= \$1,000$$

$$2\text{nd year depreciation} = (2\,/\,N) \times \text{Book Value}$$

$$= (2\,/\,5)(2,500 - 1,000)$$

$$= \$600$$

$$3\text{rd year depreciation} = (2\,/\,N) \times \text{Book Value}$$

$$= (2\,/\,5)(2,500 - 1,600)$$

$$= \$360$$

---

[21] Begins with the DDB-method and switches to the SL-method.
[22] This annual charge will change depending on when we switch over to the SL-method, as discussed later.

$$4\text{th year depreciation} = (2\,/\,N) \times \text{Book Value}$$

$$= (2\,/\,5)(2,500-1,960)$$

$$= \$216$$

$$5\text{th year depreciation} = (2\,/\,N) \times \text{Book Value}$$

$$= (2\,/\,5)(2,500-2,176)$$

$$= \$130$$

The DDB depreciation schedule is thus:

Year	1	2	3	4	5	Total:
DDB Depr.	$1,000	$600	$360	$216	$130	$2,306

Since the total DDB depreciation of $2,306 is lower than the maximum allowed ($2,400), a switch-over to the SL-method is called for[23]. This will enable the total depreciation charged to exactly equal the maximum allowable.

To do this we compare for each year the DDB-charge with the SL-charge and choose the greater of the two. For year 1, the DDB charge at $1,000 is more than the $480 SL-charge. So, we opt for DDB depreciation of $1,000 for year 1. For year 2, the DDB-charge is $600, while the SL-charge[24] will be:

$$= (1\,/\,4)(\text{total depreciation remaining to be charged})$$

$$= 0.25 \times (2,400 \ - \ 1,000)$$

$$= \ \$350$$

So, for year 2 too, we choose DDB since its $600 charge is greater than the $350 SL-charge.

We now check the third year to see whether it makes sense to switch. From the previous table, the DDB-charge for year 3 is $360. The cumulative depreciation charged so far is $1,000 + $600 = $1,600. So, the SL-method yields a depreciation charge of (1/3) × (2,400 – 1,600) = $267, which is less than $360. Thus, switching to the SL-method is not yet desirable. We thus "stick with" the DDB for the third year too.

Next check for year 4. The SL-charge for the fourth year is (½)(2,400 – 1,960) = $220, which is greater than the DDB charge of $216 for year 4. Hence, switching to the

---

[23] Had this total been greater than $2,400, we would have reduced the charges for the latter years, since tax laws do not allow total depreciation to exceed P-S. For example, if S is $500 in Example 10.10, the cumulative DDB depreciation charge of $2,306 is greater than P-S (= $2,500 – $500 = $2,000). In such a case, no composite depreciation is warranted for. With S as $500, the results of Example 10.10 would have been:

Year	1	2	3	4	5	Total:
DDB Depr.	$1,000	$600	$360	$40	$0	$2,000

Note that we charged nothing for year 5. For year 4, we reduced the charge to render the total $2,000. Since the cumulative depreciation up to year 3 is already $1,960, we charged only the remaining $40 for year 4.

[24] If we switch to the SL-method now at the end of year 1, then N is 4, and therefore depreciation rate is 1/4. Since $1,000 has already been charged for the first year by the DDB-method, depreciation that remains to be charged is $2,400 (maximum allowed) minus $1,000.

SL-method at year 4 makes sense. The switching means charging $220 for year 4, and also[25] for year 5. The composite DDB-SL depreciation schedule will thus be:

Year	1	2	3	4	5	Total:
Composite Depr.	$1,000	$600	$360	$220	$220	$2,400

The total of $2,400 being equal to P-S confirms the accuracy of the results.

Example 10.10 illustrated the combining of DDB with the SL-method. In a similar way one can combine SOYD with the SL-method, or any two or more methods, as long as only one method is used to determine the depreciation for any year. The ultimate goal of any compositing is to depreciate the asset as rapidly as possible without violating any tax law.

10.5 COMPARISON

All the methods other than the SL-method allow for accelerated depreciation in the early part of the asset life. Of the two nonlinear methods (DDB and SOYD), DDB is in general more accelerating, i.e., yields higher charges in the beginning. This is shown in Fig. 10.6, where the DDB-curve has steeper slopes[26] in the beginning than the other two. Note, as explained earlier, that the DDB method may not always lead to point S, though shown in this figure, representing the salvage value. In such a case, a composite method that switches to SL, as illustrated in Example 10.10, may have to be used to ensure that the total depreciation does not exceed P-S, the maximum allowable.

In most cases, the taxing-authority-sponsored depreciation method, such as MACRS, is practiced. Companies may be able to choose other methods in certain cases, especially for older assets under "grandfather" clauses in the tax law. For example, the DDB based on 200% of the SL-rate is used for new assets, while the 150% rate is applicable to used assets or in special cases. Non-MACRS methods may eventually disappear in the U.S., and in other countries as well which may have already adopted, or are likely to adopt, methods

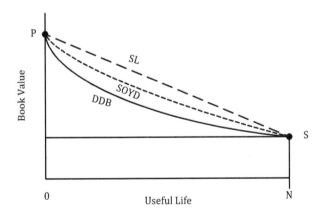

Figure 10.6 Comparison of depreciation methods.

---

[25] Once we switch to the SL-method, the depreciation for the subsequent years remains the same due to the linear nature of the method.

[26] Plural form to signify that several values are possible depending on the point selected on the curve.

similar to the MACRS. In the meantime, industries continue to practice most of the methods discussed in this chapter. We covered the non-MACRS methods too because:

1. In the U.S., properties acquired before 31 December 1980 must continue to be depreciated according to pre-1981 methods.
2. Countries other than the U.S. may not be using MACRS or a similar method. Moreover, U.S. companies operating overseas are generally required to follow the depreciation method and tax laws of the host country.

10.6 SUMMARY

Depreciation is a loss in the value of an asset. Almost all capital assets used by companies, with the exception of land, are depreciable. Depreciation is basically a means of recovering the cost of the capital invested in the assets that generate income. Assets are allowed annual depreciation deductions from profits, reducing the income tax payable. Depreciation charges modify the gross incomes from the resource. A listing of the yearly charges is called a depreciation schedule. The maximum allowable depreciation during the asset life is P-S, the loss in value, where P is the first cost and S the salvage value.

Depreciation methods commonly practiced in industry have been discussed in this chapter. These are: straight-line (SL), SOYD, declining balance such as DDB, unit-of-production, and MACRS. The SL-method is the simplest of all, charging the same depreciation for each year of the asset life. Its weakness is its inability to charge higher depreciations during the early years of the asset life; DDB and SOYD do this. The declining balance method may not ensure that the life-end book value equals the salvage value. This means that the total depreciation charged could be different from the maximum allowable P-S. The analyst must be aware of this likely situation. By opting for a composite depreciation method, it is possible to ensure that, under declining balance such as DDB, the total depreciation charge equals P-S. Sponsored by the U.S. Internal Revenue Service, MACRS is a composite DDB-SL method. Its tabular approach to depreciation costing streamlines the calculations. In the U.S., the SL-method is applicable only to pre-1981 assets; MACRS is used for depreciating recently acquired assets.

DISCUSSION QUESTIONS

10.1   Why are there so many methods for depreciation costing?
10.2   Compare the straight-line depreciation method with all the others discussed in Chapter 10 for constancy of depreciation charges and complexity of their costing.
10.3   Explain the concept behind the DDB depreciation method.
10.4   Companies account for depreciation because they pay taxes. Companies pay taxes because they account for depreciation. Which of these two statements is correct and why?
10.5   In a composite depreciation method, why do we switch to the SL-method where the slope of the declining balance curve equals that of the straight line?
10.6   Give an example from your own experience where the unit-of-production method will be appropriate.
10.7   Under MACRS, why do we charge depreciation during the sixth year for an asset of 5-year useful life?
10.8   Explain why the book value and market value of the asset may sometimes differ.
10.9   Which of the nonlinear depreciation methods do you feel presents the most realistic representation of actual depreciation?
10.10 Do you think that quickly depreciating an asset reflects actual wear and tear on a piece of equipment or merely writes the asset off the books quickly?

10.11 Why is it in the best interest of the government to limit the amount of annual record-able depreciation?

10.12 What is the initial book value of an asset and how is this value computed?

10.13 Define a situation where a company would prefer to not use the accelerated method of depreciation.

MULTIPLE-CHOICE QUESTIONS

10.14 Depreciation charges for equipment that cost $2,000 had been $600, $500, and $300. Its book value in dollars is
   a. $600
   b. $3,400
   c. $1,200
   d. $1,400

10.15 Calculation-wise the simplest depreciation method is
   a. SOYD
   b. straight line
   c. declining balance
   d. DDB

10.16 If a machine's useful life is 6 years, its SOYD value in years is
   a. 15
   b. 6
   c. 21
   d. none of the above

10.17 In SOYD depreciation, the annual charge
   a. remains the same
   b. decreases every year
   c. increases every year
   d. remains constant during the first half of useful life

10.18 The annual depreciation under the SL-method is
   a. $(S-P)/N$
   b. $(P+S)/N$
   c. $(P-S)/N$
   d. $(S+P)/N$

10.19 The depreciation method that is primarily tabular is
   a. SOYD
   b. declining balance
   c. DDB
   d. MACRS

10.20 In a composite method, usually the
   a. SL-method follows the DDB-method
   b. DDB-method follows the SL-method
   c. SOYD-method follows the SL-method
   d. DDB-method follows the SOYD-method

10.21 What is the depreciation percentage (%) under the MACRS method for a class 7-year asset in the sixth year of its useful life?
   a. 6.55
   b. 8.92
   c. 5.76
   d. 8.93

10.22 What is the depreciation percentage (%) under the MACRS method for an electrical power plant in the 13th year of its useful life?
   a. 4.462
   b. 4.461
   c. 5.90
   d. none of the above

10.23 Under the straight-line depreciation method, the annual depreciation expense for an asset valued at $45,000 with a useful life of 7 years and a salvage vale of $10,000 is:
   a. $3,500
   b. $6,429
   c. $5,000
   d. none of the above

10.24 Depreciation expense for the second year of an asset valued at $250,000 defined as a 5-year asset according to MACRS table of percentages is closest to which of the following:
   a. $90,000
   b. $50,000
   c. $80,000
   d. none of the above

10.25 What is the book value of an asset valued at $60,000 in the third year of its useful life? Assume straight-line depreciation, a 15-year useful life, and no salvage value.
   a. $32,000
   b. $60,000
   c. $44,000
   d. $48,000

10.26 A drill press is purchased for $56,000 and depreciated as a 7-year asset according to the accelerated MACRS percentage table. The drill press is sold in the fourth year of its useful life. Its book value is closest to which?
   a. $24,490
   b. $38,204
   c. $17,496
   d. $6,994

10.27 Using DDB depreciation, the book value after 3 years of a $1 million asset with a no salvage value and a recovery period of 15 years is what?
   a. $769,500
   b. $546,239
   c. $650,000
   d. none of the above

10.28 Depreciation is recorded as a(n):
   a. responsibility
   b. liability
   c. asset
   d. expense

NUMERICAL PROBLEMS

10.29 A robot has been acquired for $50,000. It is expected to remain useful for the next 5 years, at the end of which its salvage value is likely to be $1,500. What is its uniform annual depreciation charge?

10.30 The first cost of a machining center is $120,000, and its salvage value is estimated to be $10,000. During its 10-year useful life it is expected to generate a net annual income of $20,000. Prepare the machining center's cashflow table assuming constant depreciation.

10.31 Beautiful Roses buys a truck for $14,000 to deliver flowers locally. It believes that the truck could be sold for $4,000 at the end of 4 years. If the business practices straight-line depreciation, how much could it deduct each year from the profit as depreciation charge on the truck?

10.32 Premium Manufacturers acquires a fax machine for $1,350 whose estimated useful life and salvage value are 5 years and $100. Prepare the machine's depreciation schedule under the SOYD method.

10.33 Jack-of-all-Trades, a one-person home repair business, purchases a computer system to manage its accounting and customers' data. The system costs $8,000 and is expected to have a useful life of 5 years. Assuming an estimated salvage value of $500, prepare the depreciation schedule using the DDB method. Do not switch to another depreciation method, but ensure that the total depreciation does not exceed the maximum allowable.

10.34 For a networked computer system costing $45,000, determine the DDB depreciation charges and book values over 8 years of useful life, assuming a salvage value of $9,000. Without switching to another method, ensure that the total depreciation does not exceed the maximum allowable.

10.35 A tool-room lathe has been acquired by Perfect Productions for $9,000. It is likely to be useful for the next 5 years, at the end of which its salvage value is estimated to be $300. Prepare a composite depreciation schedule that begins with DDB and switches over to the SL-method at the most opportune time.

10.36 Aluminum International buys a truck for $50,000 for use in hauling bauxite. The estimated salvage value at the end of an expected 200,000-mile useful life is $2,000. The truck's use during the first 2 years had been 40,000 and 45,000 miles. What is the third-year depreciation if its use during the year has been 50,000 miles?

10.37 10.37 The engineering department of a company acquires a $12,000 workstation for designers. Prepare the workstation's depreciation schedule under MACRS.

10.38 Just-in-Time Janitors purchases an automatic floor cleaning and polishing machine for $9,000. The machine is expected to last for 10 years and to have a salvage value of $1,000. Prepare its straight-line, SOYD, DDB, and MACRS depreciation schedules.

10.39 A coal-mining company purchases a dump truck for $3 million. The equipment is classified as a 10-year asset under MACRS rules. Compose the depreciation tables for the equipment according to (a) MACRS percentages and (b) the straight-line depreciation method. What is the difference in book values at years 3 and 4 for the two methods?

10.40 A railroad company purchases two new locomotives for $400,000 each. Each locomotive is estimated to last 1 million miles of service. What is the book value of the two locomotives if in the first 4 years locomotive A travelled 100k, 150k, 175k, and 225k miles and locomotive B travelled 175k, 200k, 275k, and 175k miles respectively?

10.41 What is the DDB depreciation charge at the end of year 6 for a piece of equipment whose first cost is $650,750 and an estimates salvage value of $40,000 at the end of 9 years of useful life? Assume 200% rate.

10.42 Now using the appropriate equations, determine the cumulative DDB depreciation and the book value for Question 10.41 at the end of the third year of useful service.

10.43 A super-conducting magnet was purchased by a scientific research company for $200,000. The magnet is expected to last 7 years and have a salvage value of $29,000 at the end of its useful life. Determine the depreciation schedule based on the DDB-SL composite method, assuming rate of 150%.

10.44 Now depreciate the super-conducting magnet in Problem 10.43 using the SOYD method of depreciation.

10.45 What is the straight-line depreciation charge per year for a piece of equipment with an initial cost of $75,000, a useful life of 5 years, and a salvage value of $4,500?

10.46 An industrial embroidery machine costs $28,000, will last 4 years and have a salvage value of $4,000. Prepare the depreciation schedule using the SOYD method.

10.47 A beer bottle company purchases a piece of manufacturing equipment for $118,000. Determine the depreciation schedule as per MACRS assuming a 15-year useful life.

10.48 An oil company purchased casing for a 4-year project for $100,000. The casing's useful life is 50,000 barrels. If the barrel production for the platform is 2,000, 2,500, 3,000, 7,500, and 4,000 per year, what is the depreciation schedule for the casing for its first 5 years? Assume a salvage value of $15,000.

FE EXAM PREP QUESTIONS

10.49 Using DDB depreciation, the book value after 2 years of a $45,000 asset with a no salvage value and a recovery period of 10 years is what? Assuming a 150% rate.
   a. $25,000
   b. $28,800
   c. $39,000
   d. none of the above

10.50 The depreciation percentages for MACRS assume which of the following:
   a. straight-line switching to DDB depreciation
   b. $0 salvage value
   c. DDB switching to SL depreciation for life of asset
   d. all of the above

10.51 Using DDB depreciation, the first-year depreciation expense for an asset valued at $6 million with no salvage value and a 10-year useful life is closest to which?
   a. $1.4 million
   b. $1.2 million
   c. $345,639
   d. $450,000

10.52 When using the DDB-SL composite method:
   a. always switch from SL to DDB in year 3
   b. always switch from DDB to SL
   c. final year always use straight-line depreciation
   d. both b and c

10.53 Depreciation is reported as what:
   a. cashflow liability
   b. rental income
   c. expense
   d. all of the above

10.54 Under the straight-line depreciation method, the book value after 4 years of a $60,000 asset with a useful life of 8 years and $3,000 salvage value is:
   a. $28,500
   b. $35,600
   c. $42,750
   d. $57,000

10.55 If a company is in the 25% tax bracket on profits and has $250,000 in revenue, $75,000 in expenses, and $35,000 in depreciation, the after-tax cashflow is:
   a. $105,000
   b. $140,000
   c. $250,000
   d. none of the above

10.56 A $76,000 is depreciated over 4 years using the straight-line method. If it is sold for $45,000 at the end of year 2, what is the gain or loss on the transaction?
   a. –$12,000
   b. $10,000
   c. –$ 7,000
   d. $ 7,000

10.57 What is the depreciation percentage (%) under the MACRS method for a piece of high-tech equipment in the third year of its useful life?
   a. 19.2
   b. 14.40
   c. 17.49
   d. none of the above

10.58 Depreciation charges for equipment that cost $8,000 had been $1,000, $750, and $300. Its book value in dollars is
   a. $ 750
   b. $2,050
   c. $5,950
   d. $6,000

Chapter 11

Income Tax

LEARNING OBJECTIVES

- How to account for income tax
- Income tax calculations
- *Incremental* income tax rate
- Tax-modified cashflows
- Effect of income tax through depreciation
- After-tax ROR and other analyses
- Combined federal and state income taxes
- Other tax implications in economic analyses

To add further realism to the economic analyses discussed so far, we consider in this chapter another fact of engineering business, namely income tax obligation. The past discussions ignored the tax for the sake of simplicity. For example, we excluded in Chapter 6 the effects of income tax on rate of return, conducting what may be called before-tax ROR analysis. To render any investment decision realistic, the analyst must include tax obligations in the ROR and other analyses.

When taxes are taken into account, the procedure is appropriately called *after-tax* analysis. Such analyses reveal the effect of taxes on alternatives' attractiveness. Since taxes are a fact[1] of life in every business throughout the world, engineers invariably include the tax considerations in investment decisions.

11.1 TAXES

Most governments consider it their duty to promote an economic "climate" conducive to wealth creation. Even in the free market economies such as that of the U.S., the government is an active partner—albeit indirectly—of the business. Governments[2] participate in economic activities in various ways, one of which is through taxation. In good times when companies make profits, governments charge taxes of various types, and in bad times they help them through "soft" credits, loans, tax write-offs, and similar incentives. They sometimes set up special agencies to implement their policies of participation. For example, the

---

[1] The reality of taxes has nowhere been expressed as philosophically as by Benjamin Franklin, who said, "Two things are inevitable: death and taxes."

[2] The plural form is more appropriate since government participation could be at various levels: federal or central, state or province, county, and city.

U.S. federal government has created the Small Business Administration (SBA) to encourage small businesses to become key players in the economy.

Though various types of taxes are collected from businesses, the discussions in this chapter are limited to income tax since it has a direct bearing on investment decisions. Additionally, we consider capital gain/loss and investment credit in Section 11.8.

11.2 INCOME TAX

Income tax is payable by both individuals and corporations. The taxes are calculated using rates fixed by the governments. In the U.S., the federal government collects most of the income tax. Several states as well as city governments also levy income tax. Each government reviews its budget provisions once a year and decides the tax rates for the next fiscal (financial) year.

Income tax rates are usually variable or *graduated*, which means that the rate depends on the income level. In graduated tax systems those with higher incomes pay more taxes. Most governments of the world practice such a system. In the U.S., a move to a *flat* tax rate whereby all incomes are taxed at the same rate has been under discussion for some time.

Consider tax payers A and B. Assume that A's taxable income for the year is $50,000, while B's $80,000. Under a flat tax rate of 10%, A will pay an income tax of $5,000, while B of $8,000. Next, consider graduated taxation under the following-assumed schedule:

Taxable Income	Tax Rate
$ 0–$ 60,000	10%
$60,000–$150,000	15%

These rates mean that incomes above $60,000 are taxed at 15%. Since A's income is $50,000, all of which falls in the $0–$60,000 range, the applicable tax rate is 10%; thus, A's tax is 10% of $50,000, i.e., $5,000. In the case of B, with an income of $80,000, tax calculations are slightly complex. Referring to the above schedule, B's income of $80,000 should be considered divided in two components: $60,000 plus $20,000. For the $60,000 component the rate will be 10%, while for the $20,000 it will be 15%. Thus, B's tax will be 10% of $60,000 plus 15% of $20,000. This works out to be $6,000 + $3,000 = $9,000. Had the tax rate been flat at 10%, B would have paid only $8,000 (= 10% of $80,000). Thus, graduated taxation charged higher tax from B whose income was higher. The justification for graduated taxation is that those who can afford to pay should pay more in tax.

Income tax laws are complicated, and the rates change each year. Almost all companies, and many individuals, hire accountants or tax attorneys to handle their income tax obligations and the associated investment opportunities. The purpose of this chapter is not to learn about the intricacies of taxes, but merely to acquire skills necessary for assessing the implications of income taxes to engineering economic analyses. Such skills enable engineers and technologists to make more realistic and judicious engineering investment decisions.

Example 11.1: Income Tax

On her graduation with a BS in manufacturing engineering technology, Brenda set up a job shop called AnyMachining Inc. Since her recent use of the Internet to advertise her business through a home page, the orders have increased, resulting in a taxable income of $95,000. How much income tax is payable by AnyMachining if the applicable tax rate schedule is as follows?

Taxable Income	Rate
<$50,000	10%
$50,000–$75,000	15%
$75,000–$100,000	20%
>$100,000	25%

SOLUTION:

The given schedule comprises four tax brackets of 10%, 15%, 20%, and 25%. The taxable income of $95,000 falls in the third bracket of $75,000–$100,000. Thus, $95,000 will have three components. Since the upper cut-off for the first bracket is $50,000, the first component will be $50,000 which is taxable at 10%. This leaves $45,000 (= $95,000 − $50,000) to be taxed. The upper cut-off of the second bracket is $75,000. Therefore, the second component will be $75,000 − $50,000 = $25,000 which is taxed at 15%. The remainder of the income, i.e., $95,000 − $75,000 = $20,000, as the third component, is taxable at 20%.

On the first $50,000 component, at 10% rate, the tax is $5,000.

On the second component, the tax is 15% of $25,000 = $3,750. Since the remaining $20,000 of the income is taxed at 20%, the tax on this component is $4,000.

Thus, the total income tax[3] payable by AnyMachining

= $5,000 + $3,750 + $4,000

= $12,750

11.3 TAXABLE INCOME

Income tax is payable on *taxable* income. In the previous example, the taxable income was given; in most cases, it is determined. Taxable income is the difference between gross income and the direct expenses incurred in generating the income.

In this section, we discuss how to determine taxable income, and there from tax, for a business. Most businesses are either proprietorship or corporations. If you own a business as a proprietor, alone or in partnership, the tax payable is at individual rates. Corporations, on the other hand, pay[4] taxes at corporate rates. The rates and the other related information are contained in appropriate documents prepared by the taxing authorities. The discussions in this section are limited to taxation by the U.S. federal government[5].

---

[3] A table may be used to summarize the calculations and results, as follows.

Income Component	Rate	Tax
50,000	10%	5,000
25,000	15%	3,750
20,000	20%	4,000
Total: 95,000		12,750

[4] A company can be formed as S corporation, which is better in "infancy" when it is likely to incur losses. S corporation owners can compensate business losses by deducting them from non-business profitable incomes. They, however, run the risk of personal liability under S corporation laws. Once such a company begins to earn profit, it changes itself to the usual incorporation to avoid liability.

[5] U.S. state governments as well as governments in other countries have similar income tax provisions. Non-U.S. readers should contact the nearest office of their income taxing authorities to obtain current information on tax rates and associated matters.

11.3.1 Individual

For individual taxpayers, first determine the *total*[6] *income*, which is the sum of taxable incomes from all sources. For a sole proprietor or business partner,

Total income = Net income from business + Other incomes

For an employed person,

Total income = Taxable salary + Other incomes

Taxable salary[7] is usually lower than the gross salary by an amount invested in tax-deferred retirement and similar plans. The other incomes include interest earnings from saving accounts, dividends from investments, capital gains, etc.

The total income is next adjusted to allow for certain deductions such as moving expenses and alimony paid. Thus,

Adjusted gross income = Total income − Allowable deductions

The taxing authority may allow certain personal exemptions, and a standard or itemized[8] deduction from the adjusted gross income. This yields taxable income as:

Taxable income = Adjusted gross income

−Personal and standard (or itemized) deductions

The various deductions reflect government-sanctioned expenses likely to be incurred by an individual in earning the income, as well as other tax reliefs. They include a personal exemption for the employee, exemptions for dependents, and a lump-sum standard deduction, or itemized deductions in lieu. The itemized deductions[9] include expenses such as state and local income taxes, excessive medical costs, home-mortgage interest, donations to charities, and so on. The taxpayer has the choice of claiming either a standard deduction or itemized deductions, whichever reduces the tax. The amounts of personal exemption and standard deduction may increase each year to account for inflation.

The taxable income is taxed at rates fixed by the government for the year. The rate schedules for the US federal income tax are printed in various tables and are also available at the www.irs.ustreas.gov website. One of the tables, for example, is for individuals filing as singles (unmarried individuals); for the 1998 tax year, it was:

The table has been formatted[10] to simplify the tax calculations. For example, the dollar figures in the table's second column (under Tax) are simply the taxes on the upper cut-off

---

[6] Terminology used here is that in the Internal Revenue Service (IRS) document 1040.

[7] This information is provided by the employer to each employee at the time taxes are to be filed.

[8] Taxpayers can claim either a standard or an itemized deduction. For the latter, the taxpayer lists the claims against what is allowed. Itemized deduction is obviously claimed only if it is beneficial, i.e., if it is more than the standard deduction.

[9] The type of item and the maximum allowable deduction may vary from year to year.

[10] Table 11.1 is simply an extension of the following schedule.

Taxable Income	Rate, %	Taxable Income	Rate, %
up to $25,350	15	128,100–278,450	36
25,350–61,400	31	above 278,450	39.6
61,400–128,100	28		

One can determine the tax using this schedule, as explained in Example 11.1.

Table 11.1 Year 1998 Tax Rate Schedule for Singles

Taxable Income	Tax	Of the Amount Over
$0–$25,350	15%	$0
$ 25,350–61,400	$ 3,802.50 + 28%	$ 25,350
$ 61,400–128,100	$13,896,50 + 31%	$ 61,400
$128,100–278,450	$34,573.50 + 36%	$128,100
Above $278,450	$88,699.50 + 39.6%	$278,450

income of the previous bracket. The cut-off incomes are the income limits at which the tax rate changes; for example, at $61,400 the tax rate increases from 28% to 31%. As an illustration, the figure $13,896.50 in the second column is the tax on a $61,400 income. Any income in excess of $61,400 but less than $128,100 falls in the third bracket and, as per the rate in the table's second column, is taxed at 31%; by adding this tax to $13,896.50 the total tax is determined. The figure $13,896.50 has been obtained by charging 15% tax on $25,350 (the upper cut-off in the first bracket) and 28% on $36,050 (the income range of the second bracket = $61,400 – $25,350). In other words, $13,896.50 is the result of 0.15 × $25,350 + 0.28 × $36,050. Other dollar figures in the table's second column have been obtained the same way, i.e., based on cut-off incomes and corresponding tax rates.

If you find the concise schedule under footnote 10 easier to use than Table 11.1, then do so following the procedure explained in Example 11.1.

Example 11.2 illustrates how the income tax of a small company whose proprietor pays tax as single individual is calculated.

Example 11.2: Income Tax

On finishing her BS in computer engineering, Jasmin set up a computer repair and system configuration business as a sole proprietor. Her business income in 1998 was $180,000. The rent, utilities, and other direct expenses of the business added up to $93,000. Jasmin also had interest and dividend earnings of $5,000. How much income tax is payable if Jasmin is single with no dependents? Refer to IRS Form 1040 for any relevant data.

SOLUTION:

Jasmin's net income from the business is:

$$= \text{Business income} - \text{Expenses to earn the income}$$

$$= \$180,000 - \$93,000$$

$$= \$87,000$$

To this we add her other incomes. Thus, her

$$\text{Total income} = \text{Business income} + \text{Any other income}$$

$$= \$87,000 + \$5,000$$

$$= \$92,000$$

Assuming that this total income does not need to be modified any further as per IRS Form 1040, Jasmin's

$$\text{Adjusted gross income} = \text{Total income}$$

$$= \$92,000$$

Jasmin is entitled to only one personal exemption of $2,700 (referring to the 1998 IRS Form 1040) for herself, since she has no dependents. She is also entitled to standard deduction or itemized deductions. Assuming that Jasmin is better off[11] claiming standard deduction, she opts for it. For 1998, the standard deduction for single filers was $4,250 (IRS Form 1040). Thus, Jasmin's

$$\text{Taxable income} = \text{Adjusted gross income} - \text{Personal exemption and Standard deduction}$$

$$= \$92,000 - (\$2,700 + \$4,250)$$

$$= \$85,050$$

Noting in Table 11.1 that a taxable income of $85,050 falls in the third tax bracket of $61,400–$128,100, Jasmin's tax[12] is[13]

$$\text{Income tax} = \$13,896.50 + 31\% \text{ of income over } \$61,400$$

$$= \$13,896.50 + 31\% \text{ of } (\$85,050 - \$61,400)$$

$$= \$13,896.50 + 0.31 \times \$23,650$$

$$= \$13,896.50 + \$7,331.50$$

$$= \$21,228$$

Another group of income taxpayers is married couples filing tax jointly. For them, the 1998 U.S. tax schedule was as in Table 11.2.

Table 11.2 too has been formatted to simplify the tax calculations. It is basically an extension of the following schedule which can be used to calculate income tax in a way explained in Example 11.1.

Taxable Income	Rate
< $ 42,350	15 %
$ 42,350–102,300	28 %
$102,300–155,950	31 %
$155,950–278,450	36 %
>$278,450	39.6%

Example 11.3 illustrates the calculations for income tax obligation of a married couple filing jointly.

---

[11] Had fewer expenses resulting in itemized deductions' total less than the standard deduction.

[12] According to Table 11.1, for the third bracket, the income tax is $13,896.50 plus 31% of the taxable income over $61,400.

[13] You should get this answer also by using the schedule in footnote 10 and following the procedure of Example 11.1.

Table 11.2 1998 Tax Rates Married Filing Jointly

Taxable Income	Tax	Of the Amount Over
$ 0–$42,350	15%	$0
$ 42,350–102,300	$ 6,352.50 + 28%	$ 42,350
$102,300–155,950	$23,138.50 + 31%	$102,300
$155,950–278,450	$39,770.00 + 36%	$155,950
Above $278,450	$83,870.00 + 39.6%	$278,450

Example 11.3: Income Tax

In Example 11.2 assume that Jasmin is married to a person who, as an employee in the local video rental store, earns $18,000 as wages and $5,000 as commission. If they filed tax jointly, how much income tax would be due? Refer to IRS Form 1040 for any relevant data.

SOLUTION:

Since Jasmin is married and files jointly, she is entitled to more exemptions and deductions. However, now the

$$\text{Total income} = \text{Jasmin's business income} + \text{Spouse's income}$$

$$= \$92,000 + (\$18,000 + \$5,000)$$

$$= \$115,000$$

Assuming that this total income does not need to be modified any further as per IRS Form 1040, their

$$\text{Adjusted gross income} = \text{Total income}$$

$$= \$115,000$$

Jasmin is now entitled to two personal exemptions of $2,700 each, one for herself and one for her spouse. They are also entitled to standard deduction or itemized deductions. Assuming that they have summed up the itemized deductions and found the total to be lower than the standard deduction for couples, they claim the latter. For 1998, the standard deduction for married couples filing jointly was $7,100 (IRS Form 1040). Thus, their

$$\text{Taxable income} = \text{AGI} - \text{Personal Exemptions \& Standard Deductions}$$

$$= \$115,000 - (2 \times \$2,700 + \$7,100)$$

$$= \$115,000 - (\$5,400 + \$7,100)$$

$$= \$102,500$$

From Table 11.2, a taxable income of $102,500 falls in the third tax bracket of $102,300–$155,950; hence

$$\text{Income tax} = \$23,138.50 + 31\% \text{ of income over } \$102,300$$

$$= \$23,138.50 + 31\% \text{ of } \left(\$102,500 - \$102,300\right)$$

$$= \$23,138.50 + 0.31 \times \$200$$

$$= \$23,138.50 + \$62$$

$$= \$23,200.50$$

Each year the U.S. federal government decides upon the type and extent of *allowable deductions*. For complete details, readers should refer to the IRS document Form 1040 and its attachments, as well as the accompanying instructions. They are available free from the nearest IRS office or at their website[14]. If you already pay, or have paid, U.S. income tax as an individual, you may have found the treatment in this subsection easier to follow.

11.3.2 Corporations

Gross income is the difference between the revenue from the sale of goods and services and the operating cost incurred in their production and marketing. Thus,

$$\text{Gross income} = \text{Sales revenue} - \text{Operating cost}$$

The operating cost comprises both direct and indirect costs which include the costs of material, labor, rent, utilities, maintenance, marketing, and others. Also included are costs on non-consumable items lasting less than a year.

For incorporated businesses, taxable income is determined by deducting from the gross income expenditures on depreciable assets used in generating the sales revenue. Thus,

$$\text{Taxable income} = \text{Gross income} - \text{Depreciation expenditures}$$

Depreciation expenditures pertain to resources (equipment and machines) that last more than a year. For tax purposes, the costs of depreciable assets are not allowed as deductions from the profits in the year the assets are put to service. They must, instead, be deducted as depreciation expenses over the asset's useful lives. We have discussed the procedures of depreciation charging extensively in Chapter 10.

The taxing authorities usually provide companies information on the various tax incentives offered by the governments. Tax attorneys and accountants keep track of these matters, and are the ones businesses consult to benefit from such incentives.

Example 11.4: Income Tax Due

A small job shop's sales revenue for the last year was $782,552, while its operating costs were $458,760. The shop acquired a new CAM system for its numerically controlled machines at a cost of $125,000. The CAM system's useful life is expected to be 3 years, at the end of which it is likely to be salvaged at $20,000. For straight-line depreciation, how much income tax is due under the following schedule?

Taxable Income	Rate
<$ 50,000	10%
$ 50,000–$75,000	15%
$ 75,000–$100,000	20%
>$100,000	25%

[14] www.irs.ustreas.gov

SOLUTION:

First, we determine the gross income and taxable income as:

Gross income = Sales revenue − Operating cost

Taxable income = Gross income − Depreciation expense

For the given data, the straight-line annual depreciation charge[15] is found in cell B5.

	A	B	C	D
1	Sales Revenue	$782,552.00		
2	Operating Cost	$458,760.00		
3	Gross Income	$323,792.00	=B1-B2	
4				
5	Depr. Exp	$35,000.00		
6	$125,000-20,000/3			
7				
8	Taxable Income	$288,792.00	=B3-B5	
9				
10		Income:	Tax Rates,%:	
11		$50,000	0.1	$5,000.00
12	(first)	$25,000	0.15	$3,750.00
13	(second)	$25,000	0.2	$5,000.00
14	Remaining	$188,792	0.25	$47,198.00
15		Income Tax Due:		$60,948.00

This taxable income can be considered divided into its components to match the cut-off limits of the various tax brackets, as given in the tax schedule. Thus, $288,792 can be considered to be the result of[16] $50,000 + $25,000 + $25,000 + $188,792. The income tax due is the sum of cells D11:D14.

In the previous two sections, we discussed how to calculate income tax. While these skills are useful to engineers and technologists, the tasks of figuring out income taxes and timing the investments (whether to invest this year or next year) to reduce the tax obligations are so complex that specialists, such as tax attorneys and chartered public accountants (CPA), are hired or retained for the purpose. Engineers and technologists are usually concerned only with analyzing the impact of taxes on economic decisions.

11.4 INCREMENTAL TAX RATE

The concept of *incremental tax rate* facilitates the analysis of tax implications. Incremental tax rate is simply the tax rate that is applicable to an *incremental (additional)* income. Consider the following tax schedule.

Taxable Income	Rate
<$ 50,000	10%
$ 50,000–$ 75,000	20%
$ 75,000–$100,000	30%
>$100,000	40%

---

[15] Refer to Chapter 10 if you have difficulty following it.
[16] See Example 11.1 if necessary.

As per this schedule, for a company with taxable income of $60,000, the tax rate for the first $50,000 is 10%, and for the remaining $10,000 it is 20%. Assume now that this company can increase its gross income by $8,000 by investing in new tooling. While carrying out an economic analysis, what is important is not the amount of tax, but the tax implication of the *additional* $8,000 gross income on the decision. Assume that this additional income yields a $6,000 profit after considering the operating costs. Due to depreciation charge this profit reduces to, let us say, a taxable income of $5,000. The total taxable income will thus be $60,000 + $5,000 = $65,000, which is still in the 20% bracket. Therefore, the additional taxable income of $5,000 resulting from investment[17] in the tooling will be taxed at 20%. This effect of investment on income tax will be stated as: the *incremental tax rate* is 20%. It is this rate that is relevant in economic analyses.

Now, consider this: What if the current $60,000 income could be increased by a taxable income of $20,000 by investing in the deluxe model of the tooling? With the total taxable income becoming $80,000 ($60,000 plus $20,000), the company is in the 30% tax bracket, as evident from the above schedule. Since the lower cut-off income for 30% is $75,000, the $20,000 increment in taxable income could be considered as the total of the two components $15,000 and $5,000. The $15,000 component will raise the current $60,000 income to $75,000—the upper cut-off limit of the 20% tax bracket, and hence this $15,000 being in the same bracket will be taxed at 20%. The remaining $5,000 will, however, be taxed at the next higher rate of 30% since this component lies in that bracket. Thus, of the $20,000 additional taxable income from investment in the deluxe model, for the first $15,000 the *incremental tax rate* is 20%, and for the remaining $5,000 it is 30%. Note that two or more incremental tax rates arise only when the additional income pushes the total income to higher tax brackets.

Engineers and technologists need to know the value of incremental tax rate(s) applicable to a project. The concept explained in this section helps them in selecting the appropriate rate(s). Example 11.5 further illustrates how the incremental tax rate is assessed.

Example 11.5: Incremental Tax Rates

Kristi owns and operates a job shop that is likely to generate this year a taxable income of $95,000. She is considering investing in a personal computer that is expected to increase the taxable annual income by $15,000. What incremental tax rate or rates are appropriate for use in decision-making pertaining to this investment under the following schedule? How much additional income tax will be payable on the earnings from the investment?

Taxable Income	Rate
<$ 50,000	10%
$ 50,000–$ 75,000	20%
$ 75,000–$100,000	30%
>$100,000	40%

SOLUTION:

With the expected $95,000 taxable income being in the $75k–$100k[18] range, the job shop's tax rate in the absence of investment will be 30%. With the additional $15,000 taxable income from investment in the PC, its taxable income will be $95,000 + $15,000 = $110,000. With this income, the tax liability shifts into the next higher bracket of 40%.

---

[17] Tax obligation = 0.20 × $5,000 = $1,000. The after-tax return from the investment is thus $5,000 – $1,000 = $4,000. Note that a gross income of $8,000 was finally reduced by the operating costs, depreciation, and the tax obligation to a $4,000 after-tax cashflow.

[18] k is an abbreviation for kilo, meaning 1,000. Thus, $100k means $100,000.

But not all of the additional income will be taxed at 40%. Of the $15,000, the first $5,000 shifts the current $95k income to $100k—the upper limit of the 30% bracket. Hence this $5,000 is taxable still at 30%. The remaining $10,000 of the additional income falls in the higher bracket of 40%. Thus, an incremental tax rate of 30% is applicable to the first $5,000, while 40% to the remainder $10,000.

$$\text{Additional income tax payable} = 30\% \text{ of } \$5,000 + 40\% \text{ of } \$10,000$$

$$= \$1,500 + \$4,000$$

$$= \$5,500$$

Example 11.6: Incremental Tax Rates

Sheri's business is expected to generate a taxable income of $60,000. Her choices for income tax purposes are:

(a) Run the business as a sole proprietor and pay income tax jointly with her spouse at 40% incremental rate, or
(b) Incorporate the business and pay income tax as a C corporation for which the tax rates are:

Taxable Income	Rate
<$ 50,000	15%
Next $ 25,000	24%
Next $ 25,000	34%
Next $235,000	39%
Over $335,000	34%

From the tax obligation viewpoint, which choice is better?

SOLUTION:

As a joint income her $60,000 earnings will be taxed at the 40% incremental rate. Thus, Sheri and her spouse will together pay an additional tax = 40% of $60,000 = $24,000.

As a C corporation, referring to the tax schedule given, her $60,000 income can be considered divided into two components of $50,000 and $10,000. She would pay 15% on the $50,000 component and 24% on the $10,000. Thus,

$$\text{Tax} = 15\% \text{ of } \$50,000 + 24\% \text{ of } \$10,000$$

$$= \$7,500 + \$2,400$$

$$= \$9,900$$

She is better off incorporating, saving $24,000 − $9,900 = $14,100 in income tax.

11.5 AFTER-TAX ANALYSES

Accounting for the obligation of income tax arising from engineering investment is called *after-tax* economic analysis. Such analyses can be carried out for any of the six methods discussed in Chapters 5 through 8, and involve depreciation considerations discussed in

Chapter 10. The tax implications arising out of depreciation make the analysis complex, as discussed later in Section 11.5.2.

11.5.1 Non-Depreciable Case

For simplicity we first consider the case that does not involve depreciation. The analyst should modify the given before-tax cashflows as per the applicable incremental rate(s). The resulting after-tax cashflows are then analyzed following any of the procedures discussed in Part II of the text. In general, the following five steps are involved:

1. Gather the given (before-tax) data in a cashflow table.
2. Find out what the applicable incremental tax rate is.
3. Determine the tax payable each year.
4. Deduct the yearly tax from the before-tax (gross) cashflow to obtain after-tax (net) cashflows.
5. Analyze the after-tax cashflows under the criterion.

To minimize confusion, it is always better to implement steps 3, 4, and 5 by adding columns to the original (before-tax) cashflow table, as illustrated in Example 11.7.

Example 11.7: After-Tax Analysis

A family-owned specialty store operates ten-to-six, five days per week. Due to rising healthcare costs, the owner is in financial difficulty. One way out is to augment the income by keeping the store open on Saturdays too, which is expected to result in the following gross incomes:

Year	1	2	3	4	5
Income	$80,000	$92,000	$106,000	$112,000	$114,000

Determine the store's after-tax yearly incomes if the tax schedule is:

Taxable Income	Rate
<$ 50,000	10%
$ 50,000–$ 75,000	20%
$ 75,000–$100,000	30%
>$100,000	40%

SOLUTION:

Follow the five-step procedure outlined in Subsection 11.5.1. Since the before-tax cashflow table is already given, step 1 is complete. The given tax schedule contains information about the incremental tax rates, so step 2 is also complete. Next, we determine the tax[19] for each year.

---

[19] Note that we are assuming that these rates will not change during the next 5 years.

Let us consider year 1 first. Referring to the tax schedule, the $80,000 income falls in the third bracket. This income can be divided into three components[20] of $50,000, $25,000, and $5,000. This will result in

$$\text{Year} - 1\,\text{tax} = 10\% \text{ of } \$50,000 + 20\% \text{ of } \$25,000 + 30\% \text{ of } \$5,000$$

$$= \$5,000 + \$5,000 + \$1,500$$

$$= \$11,500$$

One can find the taxes for the other years in a similar way. But such a process will prove to be lengthy.

An efficient way is to apply the incremental tax rate on additional incomes, while being careful about the applicable rate. For example, the $92,000 income for year 2 can be looked upon as an incremental income of $12,000 over that for year 1 ($80,000). With the year-2 income of $92,000 being less than $100,000 (the upper cut-off for the third bracket), the tax rate remains at 30%. Thus, the incremental tax rate applicable to the additional $12,000 income is 30%. This gives an incremental tax of $0.3 \times \$12,000 = \$3,600$. Thus,

$$\text{Year} - 2\,\text{tax} = \text{Year} - 1\,\text{tax} + \text{Incremental tax for year 2}$$

$$= \$11,500 + \$3,600$$

$$= \$15,100$$

Other years' taxes can be determined the same way[21]. To keep track of the yearly taxes, add one more column to the given cashflow table as:

	A	B	C
1	Year	Before-Tax Income	Tax Payable[22]
2	1	$80,000	$11,500
3	2	$92,000	$15,100
4	3	$106,000	$19,900
5	4	$112,000	$22,300
6	5	$114,000	$23,100

The fourth step involves considering the effect of taxes on before-tax incomes. In this case, since the taxes are payable, they reduce the incomes. So subtract them from the corresponding incomes. Tabulate the results in another column, as follows.

	A	B	C	D	E
1	Year	Before-Tax Income	Tax Payable[22]	After-Tax Income	
2	1	$80,000	$11,500	$68,500	=B2-C2
3	2	$92,000	$15,100	$76,900	=B3-C3
4	3	$106,000	$19,900	$86,100	=B4-C4
5	4	$112,000	$22,300	$89,700	=B5-C5
6	5	$114,000	$23,100	$90,900	=B6-C6

Note that though the before-tax income for the fifth year has increased by $34,000 over that for the first year, the after-tax income has increased only by $22,400 ($90,900

---

[20] Based on the upper income limit of the tax brackets; see Example 11.1, if necessary.
[21] The incremental procedure may look to be clumsy, but with practice you will find it easier than the straightforward approach used for year 1.

– $68,500). This illustrates how taxes can distort the cashflows, and therefore must be accounted for in any economic analysis.

The fifth step of the procedure, namely analysis of the after-tax cashflows, is redundant here since no analysis is being asked for.

11.5.2 Depreciable Case

The procedure hereunder must be followed if depreciation data are given, even if the problem does not explicitly ask for the consideration of depreciation. When depreciation is to be accounted for, the before-tax cashflows should be modified *first* for depreciation. The resulting *after-depreciation incomes* are taxable, and hence the basis of the income tax obligation. The complete procedure involves the following six steps:

1. Gather the given before-tax data in a cashflow table.
2. Determine the depreciation charges and tabulate them.
3. Apply the depreciation charges to before-tax cashflows.
4. Determine taxes based on after-depreciation cashflows.
5. Apply the taxes to the before-tax cashflows.
6. Analyze the resulting after-tax cashflows.

Again, implement the steps by adding columns to the original cashflow table, as illustrated in Example 11.8.

Example 11.8: Depreciable Case

An investment of $3,900 on a special tooling is expected to result in savings of $1,000 during the first year, $1,500 during the second year, and $2,000 during the third year. At the end of the third year the tooling will be salvaged for $300. Is this a good investment considering income tax obligations if present worth (PW) is the decision criterion? Assume straight-line depreciation, 60% incremental tax rate, and 8% MARR.

SOLUTION:

The savings represent a reduction in costs which translate into increased incomes. The before-tax cashflows as given are:

Year	0	1	2	3	Salvage Value
Cashflow	–$3,900	$1,000	$1,500	$2,000	$300

The above was step 1. In step 2, depreciation charges are determined. Following the straight-line method (refer to Chapter 10, if necessary), for the given data: P = $3,900, S = $300, and N = 3 years,

$$\text{Depreciation charge} = (P - S)/N$$

$$= (\$3,900 - \$300)/3$$

$$= \$1,200$$

This yearly depreciation is charged to the before-tax incomes (step 3). Thus, the after-depreciation cashflows, which are taxable incomes[22], will be lower, as follows:

	A	B	C	D	E
1	Year	Before-Tax Cashflow	Depr. Exp	Taxable Income	
2	0	-$3,900			
3	1	$1,000	$1,200	-$200	=B3-C3
4	2	$1,500	$1,200	$300	=B4-C4
5	3	$2,000	$1,200	$800	=B5-C5
6	Salvage	$300			

Next, as step 4, we determine the tax obligations for each year. The incremental tax rate is given to be 60%. The calculated taxes are tabulated in yet another column, as follows[23].

	A	B	C	D	E	F
1	Year	Before-Tax Cashflow	Depr. Exp	Taxable Income	Tax @ 60%	
2	0	-$3,900				
3	1	$1,000	$1,200	-$200	-$120.00	=D3*0.6
4	2	$1,500	$1,200	$300	$180.00	=D4*0.6
5	3	$2,000	$1,200	$800	$480.00	=D5*0.6
6	Salvage	$300				

As step 5, the effect of taxes on incomes (before-tax cashflows) is accounted for, generating after-tax cashflows, as follows, in another column.

	A	B	C	D	E	F	G
1	Year	Before-Tax Cashflow	Depr. Exp	Taxable Income	Tax @ 60%	After-Tax Cashflow	
2	0	-$3,900				-$3,900	=B2-E2
3	1	$1,000	$1,200	-$200	-$120.00	$1,120	=B3-E3
4	2	$1,500	$1,200	$300	$180.00	$1,320	=B4-E4
5	3	$2,000	$1,200	$800	$480.00	$1,520	=B5-E5
6	Salvage	$300					

The after-tax cashflows in the last column have thus accounted for the tax benefits of asset depreciation. As the final (sixth) step, we analyze them for their PW to judge the project's investment worthiness.

$$PW = -3,900 + 1,120(P/F,8\%,1) + 1,320(P/F,8\%,2) + 1,820(P/F,8\%,3)$$

$$= -3,900 + 1,120 \times 0.9259 + 1,320 \times 0.8573 + 1,820 \times 0.7938$$

$$= -3,900 + 1,037 + 1,132 + 1,445$$

$$= -\$286$$

Since the PW of the after-tax cashflows is –ve, the investment in the special tooling is not recommended.

---

[22] We are assuming that the before-tax cashflows represent net savings (incomes) that have already taken the operating costs into account.

[23] In the tables in this example, the first row corresponding to year 0 and the salvage value can initially be excluded, if found confusing. The cashflow in this row represents the investment—cost incurred now—which is not affected by depreciation or income tax. Likewise, salvage value remains unaffected by depreciation and tax considerations. However, these must be added later to complete the after-tax cashflow table, just prior to applying the decision criterion (PW in this case).

After-tax analyses render economic decisions more realistic and reliable. Had we ignored the tax obligation in Example 11.8, we would have erred[24] by investing in the special tooling.

11.6 AFTER-TAX ROR ANALYSIS[25]

Though the after-tax analysis can be carried out under any of the decision criteria discussed in Part II, it is common with the ROR.

The procedure for after-tax ROR analysis is very similar to that explained previously in Subsection 11.5.2. The underlying principle is a two-part process. First, modify the before-tax cashflows to account for depreciation and the associated income tax. This yields after-tax cashflows. Second, analyze the after-tax cashflows to judge the project's worthiness based on the rate of return. The ROR analysis follows the usual procedure discussed in Chapter 7, as illustrated in Example 11.9.

Example 11.9: ROR Analysis

Woman Wardrobe, a garment manufacturer, plans to buy an old truck for business use for $13,000. The truck is expected to increase company earnings annually (through savings) by $4,000. The company will sell the truck for an estimated price of $7,000 at the end of 3 years. Assuming straight-line depreciation, should the company invest in the truck if the incremental income tax rate is 40% and MARR is 12%?

SOLUTION:

Note that in the third year there will be two cash inflows; one is the $4,000 annual benefit and the other is the $7,000 salvage. The before-tax cashflow table for the project is:

Year	0	1	2	3	Salvage Value
Cashflow	−$13,000	$4,000	$4,000	$4,000	$7,000

Though the criterion of decision-making is not explicit, there is a hint to use ROR-criterion since MARR is given. With a given MARR, we can avoid having to evaluate ROR exactly. We can instead decide on the basis of PW's sign corresponding to MARR.

We first determine the after-tax cashflows by accounting for depreciation and the associated income tax. Let us follow the six-step procedure explained earlier in Section 11.5.2

---

[24] Had the tax obligations not been considered in Example 13.8, PW would have been +ve, as shown as follows, recommending investment in the tooling.

$$PW = -3,900 + 1,000(P / F, 8\%, 1) + 1,500(P / F, 8\%, 2)$$

$$+2,300(P / F, 8\%, 3)$$

$$= -3,900 + 1,000 \times 0.9259 + 1,500 \times 0.8573 + 2,300 \times 0.7938$$

$$= -3,900 + 926 + 1,286 + 1,826$$

$$= \$138$$

[25] Though we discuss only the ROR-analysis in this section, similar procedure is applicable to analyses under on any of the other decision criteria covered in Part II.

and applied in Example 11.8. We begin by determining the yearly straight-line depreciation charge as

$$\text{Depreciation charge} = (P - S) / N$$

$$= (\$13,000 - \$7,000) / 3$$

$$= \$2,000$$

The analyses for depreciation and taxes yield the following[26] table (see Example 11.8 for details on how to carry out the calculations).

	A	B	C	D	E	F
1	Year	Before-Tax Cashflow	Depr. Exp	Taxable Income	Tax @ 40%	After-Tax Cashflow
2	0	-$13,000				-$13,000
3	1	$4,000	$2,000	$2,000	$800	$3,200
4	2	$4,000	$2,000	$2,000	$800	$3,200
5	3	$4,000	$2,000	$2,000	$800	$3,200
6	Salvage	$7,000				$7,000

Now, determine[27] the PW of the after-tax cashflows corresponding to 12% MARR, which is

$$PW = -13,000 + 3,200(P / A, 12\%, 3) + 7,000(P / F, 12\%, 3)$$

$$= -13,000 + 3,200 \times 2.402 + 7,000 \times 0.7118$$

$$= -13,000 + 7,686 + 4,983$$

$$= -\$331$$

Since the PW is negative, after-tax ROR will be lower than the 12% MARR. Therefore, the company should not invest[28] in the old truck.

---

[26] Some analysts post the payable tax data (column 5 in the table) with a negative sign, considering them as costs. For year 1, their entry will be –$800, instead of $800. The signed tax data (column 5) is added algebraically to the before-tax data (column 2) to get the after-tax data (column 6). While this concept is mathematically sounder, we practice in this text what is easier to comprehend. We post the payable taxes with no sign (i.e., a +ve sign) and subtract it from the before-tax data. That taxes are paid out and should therefore be subtracted from the income which is easier to understand. Situations may arise where tax implications of an investment are beneficial, for example when depreciation cost exceeds the before-tax income (Example 11.8 data for year 1). In such cases we sign the income tax as negative, meaning a refund rather than payment. When subtracted from the before-tax cashflow the negative tax renders the after-tax cashflow larger than the before-tax cashflow.

[27] It may be better to keep the salvage value separate from the last period cashflow. This avoids charging any tax on the salvage value. Moreover it can keep the evaluation simpler. In here, for example, the evaluation of PW is simplified by considering the $3,200 after-tax data for period 3 with the annual cash inflows.

[28] Were the income-tax implications not considered, PW would have been

$$PW = -13,000 + 4,000(P / A, 12\%, 3) + 7,000(P / F, 12\%, 3)$$

$$= -13,000 + 4,000 \times 2.402 + 7,000 \times 0.7118$$

$$= -13,000 + 9,608 + 4,983$$

$$= \$1,591$$

The +ve PW would have led the engineer to invest in the truck—obviously an erroneous decision. After-tax analysis may decline funding a project that would have been approved otherwise. This is so because taxation usually reduces the benefits, while the first cost remains unchanged.

11.7 COMBINED INCOME TAX

Some state governments[29] also charge income tax from individuals as well as from businesses. In such cases the implications of both state and federal income taxes should be considered in the analysis. The procedures discussed in the earlier sections, including the concept of incremental tax rate, remain applicable when taxes are due to both federal and state governments. The simplest approach is to create another column in the cashflow table and account for state tax as we did for the federal tax. However, the usual practice is to determine a *combined incremental tax rate* that takes both the taxes into account, and use this rate in the analysis.

The combined incremental tax rate is based on the common practice that the federal government allows state income tax as an itemized deduction. Under this practice,

$$\text{Combined incremental tax rate} = r_s + r_f\left(1 - r_s\right)$$

where
r_s = *state incremental tax rate*
r_f = *federal incremental tax rate*

The basis of this relationship[30] can easily be explained. Since the state income tax is deducted from the taxable income for federal tax purposes, the factor $(1 - r_s)$ normalizes the federal tax obligation from rate r_f to $r_f(1 - r_s)$. The normalized rate is then added to the state rate r_s to yield the combined incremental tax rate, as illustrated in Example 11.10.

Example 11.10: Combined Income Tax

RST Company operates in a state that charges no income tax. In investment analyses the company uses an incremental rate of 40% for federal income tax. It is considering moving its operations to a neighboring state which charges income tax. If the incremental rate for state income tax is 10%, what will the combined rate be to include both federal and state income tax obligations?

SOLUTION:

Given, $r_s = 10\% = 0.10$
$r_f = 40\% = 0.40$

$$\text{Combined income tax rate} = r_s + r_f\left(1 - r_s\right)$$

$$= 0.10 + 0.40\left(1 - 0.10\right)$$

$$= 0.10 + 0.36$$

$$= 0.46$$

$$= 46\%$$

Note that the combined rate of 46% is higher than the federal rate (40%), but less than the algebraic sum of the federal and state tax rates (40% + 10% = 50%).

---

[29] In the U.S., even some city governments levy income tax.
[30] For every dollar of taxable income, only (1 – rs) dollar is taxed at federal rate rf. Thus, the federal tax is rf(1 – rs) while the state tax is rs. This yields a total tax of rs + rf(1 – rs) for every taxable dollar.

If the state tax rate is negligible in comparison to the federal tax rate, one can ignore the effect of state income tax on economic analysis. This is obvious in the above relationship, where the combined tax rate $r_s + r_f(1 - r_s)$ reduces to r_f when r_s is zero. As an illustration, if the state rate is 3% and the federal rate is 50%, the combined rate from the above relationship works out to be 51.5% which is close to the 50% federal rate. Ignoring the state tax implications in such a case may be acceptable.

Example 11.11: Combined Income Tax

The RST company of Example 11.10 is considering a project whose data are in Example 11.9. Is the project worthy of investment?

SOLUTION:

Given, $r_s = 10\% = 0.10$
$\quad r_f = 40\% = 0.40$

Since the state rate is not negligible in comparison to the federal rate, consider them both. From Example 11.10, the combined income tax rate is 46%. Redo Example 11.9 with 46% tax rate. The results are summarized below.

	A	B	C	D	E	F
1	Year	Before-Tax Cashflow	Depr. Exp	Taxable Income	Tax @ 46%	After-Tax Cashflow
2	0	-$13,000				-$13,000
3	1	$4,000	$2,000	$2,000	$920	$3,080
4	2	$4,000	$2,000	$2,000	$920	$3,080
5	3	$4,000	$2,000	$2,000	$920	$3,080
6	Salvage	$7,000				$7,000

Thus, the present worth of the after-tax cashflows, considering both the federal and state taxes, is:

$$PW = -13,000 + 3,080(P/A, 12\%, 3) + 7,000(P/F, 12\%, 3)$$

$$= -13,000 + 3,080 \times 2.402 + 7,000 \times 0.7118$$

$$= -13,000 + 7,398 + 4,983$$

$$= -\$619$$

Since PW is –ve, the project is not worthy of investmentf[31].

11.8 OTHER CONSIDERATIONS

As mentioned earlier, a major role of governments throughout the world is to actively participate[32] in the national economy. The income tax, discussed so far in this chapter, is one

---

[31] The decision has not unchanged in this case. But in cases where the PW is marginally +ve, consideration of state income tax may render it –ve, thus changing the decision.

[32] Though the participation is good-intentioned, it may hurt the economy, rather than help it, if the government policies are not implemented judiciously. The U.S. is one of the few countries that seems to understand the danger of playing a role bigger than necessary.

of the modes of direct participation. Another one is through taxation on capital gains and write-offs on capital losses.

11.8.1 Capital Gain and Loss

Companies may earn a profit or incur loss when a capital asset is disposed off. If a profit is realized, the resulting *gain* is taxable. A capital gain occurs when the asset is sold for a price above its book value. Thus,

Capital gain = Selling price − Book value

The capital gain is treated as an income in the year it is realized. It is accounted for by considering its impact on the cashflow, as illustrated in Example 11.12.

Example 11.12: Capital Gains and Losses

If in Example 11.8, the tooling is sold for $1,500 at the end of its 3-year life, how much is the capital gain? What effect does this gain have on the net cashflows and what is the present worth?

SOLUTION:

Note that the capital consumed (= P-S) in the tooling investment has been recovered over the years through the annual $1,200 depreciation. Thus, the book value equals the $300 salvage value. Since the tooling is sold for $1,500—more than the book value—there is a capital gain.

Capital gain = $1,500 − $300

$$= \$1,200$$

We follow the same procedure as in Example 11.8, but account for this capital gain by posting it in the year it is realized, i.e., year 3. This has been done below by adding columns (5) and (6).

	A	B	C	D	E	F	G	H
1	Year	Before-Tax Cashflow	Depr. Exp	After Depr. Cashflow	Capital Gain	Taxable Income	Tax @ 60%	After-Tax Cashflow
2	0	-$3,900						-$3,900
3	1	$1,000	$1,200	-$200		-$200	-$120	$1,120
4	2	$1,500	$1,200	$300		$300	$180	$1,320
5	3	$2,000	$1,200	$800	$1,200	$2,000	$1,200	$2,000
6	Salvage	$300						$300

The effect of capital gain is reflected in the increased taxable income for the third year from $800 to $2,000. This has increased the tax for the year from $480 (Example 11.8) to $1,200; the $720 increase is 60% of the $1,200 capital gain. The net effect is the reduced third-year after-tax cashflow, from $1,520 to $800. Thus, the cashflows in the last column account for depreciation, capital gain, as well as income taxes.

Once the cashflows have been modified to account for all the applicable tax obligations, the net after-tax cashflows are analyzed the usual way. For this case,

$$PW = -3,900 + 1,120(P/F,8\%,1) + 1,320(P/F,8\%,2) + (800 + 300)(P/F,8\%,3)$$

$$= -3,900 + 1,120 \times 0.9259 + 1,320 \times 0.8573 + 1,100 \times 0.7938$$

$$= -3,900 + 1,037 + 1,132 + 873$$

$$= -\$858$$

Sometimes, a capital loss may occur when the asset is sold for less than its book value, i.e.,

Capital loss = Book value – Selling price

The capital loss is treated as a cost in the year it occurs, reducing the profit and hence the income tax. It is accounted for in the analysis by considering it accordingly, as illustrated in Example 11.13.

Example 11.13: Capital Loss

If in Example 11.8, the tooling is sold for $100 at the end of its 3-year life, what is the present worth of the project considering capital loss?

SOLUTION:

The tooling's book value is $300. Since it is sold for $100, there occurs a capital loss.

Capital loss = $300 – $100

$$= \$200$$

We follow the procedure of Example 11.8 and account for this loss by posting it for the year it occurred, i.e., for year 3. This has been done below by adding columns (5) and (6).

	A	B	C	D	E	F	G	H
		Before-Tax	Depr.	After Depr.	Capital	Taxable		After-Tax
1	Year	Cashflow	Exp	Cashflow	Loss	Income	Tax @ 60%	Cashflow
2	0	-$3,900						-$3,900
3	1	$1,000	$1,200	-$200		-$200	-$120	$1,120
4	2	$1,500	$1,200	$300		$300	$180	$1,320
5	3	$2,000	$1,200	$800	$200	$600	$360	$1,440
6	Salvage	$300						$300

The capital loss has decreased the third-year taxable income by $200. This has the effect of reduced income tax for that year from $480 (Example 11.8) to $360. The $120 decrease represents 60% of the $200 capital loss. The ultimate effect is increased third-year net cashflow in the last column. Once the cashflows have been modified to account for all the applicable tax implications, the net after-tax cashflows are analyzed the usual way. For this case,

$$PW = -3,900 + 1,120(P/F,8\%,1) + 1,320(P/F,8\%,2) + (1,640 + 300)(P/F,8\%,3)$$

$$= -3,900 + 1,120 \times 0.9259 + 1,320 \times 0.8573 + 1,940 \times 0.7938$$

$$= -3,900 + 1,037 + 1,132 + 1,540$$

$$= -\$191$$

Taxation rules governing capital gains and losses change continually, depending on the extent and frequency of government intervention in the economy. Engineers and technologists need to keep up-to-date[33] with these changes so as to incorporate them in economic analyses. The best information resource on the changes is usually the taxing authority and the company tax attorney or accountant.

---

[33] For example, as an incentive to small companies, the tax laws sometimes allow equipment within a certain cost limit—currently $17,500 in the U.S.—to be written off in the year of the acquisition. In other words, the entire cost can be depreciated in the first year.

11.8.2 Investment Credit

When there is a slowdown in the economy, governments encourage investment by offering industry incentives of various types. One such incentive is *investment tax credit*, which indirectly pays companies a certain percentage of the investment in equipment. Investment tax credits may vary from year to year; the relevant information can be obtained from the taxing authority or company attorney.

The investment tax credit must be claimed in the year the equipment is bought. It does not affect the initial cost of the equipment, and therefore depreciation charges remain unaffected. Again, the cashflows should be modified to include the effect of tax credit, as illustrated in Example 11.14.

Example 11.14: Investment Credit

If an investment tax credit of 10% is allowed in Example 11.9, how is the decision affected?

SOLUTION:

The overall procedure and the discussions of Example 11.9 remain applicable. The investment tax credit (= $1,300) posted in the following *modified* cashflow table in the row[34] for year 1 under column 6 affects the after-tax cashflow (column 7). This cashflow is higher by $1,300, from $3,200 to $4,500. Note that the investment tax credit does not affect depreciation.

	A	B	C	D	E	F	G	H
1	Year	Before-Tax Cashflow	Depr. Exp	Taxable Income	Tax @ 40%	Investment Credit	After-Tax Cashflow	
2	0	-$13,000					-$13,000	
3	1	$4,000	$2,000	$2,000	$800	$1,300	$4,500	=B3-E3+F3
4	2	$4,000	$2,000	$2,000	$800		$3,200	
5	3	$4,000	$2,000	$2,000	$800		$3,200	
6	Salvage	$7,000					$7,000	

The present worth with capital investment credit considered should be higher, since after-tax cash inflow has increased for year 1. As seen below, it is so (compared to Example 11.8). In fact, PW is now positive, rendering the project worthy of investment.

$$PW = -13{,}000 + 1{,}300 \left(P/F, 12\%, 1\right) + 3{,}200 \left(P/A, 12\%, 3\right) + 7{,}000 \left(P/F, 12\%, 3\right)$$

$$= -13{,}000 + 1{,}300 \times 0.8929 + 3{,}200 \times 2.402 + 7{,}000 \times 0.7118$$

$$= -13{,}000 + 1{,}161 + 7{,}686 + 4{,}983$$

$$= \$830$$

As seen in this example, the investment tax credit can turn an otherwise unworthy project into a worthy one. Engineers should be on the lookout for any investment opportunity offered by the government, and reassess the projects that were denied funding in the past.

---

[34] The investment tax credit is usually taken at the end of the year the asset is acquired.

11.9 SUMMARY

Income taxes and other tax obligations may have significant implications to engineering investment decisions. To boost economic growth, governments throughout the world use income tax and other incentives, such as capital gains and losses, and investment tax credits, to encourage industry to invest in new equipment. These incentives may vary from year to year. Engineers should keep in touch with company management which tracks such incentive programs. The company tax attorney and/or accountant are also useful resources on such matters.

Income taxes are easier to calculate based on the concept of incremental tax rate. The federal and state income tax rates can be considered jointly through a combined tax rate. The basic approach to accounting for taxes and tax incentives is to modify the original (before-tax) cashflows. Modifications are better carried out by widening the cashflow table. The modified after-tax cashflows are then analyzed the usual way under any of the decision criteria discussed in Part II of the text. Decisions can change when tax implications are considered. Most engineering economics analyses include depreciation, income taxes, and other factors that may affect the investment decision.

DISCUSSION QUESTIONS

11.1 Why should engineers be conversant with current laws on income tax and other government investment incentives?

11.2 What roles can company tax attorneys play in decision-making by engineers and technologists pertaining to engineering projects?

11.3 Discuss in not more than 100 words how engineering economic analysis is carried out when income tax is to be accounted for.

11.4 Does the investment tax credit affect depreciation charges? Explain.

11.5 If the investment tax credit is spread over the asset life, rather than taken in the year of investment, will it make the investment more attractive or less attractive? Explain your answer.

11.6 Explain the concept of incremental income tax rate.

11.7 Why is the combined income tax rate lower than the algebraic sum of the federal and state tax rates?

11.8 Would pre-tax cashflows or after-tax cashflows provide a more accurate representation of the performance of a company in a given period?

11.9 Why is the depreciation expense deducted from the before-tax income of a company rather than the after-tax income?

11.10 Explain the treatment of salvage value and its taxability.

11.11 Explain the treatment of an investment credit on the purchase of an asset.

11.12 How do taxes affect analysis of capital investment opportunities?

MULTIPLE-CHOICE QUESTIONS

11.13 Taxable income is usually _____ the adjusted gross income.
   a. greater than
   b. same as
   c. less than
   d. none of the above

11.14 In computing personal income tax, you can deduct the
    a. state income tax from the federal taxable income
    b. federal income tax from the state taxable income
    c. either a or b, whichever reduces your total income tax
    d. neither a nor b

11.15 While accounting for both depreciation and income tax, modify the cashflows
    a. first for income tax and then for depreciation
    b. first for depreciation and then for income tax
    c. either a or b; it does not matter
    d. either a or b, whichever reduces the tax obligation

11.16 Capital loss on an equipment, if allowed, must be accounted for in the year
    a. of the investment
    b. the asset undergoes its first major overhaul
    c. half-way through its useful life
    d. it is disposed off

11.17 The best resource for information on the current income tax rules, while analyzing an engineering project, is the
    a. county's property tax collector
    b. boss who is a technical manager
    c. company tax attorney
    d. last year's file on the subject

11.18 Incremental income tax rate is the
    a. increase this year in the tax rate over last year's
    b. difference between federal and state income tax rates
    c. difference in individual and corporation tax rates
    d. none of the above

11.19 Which of the following would not affect after-tax cashflows?
    a. loans received
    b. revenues
    c. expenses
    d. all of the above

11.20 One benefit of income tax credit for investments is:
    a. decrease in annual expenses
    b. increased revenues in given period
    c. rapid depreciation of assets
    d. none of the above

11.21 A capital loss on an asset is treated as which of the following?
    a. depreciation expense
    b. rare and infrequent liability
    c. revenue
    d. cost

11.22 Which of the following result in an increased taxable income?
    a. annual depreciation expense
    b. capital gain on investment
    c. capital loss on investment
    d. none of the above

11.23 Income tax is paid on:
    a. gross income
    b. gross income – depreciable expenditures
    c. operating cost less gross income
    d. all of the above

11.24 Indirect operating costs include:
  a. labor
  b. marketing
  c. maintenance
  d. all of the above

NUMERICAL PROBLEMS

11.25 Universal Robotics' gross income for the current year is $265,000. It has invested in a special hand gripper costing $15,000, whose estimated useful life and salvage values are 3 years and $3,000. Its operating expense for the year has been $70,000. Estimate the payable income tax if applicable tax rate schedule is:

Taxable Income	Tax Rate
Below $100k	30%
$100k–$150k	40%
$150k–$200k	50%
Above $200k	60%

11.26 In Problem 11.25, if Universal Robotics can increase its income by $5,000 by investing in a special tooling, what is the incremental tax rate?

11.27 A used truck is purchased for $12,000. It is estimated to have a useful life of 5 years during which MACRS-based depreciation is charged. The truck is expected to annually save the company $2,000 in material-handling costs. Assuming zero salvage value and a tax rate of 50%, is it a good investment? Assume MARR = 12%.

11.28 A company operating in Mississippi pays 5% state income tax. If its federal income tax rate is 50%, what is the combined income tax rate?

11.29 Sam's Salvage Store (3S) is considering buying a van for $35,000 so that it can offer free local delivery. The free delivery is likely to increase 3S's sale by $16,000 each year. The truck's annual operating cost will be $6,000. Assuming zero salvage value and a useful life of 7 years, is this investment sound on the basis of present worth? The company uses SL-depreciation and pays income tax at a combined incremental rate of 50%. Assume MARR = 8%.

11.30 If MARR is 10%, is the investment in Problem 11.29 sound on the basis of after-tax rate of return?

11.31 What is the expected after-tax ROR in Example 11.29?

11.32 For an initial investment of $100k in R&D, Fancy Products expects its new solar-powered hair dryer to bring in annual revenue of $30k for the next 10 years. The production and marketing costs will be $5k per year. Assuming 50% income tax rate and SL-depreciation, determine the after-tax ROR if salvage value is zero?

11.33 Sam's Snowmobiles is considering investing in an automatic tool changer that will reduce the setup time on one of its CNC mills. The investment cost and the resulting benefits are summarized in the following cashflow table. Note that the $300 cashflow for year 5 includes a salvage value of $200. If the company practices DDB depreciation and its income tax rate is 45%, should the project be funded? Assume MARR = 10%.

Year	Before-Tax Cashflow
0	$1,500
1	$ 750
2	$ 500
3	$ 350
4	$ 150
5	$ 350

11.34 Precision Machinists believes in keeping its resources technologically updated. It is considering investing $250,000 in a new machining cell which at the end of its 5-year useful life is likely to fetch 20% of the initial cost. The benefits from the cell are estimated to be $50k in the first year, $70k in the second year, and $90k in each of the remaining 3 years. The cell is eligible for 10% investment tax credit. Any capital gain is taxed, or loss tax-credited, at a rate of 25%. For a 10% MARR and 50% tax rate, is the investment desirable?

11.35 An asset valued at $139,000 is depreciated over 8 years using the straight-line method. If it is sold for $98,000 at the end of year 2, what is the gain or loss on the transaction?

11.36 UnaFanta bottling company built a manufacturing plant in Botswana in 2006 at a cost of $4 million. The plant has a capacity of 20 million cases per year. Make the following assumptions: (1) the plant in classified as a 15-year property according to MACRS. Prepare an after-tax diagram of the investment over a 5-year period if there is no salvage value at the end of the useful life. Starting in year 0, the sales are expected to be 400,000 at $6/case, which is expected to grow by 5% per year. Variable costs are expected to be $3/case. The fixed costs are $250,000 per year. The effective tax rate is 35% and all production is sold.

11.37 Max Q owns "Mad Max" Dry Cleaners as well as a number of other investments. Last year the cleaners grossed $95,000. From his other investments Max made an additional $130,000. What is the total taxable income Max Q earned last year?

11.38 Jessie runs a successful outdoor recreation company in the Northwest. Last year her business income was $200,000. The rent, utilities, and other expenses of her business equaled $75,000. She made an additional $7,500 in dividends and interest earnings. How much income tax is payable if Jessie is single with no dependents? Refer to Table 11.1 for relevant data.

11.39 The Big Black River Brewery's sales revenue for last year was $375,000, while its operating costs were $237,000. The microbrewery was able to procure an antique brew kettle for traditional brewing purposes at a cost of $56,000. The useful life of the kettle is expected to be 8 years with a salvage value of $13,000. For straight-line depreciation, how much income tax is due under the following schedule?

Taxable Income	Rate
<$ 50,000	20%
$ 50,000–$ 75,000	30%
$ 75,000–$100,000	40%
>$100,000	50%

11.40 Now suppose that the Big Black River Brewery is considering expanding the production capabilities of their plant by adding a couple of new delivery trucks. This decision would boost their revenue by $45,000 per year. What additional income tax will be due on the earnings of the investment? Refer to Problem 11.39 as needed.

11.41 Craig runs an oil consulting firm that is expected to generate $95,000 in taxable income. Craig wants to save as much money on taxes as possible. He can either file jointly with his wife ($15,000 taxable income) or pay income taxes as an S Corporation for which the tax rates are found in the following table. From the tax obligation viewpoint, which choice is better? Refer to Table 11.2.

Taxable Income	Rate
<$ 50,000	15%
$ 50,000–$75,000	24%
$ 75,000–$100,000	29%
>$100,000	35%

11.42 Due to a tremendous spike in sales of a new stylish bathrobe design, the Super Sweet Robe Company is considering operating six days a week instead of the traditional five. The additional operating shift is expected to result in the following gross incomes:

Year	Before-Tax Cashflow
0	$60,000
1	$75,000
2	$35,000
3	$55,000

Estimate the company's extra tax obligation on the additional revenue if the tax schedule is:

Taxable Income	Rate
<$ 50,000	15%
$ 50,000–$ 75,000	24%
$ 75,000–$100,000	29%
>$100,000	35%

11.43 WorldWide Logistics company plans to purchase five new trailers for $80,000. These new state of the art trailers are expected to increase revenues by $50,000 per year and require $10,000 in operating expense. The trailers will be sold after 6 years for $15,000. Assuming straight-line depreciation, should the trailers be purchased if the MARR is 8%?

11.44 If an investment tax credit of 12% of purchase price for the first 2 years after the purchase is allowed for Problem 11.43, how is the decision affected?

FE EXAM PREP QUESTIONS

11.45 If 10% of taxes are paid on the first $75,000, 15% on the next $100,000, and 20% on the next $100,000; the amount of taxes paid on $200,000 of profits is closest to:
a. $ 45,000
b. $ 27,500
c. $108,000
d. $ 85,000

11.46 If the tax rate is a flat rate of 25% on profits and a company generates $500,000 of revenues, $200,000 of operating expenses, and $45,000 of depreciation, the amount of taxes paid is:
   a. $63,750
   b. $55,000
   c. $0
   d. none of the above

11.47 If 10% of taxes are paid on the first $75,000, 15% on $75,000–$150,000, and 20% on $150,000–$250,000; the effective tax rate for $190,000 is closest to:
   a. 14%
   b. 12%
   c. 15%
   d. none of the above

11.48 If the MARR (used when evaluating projects) is 12% per year for before-tax calculations and the effective tax rate is 34%, then the after-tax MARR to be used for analysis is closest to:
   a. 21%
   b. 22%
   c. 12%
   d. none of the above

11.49 Gross income is equal to:
   a. sales revenue + depreciation expense
   b. total income + depreciation expense
   c. total income-allowable deductions
   d. sales revenue-operating expense

11.50 Which of the following result in a decreased taxable income?
   a. annual depreciation expense
   b. investment tax credit
   c. capital loss on investment
   d. all of the above

11.51 Capital gain on equipment, if allowed, must be accounted for in the year
   a. of the investment
   b. the asset undergoes its first major overhaul
   c. at the time the gain is realized
   d. it is disposed off

11.52 If the tax rate is a flat rate of 20% on profits and a company generates $250,000 of revenues, $70,000 of operating expenses, and $25,000 of depreciation, the amount of taxes paid is:
   a. $15,000
   b. $25,550
   c. $31,000
   d. none of the above

11.53 A $64,000 asset is depreciated over 6 years using the straight-line method. If it is sold for $17,500 at the end of year 4, what is the gain or loss on the transaction?
   a. –$10,000
   b. $ 8,000
   c. –$ 3,832
   d. $ 4,572

11.54 If a company is in the 35% tax bracket on profits and has $450,000 in revenue, $80,000 in expenses, and $45,000 in depreciation, the after-tax cashflow is:
  a. $129,546
  b. $113,750
  c. $ 76,452
  d. none of the above

Chapter 12

Replacement Analysis

LEARNING OBJECTIVES

- Defender-challenger concept
- Analysis of the defender
- Defender's economic worth
- Remaining life of the defender
- Analysis of the challenger
- Challenger's economic life
- EUAC-based analysis
- Marginal-cost-based analysis
- MAPI method

In the economic analyses discussed in the text, we implicitly assumed a "green field" scenario. Under such a scenario, the plant is new and resources are procured for the first time. In other words, we assumed that the resources did not exist. While this scenario prevails occasionally, engineers and technologists often make economic decisions to replace resources that already exist and are in use. Such a decision is called replacement analysis.

In replacement analyses, the reference for comparison is the existing resource—a machine or equipment. The basic question is: When to replace the resource[1]? In fact, the more focused question is: Should we replace the equipment now (during the next fiscal year) or sometime later? The replacement analysis offers answers to such a question. While this question can be raised anytime, it is usually entertained once a year at the time of capital budgeting. Thus, the precise question is: *Should we budget now to replace the resource during the next fiscal year?*

12.1 DEFENDER-CHALLENGER CONCEPT

The existing resource is analyzed in relation to what the market has to offer as a replacement. If the market has nothing to offer, the replacement question does not arise. The existing resource is continued as long as its operational cost is acceptable, or else it is retired.

Replacement analysis can be conceptualized better by considering the existing resource as a *defender*—as if it were trying to defend its continued use. The one being considered to replace the defender is appropriately called a *challenger*. The replacement analysis can thus

---

[1] The term resource is used in a generic way to mean anything that is used in business such as machines, equipment, buildings, or tools.

be thought of as a "match" between the defender and the challenger, with the engineering economist as the referee.

The challenger is selected from among the alternatives by any of the methods discussed in Part II, for example present worth (Chapter 6) or rate-of-return (Chapter 7). Usually, the challenger is economically the best the market has to offer at the time of replacement. Engineering economists sometimes postpone the replacement decision to wait for a better challenger likely to be available in the near future.

A complete replacement analysis involves three tasks:

1. selection of the defender and its analysis
2. selection of the challenger and its analysis
3. defender-challenger comparison

We discuss these tasks, illustrating them through examples where appropriate, in the next five sections (12.2 through 12.6).

12.2 THE DEFENDER

At any time, there may be more machines needing replacement than the capital budget would permit. Companies usually look into replacement needs once a year as part of the budget process. Department managers request funds for replacing the troublesome resources. Although they may like to replace all the resources impeding productivity, the company's limited capital budget does not allow it. Budget limitations and the severity of productivity bottlenecks created by the failing resources largely dictate the selection of the defender(s). These bottlenecks are measured by the frequency of past breakdowns, repair costs, average repair time, mean time between failures, effect on product quality, and so on, which eventually impact the cost of production, and hence the profit.

Once a defender has been selected for replacement, the next relevant question is: What is its economic worth?

12.2.1 Economic Worth

Several types of costs or values are associated with a resource, as discussed below.

Initial Cost

The initial cost is the sum spent when the defender was acquired. It usually includes shipping, installation, and other *one-time costs* incurred at the time of procurement. It is also called *original cost* or *first cost*.

Trade-In Value

The trade-in value is the price offered for the defender when the challenger is being negotiated for acquisition. This value is usually "inflated," i.e., higher than the real value. You may be aware of the high price car dealers offer for an old car as trade-in.

Book Value

Since the defender has been in use for some time, depreciation may have been charged against it. The book value of an asset is the difference between the first cost and the cumulative depreciation. For example, if the first cost of a 3-year-old defender was $75,000, and the annual depreciation charge had been $15,000, then the book value at the end of third year of its life is $75,000 – (3 × $15,000) = $30,000.

Market Value

The defender's market value is what the defender will fetch if sold in the "open" market. It is based on free sale, with no "strings attached." If the dealer is willing to pay you $1,000 for the trade-in which is worth no more than $400 (market value), the "attached string" is represented by the profit they are trying to make on the new car you are buying. The price of the new car has certainly been marked up, at least by $600—the loss the dealer is ready to suffer on the trade-in.

Present Cost

The present cost of the defender is the current price of the same type and model if bought new. Due to inflation, it is usually higher than the defender's first cost. At times, it may be the same or even lower due to technological advances that reduce the price by more than the inflationary increase. For products undergoing rapid technological changes, such as personal computers, prices of new models may be lower even with enhanced capabilities.

Which of the aforementioned costs or values is the *real* economic worth of the defender, especially for the purpose of replacement analysis? The answer is: *market value*, in most cases[2].

The market value of an asset may at times be lower than its book value, the difference being the capital loss. Thus,

Capital loss = Book value – Market value

Oftentimes, a capital gain may be realized since the market value is more than the book value, where

Capital gain = Market value – Book value

Capital loss and other such losses are collectively called *sunk cost*. Analysts may be tempted to recover the sunk cost by including it in the replacement analysis. But that is erroneous. Economic analysis should never include the sunk cost (past losses); only the present and estimated future costs are relevant. Accordingly, *all sunk costs are ignored in economic analyses*. Though no attempt should be made to recover the sunk cost, the capital loss may be partially recovered through reduced taxes (see Chapter 11).

Example 12.1: Economic Analysis

A replacement analysis of *Bahadur*, a pick-and-place robot, purchased 3 years ago for $20,000 has been initiated by the job shop. At the time of purchase Bahadur's useful life

[2] The trade-in may at times be attractive. Consider that the market value of a defender is $3,000 and that the challenger costs $10,000 if bought in the open market. A distributor offers trade-in of $5,000, with the challenger at $11,500. Trading-in is thus preferable since the challenger's additional cost of $1,500 over the open market price is offset by the additional $2,000 on the trade-in. The *real* economic worth of the defender, however, remains the same—its market value of $3,000.

was estimated to be 5 years, with an expected salvage value of $5,000. During the past 3 years, SL-depreciation has been charged.

Due to recent innovations in microprocessor technology, a new robot is available in the market for $15,000. The robot manufacturer is offering $7,000 for the defender as trade-in. The job shop has researched to conclude that Bahadur can be sold in the open market for $5,000. Doing so will enable the job shop to buy the new versatile robot that will not only do Bahadur's job but also facilitate integration among the other computer-controlled machines of the plant. Determine Bahadur's

(a) book value,
(b) economic worth, and
(c) sunk cost

SOLUTION

Bahadur has already been in use for 3 years of its 5-year useful life. Let us assume that its analysis is being conducted during the end of the third year for possible replacement in the fourth year.

(a) The book value is the difference between the first cost and the cumulative depreciation over 3 years[3].

Annual straight-line depreciation

$$= (P - S) / N$$

$$= (\$20,000 - \$5,000) / 5$$

$$= \$3,000$$

Cumulative depreciation over 3 years

$$= \$3,000 \times 3$$
$$= \$9,000$$

Thus,

$$\text{Book value} = \text{First cost} - \text{Cumulative depreciation}$$

$$= \$20,000 - \$9,000$$

$$= \$11,000$$

(b) The economic worth of the defender is its market value, which is $5,000 (given).

(c) If the jobshop replaces Bahadur, it will suffer a capital loss since its market value is lower than the book value, given by

$$\text{Capital loss} = \text{Book value} - \text{Market value}$$

$$= \$11,000 - \$5,000$$

$$= \$6,000$$

This $6,000 loss in capital at the time of replacement is the sunk cost.

---

[3] It is being assumed that the third-year depreciation will be charged by the year-end.

12.2.2 Remaining Life

The defender being considered for replacement can in most cases be continued in use, at least theoretically[4], provided we do not mind its operational cost of repairs, breakdowns, productivity loss, etc. In reality, however, the operational cost becomes prohibitive.

How long a defender can be continued is decided upon usually on the basis of the cost of its use. This cost comprises two basic elements: capital cost and maintenance cost.

Capital Cost

Capital cost is the cost of current investment in the existing resource[5], which is based on its economic worth and the prevailing interest rate or MARR. By not selling off the defender, a capital equal to its market value remains "tied up." This realizable capital has the potential of use elsewhere in the company.

Maintenance Cost

The existing machine might break down, resulting in cost of repair and lost production. Even when it is operational, its age may contribute to higher scrap, rework, or customer returns. It also costs money for its usual supplies. All such costs, along with those on insurance, space, utilities, and others, are grouped together as *maintenance or operational cost*.

The cost to keep the existing resource in use is thus the total of the capital cost and maintenance cost, as

$$\text{Defender's Cost} = \text{Capital Cost} + \text{Maintenance Cost}$$

Noting that the time period in replacement analysis is usually year, these costs are considered annual unless mentioned otherwise. Example 12.2 illustrates the discussions of this subsection.

> ### Example 12.2: Economic Comparison
>
> Make an economic comparison between the following two options pertaining to Bahadur, the pick-and-place robot of Example 12.1:
>
> (a) Sell it off.
> (b) Keep it operational for one more year, during which its maintenance cost is estimated to be $200. The salvage value at the end of the next year is estimated to be $4,000. Assume MARR = 10%.
>
> ### SOLUTION
>
> (a) By selling the robot, the company can recover the "tied up" capital[6], which is Bahadur's market value of $5,000. This capital can be used elsewhere.

---

[4] The defender may become completely useless, for example what it was doing is no longer required, or it may become redundant due to technological breakthroughs. In such cases, the resource is simply *retired*, rather than replaced. The analysis for retirement or replacement is conceptually similar and hence most discussions in this chapter are valid for both. Section 12.7 briefly discusses retirement analysis.

[5] The terms *resource, machine*, and *equipment* are used synonymously.

[6] The term *capital* to denote the market value is appropriate since, once recovered by selling off the robot, this $5,000 becomes capital for the company. Note that the same sum may be described by different terms—market value, salvage value, or capital—depending on the context.

(b) The second option of keeping the robot operational for one more year will cost money, which is the total of capital and maintenance costs for the year. The maintenance cost is given to be $200. We need to determine the capital cost for the year.

By not selling off the robot, the job shop keeps $5,000 of its capital "tied up," which would have earned interest at the (given) 10% MARR. Thus, an interest income of $500 (10% of $5,000) is foregone for the year. Moreover, the robot's salvage value of $4,000 at the end of the next year is lower than this year's salvage value of $5,000. So, by not selling off this year, a loss of $1,000 in the value of the robot (capital) is also incurred. Hence, the cost of capital in keeping the robot in use for one more year

$$= \text{Loss in capital} + \text{Loss in interest income}$$

$$= \$1,000 + \$500$$

$$= \$1,500$$

Thus, the total cost of keeping the robot for one more year

$$= \text{Capital cost} + \text{Maintenance cost}$$

$$= \$1,500 + \$200$$

$$= \$1,700$$

A comparison of the two options shows that option (a) generates a capital of $5,000 now (this year) with the loss of Bahadur as a resource, while option (b) keeps Bahadur in use for a year at a cost of $1,700, and releases a capital of $4,000 at the end of the year as salvage value.

The $1,700 cost in option (b) can also be explained through the cashflow diagram in Figure 12.1. The upper diagram represents option (a). The selling-off of the robot generates a cash inflow of $5,000 now (year 0), shown by its upward vector. Option (b) is represented by the lower diagram, where not selling it this year is equivalent to an investment of $5,000. This is represented by the downward vector at year 0. The $200 downward vector at year 1 represents the maintenance cost for the year (following the period-end convention of cashflow posting), while the $4,000 upward vector denotes the cash inflow from salvaging the robot next year.

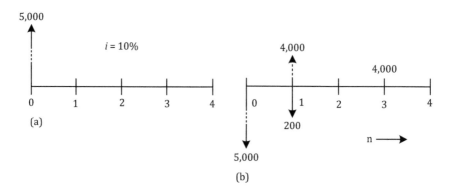

Figure 12.1 Cost of keeping the defender for the year.

Referring to the lower diagram for option (b), the $5,000 investment at year 0 is equivalent to its future value F at year 1 as

$$F = P(F/P, i, n)$$

$$= \$5,000(F/P, 10\%, 1)$$

$$= \$5,000 \times 1.1$$

$$= \$5,500$$

Thus, instead of the $5,000 vector at year 0, we can consider a downward vector of $5,500 (not shown) at year 1. The net effect of this transferred $5,500 vector and the other two vectors already existing at year 1 is (following the sign convention learned in Chapter 2)

$$= -\$5,500 - \$200 + \$4,000$$

$$= -\$1,700$$

Being negative, this $1,700 represents a cost at year 1. Thus, the cost of keeping the robot in use for a year[7] is $1,700.

We can determine the present worth of the $1,700 cost by transferring it to year 0 as:

$$P = \$1,700(P/F, 10\%, 1)$$

$$= \$1,700 \times 0.9091$$

$$= \$1,545$$

This $1,545 is the present value of the cost of keeping the robot for one more year. In contrast, option (a) results in present benefit of $5,000, but with no resource available for use.

The discussions in Example 12.2 have illustrated that while analyzing the defender for its cost implications (as well as other considerations of the replacement analysis) the time value of money must be accounted for. Rather than the asset's present or future worth of the costs, its annual worth is usually the basis of decision in replacement analysis. The *equivalent uniform annual cost* (EUAC)[8] serves this purpose. Example 12.3 illustrates the evaluation and role of the EUAC in analyzing a defender.

Example 12.3: EUAC

Determine the annual cost (EUAC) of operating the pick-and-place robot of Example 12.1 for each of the next 5 years. The year-end salvage values and the maintenance costs are:

Year	Year-End Salvage Value	Maintenance Cost
1	$4,000	$200
2	$3,500	$300
3	$3,000	$400
4	$2,000	$500
5	$1,000	$600

---

[7] Based on year-end cashflow convention, the $1,700 cost at year 1 represents the cost from now (year 0) until the end of the year (year 1).

[8] See Chapter 8 if you find EUAC unfamiliar.

SOLUTION:

The given data may be difficult to comprehend. What they mean, for example for year 1, is that if we keep the robot for one more year, then its maintenance cost for the year will be $200, and at the end of the year its salvage value will be $4,000.

On the other hand, the data for year 3 say that if we keep the robot for 3 years, then its maintenance costs will be $200 for year 1, $300 for year 2, and $400 for year 3, and at the end of the third year its salvage value will be $3,000. That means, as far as the data for year 3 are concerned, the $4,000 and $3,500 data in column 2 for years 1 and 2 are irrelevant.

Thus, the appropriate cashflow table, after posting the market value as year-0 cash outflow, for analyzing the robot to keep it for 3 years is:

Year	0	1	2	3	Salvage Value
Cashflow	−$5,000	−$200	−$300	−$400	3,000

Similarly, the cashflow table appropriate for the 5-year analysis is

Year	0	1	2	3	4	5	Salvage Value
Cashflow	−$5,000	−$200	−$300	−$400	−$500	−$600	$1,000

Thus, the given data in this example are concise, wherefrom the cashflow table appropriate to the particular analysis should be derived.

We begin by creating a modified table based on the previous discussions. The table will also contain results of the analysis for each year under the heading EUAC.

Remaining Life n (yrs)	Year-end Value	Maintenance Cost	EUAC
0	$5,000		
1	$4,000	$200	
2	$3,500	$300	
3	$3,000	$400	
4	$2,000	$500	
5	$1,000	$600	

Note that we have changed the title of column 1 to reflect more accurately the meaning of "year" by calling it the *remaining life* of the robot. The second column entitled *year-end value* denotes the salvage value at the end of the year, while the maintenance costs appear in column 3. Also note the introduction of a new row for year 0, containing the robot's current market value of $5,000. Note that this sum, and all the others under column 2, represents the investment, for the corresponding remaining life, in the defender if it is kept in use (not sold off).

We now analyze the data, following the procedure explained in Example 12.2, to determine the EUAC for each of the five options—keeping the robot for 1 year, keeping the robot for 2 years ... and keeping the robot for 5 years.

The calculations to determine[9] the EUAC can be approached by one of the following two ways.

Sketch and Analyze the Appropriate Cashflow Diagram

We illustrate this approach for the option of keeping the robot for 2 years, for which the appropriate cashflow diagram is Figure 12.2. In (a), the downward vector at year 0 represents the current investment. The maintenance costs are shown at the appropriate time

---

[9] Refer to Chapter 6 if you can't follow the discussions.

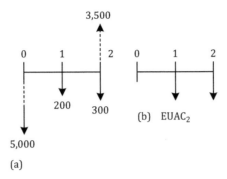

Figure 12.2   Diagram for Example 12.3.

periods, while the $3,500 salvage value is sketched as an upward vector at year 2 when the robot will be sold off.

Our task is to determine an equivalent uniform annual cost $EUAC_2$ for the diagram in (a). In other words, what value[10] of $EUAC_2$ in diagram (b) will render the two diagrams (a) and (b) equivalent. Thus,

$$EUAC_2 = 5,000(A/P,10\%,2)+200(P/F,10\%,1)(A/P,10\%,2)$$

$$-(3,500-300)(A/F,10\%,2)$$

$$= 5,000\times0.5762+200\times0.9091\times0.5762-3,200\times0.4762$$

$$= 2,881+105-1,524$$

$$= \$1,462$$

In a similar way the EUAC for the other four options[11] can be calculated[12] by referring to their cashflow diagrams, and the results can be summarized in the table as:

Remaining Life n (yrs)	Year-end Value	Maintenance Cost	EUAC
0	$5,000		
1	$4,000	$200	$1,700
2	$3,500	$300	$1,462
3	$3,000	$400	$1,398
4	$2,000	$500	$1,485
5	$1,000	$600	$1,536

---

[10] Note that the EUAC relates to 2 years only, not 5 years, since the plan is to keep the robot for 2 years only. For a 4-year analysis, i.e., the option of keeping the robot for 4 years, the EUAC will be spread over 4 years.

[11] A separate diagram is required for each option, with different timeline lengths. For example, for the option of keeping the robot for 4 years, there will be four periods in the diagram. While calculating the EUAC, exploit the arithmetic-series pattern in the maintenance costs. For example, for year 4, the most efficient solution is:

$$EUAC_4 = 5,000(A/P,10\%,4)+200+100(A/G,10\%,4)-2,000(A/F,10\%,4)$$

$$= 5,000\times0.3155+200+100\times1.381-2,000\times0.2155$$

$$= 1,578+200+138-431$$

$$= 1,485$$

[12] The value for year 1 is already worked out in Example 12.2.

What do the EUAC values mean in the table? They are simply the total annual cost for each of the five options (keeping for 1 year, keeping for 2 years ... keeping for 5 years). For example, if the robot is kept for 4 years, then the total cost of doing so will be $1,485 per year during the next 4 years. That means, for this option, the entries $1,700, $1,462, and $1,398 in the table under EUAC are irrelevant, and so is the $1,536 entry for year 5.

Equation-Based Calculations (Cashflow Diagrams Not Essential)

The equation-based approach attempts to do away with the need for sketching the cashflow diagrams. In here, the calculations for the EUAC are done in two parts: one for the capital cost or recovery, and the other for the maintenance cost. The two parts are then added together to get the EUAC for the year under consideration, as in part (b) of Example 12.2.

The capital cost is evaluated by determining an equivalent annual cost of the capital P tied up in the asset and the salvage cash inflow S. It is the annual recovery of the net "used up" capital as represented[13] by P-S.

Refer to the cashflow diagram in Figure 12.3 showing an investment P (now) and a salvage S at year n. Thus, the capital used up over n years is P-S. The annual cost of this amount of used capital in terms of the EUAC can be obtained by subtracting the annual worth of S from that of P, as

$$EUAC_n = P(A / P, i, n) - S(A / F, i, n)$$

The use of this relationship requires evaluating two functional notations. For efficient solution, this can be reduced to one by modifying it as follows, since $(A/F, i, n) = (A/P, i, n) - i$ (see Chapter 3).

$$
\begin{aligned}
EUAC_n &= P(A/P, i, n) - S(A/F, i, n) \\
&= P(A/P, i, n) - S\{(A/P, i, n) - i\} \\
&= (P - S)(A/P, i, n) - iS
\end{aligned}
\tag{12.1}
$$

Equation (12.1) can be used to determine the EUAC—the annual cost[14] of using up the capital (P-S)—corresponding to any period by substituting for n.

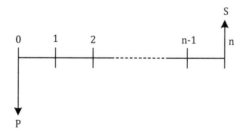

Figure 12.3 Effect of P and S on the EUAC.

---

[13] Note that P and S occur at two different periods, necessitating a consideration of their time values.

[14] When equipment is so old that its salvage value depends mostly on the scrap value of the material it is made of, the salvage value does not change much with its age. In other words, S does not change, and the investment P (the market value) in the defender almost equals S; thus, (P-S) tends to zero. In such cases, from Equation (12.1), the EUAC of capital recovery is simply iS, since the first term is almost zero.

To facilitate the posting of the two parts of EUAC, the previous table has been modified, as follows, by having three sub-columns under EUAC.

Remaining Life n, Years	Year-End Value, S	Maintenance Cost	EUAC Capital Recovery	Maint.	Total
0	P = $5,000				
1	4,000	$200			
2	3,500	300			
3	3,000	400			
4	2,000	500			
5	1,000	600			

In the first row, column 2, note that the robot's market value of $5,000 has been denoted
by P. The salvage value S corresponding to the remaining lives (years) can be denoted by
$S_1, S_2, \ldots$ to distinguish one from the other. These notations avoid any confusion in using
Equation (12.1). So, from the table $S_3 = \$3,000$ while $S_5 = \$1,000$.

Let us now illustrate the calculations for year 2, as an example, based on the equation
approach. For year 2, the salvage value $S_2 = \$3,500$. The EUAC of capital recovery from
Equation (12.1) is

$$EUAC_2 = (P - S_2)(A/P, i, 2) + iS_2$$

$$= (P - S_2)(A/P, 10\%, 2) + 0.1\,S_2$$

$$= (5,000 - 3,500) \times 0.5762 + 0.1 \times 3,500$$

$$= 864 + 350$$

$$= \$1,214$$

In a similar way, the EUAC of capital recovery for the other years can be calculated, as
follows.

$$EUAC_1 = (P - S_1)(A/P, 10\%, 1) + 0.1S_1$$

$$= (5,000 - 4,000) \times 1.1 + 0.1 \times 4,000$$

$$= 1,100 + 400$$

$$= \$1,500$$

$$EUAC_3 = (P - S_3)(A/P, 10\%, 3) + 0.1S_3$$

$$= (5,000 - 3,000) \times 0.4021 + 0.1 \times 3,000$$

$$= 804 + 300$$

$$= \$1,104$$

$$EUAC_4 = (P - S_4)(A/P, 10\%, 4) + 0.1S_4$$

$$= (5,000 - 2,000) \times 0.3155 + 0.1 \times 2,000$$

$$= 947 + 200$$

$$= \$1,147$$

$$EUAC_5 = (P - S_5)(A/P, 10\%, 5) + 0.1S_5$$

$$= (5,000 - 1,000) \times 0.2638 + 0.1 \times 1,000$$

$$= 1,055 + 100$$

$$= \$1,155$$

On posting these capital recovery parts of the EUAC the table looks like:

			EUAC		
			Capital	Maint.	Total
Remaining Life n, Years	Year-End Value, S	Maintenance Cost	Recovery		
0	P = $5,000				
1	4,000	$200	$1,500		
1	3,500	300	1,214		
3	3,000	400	1,104		
4	2,000	500	1,147		
5	1,000	600	1,155		

Next, the EUACs of the maintenance are calculated. Illustrated as follows is the calculation for year 2. Looking at the pattern[15] in the maintenance costs, they follow an arithmetic series for which we have a functional notation (Chapter 4). Thus[16][17][6],

$$EUAC_{2 \text{ maintenance}} = 200 + 100(A/G, 10\%, 2)$$

$$= 200 + 100 \times 0.476$$

$$= 200 + 48$$

$$= \$248$$

---

[15] If the maintenance costs do not follow any pattern—either as given or as "tailored"—then the EUAC is determined by the straightforward method, i.e., by considering the time value of each cost individually.

[16] We could have done

$$EUAC_{2 \text{ maintenance}} = \{200(P/F, 10\%, 1) + 300(P/F, 10\%, 2)\}(A/P, 10\%, 2)$$

$$= (200 \times 0.9091 + 300 \times 0.8264) \times 0.5762$$

$$= (181.82 + 247.92) \times 0.5762$$

$$= 248$$

Or,

$$EUAC_{2 \text{ maintenance}} = 200(P/F, 10\%, 1)(A/P, 10\%, 2) + 300(A/F, 10\%, 2)$$

$$= (200 \times 0.9091 \times 0.5762 + 300 \times 0.4762)$$

$$= 104.76 + 142.86$$

$$= 248$$

Or,

$$EUAC_{2 \text{ maintenance}} = \{200(F/P, 10\%, 1) + 300\}(A/F, 10\%, 2)$$

$$= (200 \times 1.1 + 300) \times 0.4762$$

$$= 248$$

In a similar way, the maintenance EUACs for the other years are calculated and posted in the table.

Finally, by adding the EUAC of the capital recovery to that of the maintenance, the total EUAC for the year is obtained, as shown in the last column below.

Remaining Life n, Years	Year-End Value, S	Maintenance Cost	EUAC Capital Recovery	EUAC Maint.	EUAC Total
0	P = $5,000				
1	4,000	$200	$1,500	$200	$1,700
2	3,500	300	1,214	248	1,462
3	3,000	400	1,104	294	1,398
4	2,000	500	1,147	338	1,485
5	1,000	600	1,155	381	1,536

As expected, the EUAC results based on the equation are the same as the previous ones based on cashflow diagrams. You can use either of these two approaches you feel comfortable with to evaluate the EUAC in replacement analysis.

12.2.3 Economic Life

The various costs of keeping a defender in use can be expressed as an equivalent uniform annual cost (EUAC), as explained in the previous subsection. The variation in the EUAC with the remaining life of the defender shows the cost implication of keeping it in use. Obviously, the best time to replace a defender is when its EUAC is at its minimum. In some cases, this minimum occurs at year O (now) since the EUAC increases with n, as depicted by variation (a) in Figure 12.4. In such a case the equipment is replaced now.

In other cases, the EUAC reaches a minimum at some future time, as depicted by variation (b) in Figure 12.4, which is when the defender should be replaced. The remaining life for which EUAC is minimum is called the *economic life* of the defender. In Example 12.3 earlier, browsing through the total EUAC in the last column of the final table, Bahadur's economic life is 3 years, when its EUAC is at its minimum of $1,398. Rather than browse through, one can plot the EUAC-values to generate a curve like Figure 12.4(b) and read the

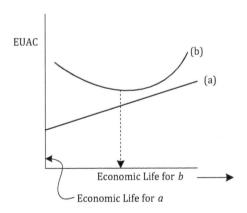

Figure 12.4 EUAC as a function of remaining life.

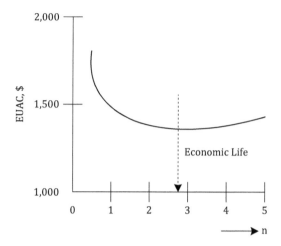

Figure 12.5 Economic life corresponds to minimum EUAC.

economic life off the x-axis. This is illustrated in Figure 12.5 for the results of Example 12.3, wherefrom Bahadur's economic life is 2.9 years.

Whether the EUAC is minimum now or at some time later depends on the rates at which the asset salvage value decreases and maintenance cost increases as it ages.

12.3 THE CHALLENGER

Replacement analysis begins with separate analyses of the defender and the challenger, and ends with their comparison. In the previous section we discussed the analysis of the defender. In this section, we do that for the challenger.

12.3.1 Selection

The selection of the challenger involves deciding the best resource available in the market as a possible replacement of the defender. In the simplest scenario, the manufacturer of the existing equipment offers a functionally equivalent or better replacement. If the equipment is specialized, then there may not be a choice in the market, and what the manufacturer offers becomes the challenger.

In cases where the defending equipment is of the general-purpose type, several alternatives may be available in the market, some of which may be from other manufacturers. In such a case, all the possible[18] alternatives are analyzed by one or more methods discussed in Part II of the text to select the best one. This best becomes the challenger and is analyzed further as discussed next.

---

[18] One of the options in economic analysis, as pointed out earlier in other chapters, is *do nothing*, i.e., maintain the status quo. In replacement analysis, this option does not exist since the defender has already signaled the need of replacement through unacceptable breakdowns and the associated costs. In fact, the analysis of the defender is triggered by the absence of the do-nothing option.

12.3.2 Analysis

The analysis of the challenger is similar to that of the defender discussed in Section 12.2.2. Its objective is to evaluate the challenger's economic life, which may not be the same as useful life, as illustrated in Example 12.4.

Example 12.4: Economic Life

The manufacturer of Bahadur, the pick-and-place robot, now markets an updated version *Virbahadur*. *Virbahadur* costs $15,000 which is $5,000 less than what Bahadur cost 3 years ago. It comes with free maintenance for the first 2 years. For the salvage values and maintenance costs given as follows, determine Virbahadur's economic life, assuming *i* to be 10%.

Year	Year-End Salvage Value	Maintenance Cost
1	$13,000	$0
2	$10,000	$0
3	$ 6,000	$200
4	$ 4,500	$300
5	$ 3,000	$800
6	$ 1,500	$200
7	$0	$300

The high maintenance cost of $800 at year 5 is due to an overhaul which keeps the subsequent years' costs low.

SOLUTION:

The analysis of the challenger is similar to that of the defender; hence the procedure discussed in Example 12.3 is applicable. Let us follow the equation-based approach. However, keep the appropriate cashflow diagrams in mind while determining the maintenance EUACs.

The economic life corresponds to the minimum EUAC. Each year's EUAC is obtained by determining the EUAC of the capital recovery using Equation (12.1) and adding it to the corresponding maintenance EUAC.

EUACs of capital recovery:

$$\text{EUAC}_{1\,\text{capital recovery}} = (P - S_1)(A/P, 10\%, 1) + 0.1 S_1$$

$$= (15,000 - 13,000) \times 1.1 + 0.1 \times 13,000$$

$$= 2,200 + 1,300$$

$$= \$3,500$$

$$\text{EUAC}_{2\,\text{capital recovery}} = (P - S_2)(A/P, 10\%, 2) + 0.1 S_2$$

$$= (15,000 - 10,000) \times 0.5762 + 0.1 \times 10,000$$

$$= 2,881 + 1,000$$

$$= \$3,881$$

$$\text{EUAC}_{3 \text{ capital recovery}} = (P - S_3)(A/P, 10\%, 3) + 0.1 S_3$$

$$= (15,000 - 6,000) \times 0.4021 + 0.1 \times 6,000$$

$$= 3,619 + 600$$

$$= \$4,219$$

$$\text{EUAC}_{4 \text{ capital recovery}} = (P - S_4)(A/P, 10\%, 4) + 0.1 S_4$$

$$= (15,000 - 4,500) \times 0.3155 + 0.1 \times 4,500$$

$$= 3,313 + 450$$

$$= \$3,763$$

$$\text{EUAC}_{5 \text{ capital recovery}} = (P - S_5)(A/P, 10\%, 5) + 0.1 S_5$$

$$= (15,000 - 3,000) \times 0.2638 + 0.1 \times 3,000$$

$$= 3,166 + 300$$

$$= \$3,466$$

$$\text{EUAC}_{6 \text{ capital recovery}} = (P - S_6)(A/P, 10\%, 6) + 0.1 S_6$$

$$= (15,000 - 1,500) \times 0.2296 + 0.1 \times 1,500$$

$$= 3,100 + 150$$

$$= \$3,250$$

$$\text{EUAC}_{7 \text{ capital recovery}} = (P - S_7)(A/P, 10\%, 7) + 0.1 S_7$$

$$= (15,000 - 0) \times 0.2054 + 0.1 \times 0$$

$$= 3,081 + 0$$

$$= \$3,081$$

EUACs of maintenance[19]:

$$\text{EUAC}_{1 \text{ maintenance}} = 0 \,(\text{given})$$

$$\text{EUAC}_{2 \text{ maintenance}} = 0 \,(\text{given})$$

$$\text{EUAC}_{3 \text{ maintenance}} = 200 \,(A/F, 10\%, 3)$$

$$= 200 + 0.3021$$

$$= \$60$$

$$\text{EUAC}_{4 \text{ maintenance}} = \left[200 (F/P, 10\%, 1) + 300 \right] (A/F, 10\%, 4)$$

$$= (200 \times 1.1 + 300) \times 0.2155$$

$$= 520 \times 0.2155$$

$$= \$112$$

---

[19] Be careful while evaluating maintenance EUACs for fourth through seventh year. What is within the bracket [ ] yields the P-value at year 0, which is then converted into annual value.

$$\text{EUAC}_{5 \text{ maintenance}} = \left[200(\text{F/P},10\%,2)+300(\text{F/P},10\%,1)+800\right](\text{A/F},10\%,5)$$

$$= (200\times1.21+300\times1.1+800)\times0.1638$$

$$= 1,372\times0.1638$$

$$= \$225$$

$$\text{EUAC}_{6 \text{ maintenance}} = [200(\text{F/P},10\%,3)+300(\text{F/P},10\%,2)$$

$$+800(\text{F/P},10\%,1)+200](\text{A/F},10\%,6)$$

$$= (200\times1.331+300\times1.21+800\times1.1+200)\times0.1296$$

$$= 1,709\times0.1296$$

$$= \$221$$

$$\text{EUAC}_{7 \text{ maintenance}} = [200(\text{F/P},10\%,4)+300(\text{F/P},10\%,3)+800(\text{F/P},10\%,2)$$

$$+200(\text{F/P},10\%,1)+300](\text{A/F},10\%,7)$$

$$= (200\times1.464+300\times1.331+800\times1.21+200\times1.1+300)\,0.1054$$

$$= 2,180\times0.1054$$

$$= \$230$$

Carefully post these EUACs in their columns and add them for each year to determine the total EUAC, as follows.

Remaining Life n, Years	Year-End Value, S	Maintenance Cost	EUAC Capital	EUAC Maint. Recovery	EUAC Total
0	P =$15,000				
1	13,000	$0	$3,500	$0	$3,500
2	10,500	0	3,881	0	3,881
3	6,000	200	4,219	60	4,279
4	4,500	300	3,763	112	3,875
5	3,000	800	3,466	225	3,691
6	1,500	200	3,250	221	3,471
7	0	300	3,081	230	3,311

The economic life is obtained by browsing through the data in the last column and locating the minimum EUAC. This is evidently 7 years corresponding to the EUAC of $3,311.

12.4 EUAC-BASED ANALYSIS

As mentioned earlier, replacement analysis involves three tasks:

1. Defender analysis
2. Challenger analysis
3. Defender-Challenger comparison

We discussed the first two tasks in the previous two sections. We now embark on the third task, namely comparison of the defender with the challenger to decide whether to replace or not. There are two approaches to carrying out this task. In this section we discuss the first approach based on the EUAC. The second approach based on marginal cost is presented in the next section.

The defender-challenger comparison can proceed only after both have been analyzed individually for their economic lives, as discussed in Sections 12.2 and 12.3. Two different situations can arise, as discussed in the next two subsections. In the first one, defenders' and challengers' economic lives are equal, whereas in the other, which is relatively more difficult to analyze, their economic lives are unequal.

12.4.1 Equal Lives

When the defender's remaining economic life is equal to the challenger's economic life, the comparison is simpler. In fact, the problem reduces to the analysis of two alternatives. Any of the analyses discussed in the text, including the after-tax ROR, can be carried out. Examples 12.5 and 12.6 illustrate two of the possible replacement scenarios.

Example 12.5: EUAC Analysis

A 5-year-old lathe has been analyzed as a defender and found to have an economic life of 5 years. Its current market value is $5,000, and maintenance cost is $500 per year. Its salvage value at the end of economic life is estimated to be $500.

The challenger is a modern CNC lathe whose analysis indicated its economic life to be 5 years. The CNC lathe is capable of operating as a turning cell due to its automatic tool changer, enhanced integratability, and automatic palletizing. Its first cost is $55,000, while the annual maintenance cost is $400. It is expected to save $14,000 per year through higher productivity. Its estimated salvage value at the end of 5 years is $5,000. Should the defender be replaced, if MARR = 12%?

SOLUTION:

Note that the defender's economic life of 5 years is the same as that of the challenger. Since no criterion of comparison has been specified, we can choose any of those discussed in Part II of the text.

Let us use EUAC as the basis of comparison. Our task then is to determine the EUAC of the defender and compare it with that of the challenger. The one with the lower EUAC should win. For both, i = MARR = 12%, and n = 5 years.

$$\text{Defender (lathe):} \quad P = \text{current market value} = \$5,000$$

$$S = \$500$$

$$AM = \text{annual maintenance} = \$500$$

From the defender's cashflow diagram in Figure 12.6[20],

---

[20] Since the two cashflows at period 5 cancel each other out, you may be tempted to evaluate

$$\text{EUAC}_{\text{defender}} = 500 + 5,000(A/P, 12\%, 4)$$

But this is incorrect since this $\text{EUAC}_{\text{defender}}$ covers only 4 years. What we need to determine is the value of $\text{EUAC}_{\text{defender}}$ spread over the 5-year useful life.

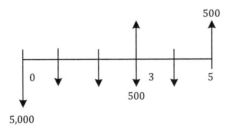

Figure 12.6 Diagram for Example 12.5.

$$\text{EUAC}_{\text{defender}} = 500 + 5,000\left(\text{A/P},12\%,5\right) - 500\left(\text{A/F},12\%,5\right)$$

$$= 500 + 5,000 \times 0.2774 - 500 \times 0.1574$$

$$= 500 + 1,387 - 79$$

$$= \$1,808$$

Alternatively, we can use Equation (12.1), from which

$$\text{EUAC}_{\text{defender}} = \text{Capital recovery cost} + \text{Maintenance cost}$$

$$= \left\{(P-S)\left(\text{A/P},12\%,5\right) + iS\right\} + \text{AM}$$

$$= (5000 - 500) \times 0.2774 + 0.12 \times 500 + 500$$

$$= 4,500 \times 0.2774 + 60 + 500$$

$$= \$1,808$$

Challenger (CNC lathe):   P = \$55,000

S = \$5,000

AM = annual maintenance = \$400

AB = annual benefit = \$14,000

Again, we can follow either the first-principles approach based on cashflow diagram, or use Equation (12.1). From the equation, we get

$$\text{EUAC}_{\text{challenger}} = \text{Capital recovery cost} + \text{Maintenance} - \text{Benefit}$$

$$= \left[(P-S)\left(\text{A/P},12\%,5\right) + iS\right] + \text{AM} - \text{AB}$$

$$= (55,000 - 5,000)(0.2774) + 0.12 \times 5,000 + 400 - 14,000$$

$$= 50,000 \times 0.2774 + 600 - 13,600$$

$$= \$870$$

Since the challenger's EUAC is lower, replace the defender with the CNC lathe.

Example 12.6: EUAC Analysis

The replacement analysis of a welding robot purchased 3 years ago at an initial cost of $20,000 has determined its economic life to be 5 years and its current market value to be $5,000. The robot began straight-line depreciation over its then-estimated useful life of 5 years, with zero salvage value. Its future maintenance cost is predicted to be $200 per year.

It (defender) is being considered for replacement by a more accurate deluxe model (challenger), whose analysis yields an economic life of 5 years. The challenger is priced at $25,000 and is estimated to cost $300 annually in maintenance. It will also be straight-line depreciated over its useful life, with zero salvage value. An analysis has determined the challenger's useful life to be its economic life. In relation to the defender, the deluxe model is estimated to generate an annual income of $10,000 through savings. With tax obligation at the 50% rate, should the deluxe model replace the defender if $i = 15\%$?

SOLUTION:

Note that the defender's 5-year economic life equals that of the challenger. The analysis period should therefore be 5 years. Also note that the defender has already been depreciated for 3 years.

Since no comparison criterion is specified, we can choose any of those discussed in the text. Let us opt for EUAC. Our task is to determine the defender's EUACs and compare them with those of the challenger, taking depreciation and income tax into account. The one with the lower EUAC should be selected. For both, $i = 15\%$, and n = 5 years.

Since the redundant data in the problem statement can be confusing, it is preferable to summarize the relevant data.

Defender (existing welding robot):

$$\text{Initial Cost} = \$20,000$$

$$\text{Then-estimated Life} = 5\,\text{years}$$

$$\text{Then-estimated Salvage} = \$0$$

$$\text{Annual Maintenance} = \$200$$

$$\text{Current Market Value} = \$5,000$$

$$\text{Annual Depreciation}\,(\$20,000/5) = \$4,000$$

Since depreciation has already been charged for 3 years,

$$\text{Book value} = \text{First cost} - \text{Cumulative depreciation}$$

$$= \$20,000 - 3 \times \$4,000$$

$$= \$8,000$$

Since the defender's market value of $5,000 is lower than its $8,000 book value, a capital loss occurs in replacing the defender now.

$$\text{Capital loss} = \text{Book value} - \text{Current market value}$$

$$= \$8,000 - \$5,000$$

$$= \$3,000$$

We can summarize the results in a table as follows. An explanation of the entries is provided following the table. Note that the columns have been selected on the basis of what we learned in Chapters 10 and 11. The $3,000 capital loss can be used to reduce income tax now (year 0 in the table).

(1) Year	(2) Before-Tax Cashflow	(3) Depr. Loss	(4) Capital Loss	(5) Taxable Income	(6) Tax @ 50%	(7) After-Tax Cashflow
0	−$5,000		$3,000		$1,500	$6,500
1	−$ 200	$4,000		$4,200	−$2,100	$1,900
2	−$ 200	$4,000		$4,200	−$2,100	$1,900
3	−$ 200	$0		−$ 200	−$ 100	−$ 100
4	−$ 200	$0		−$ 200	−$ 100	−$ 100
5	−$ 200	$0		−$ 200	−$ 100	−$ 100

Explanatory Notes:

1. In the first column, year 0 means now. The past 3 years of the defender are of no relevance. The $5,000 entry in the first row second column represents the defender's current market value.
2. In the fourth column, the row for year 0, there is an entry[21] of $3,000. Since replacing the defender now would result in a capital loss of $3,000, saving the company $1,500 in income tax, by not doing so an opportunity to reduce tax is forfeited. Thus, the defender represents an investment[22] of $5,000 + $1,500 = $6,500.
3. In the third column, the $4,000 SL-depreciation for the remaining 2 years has been charged. For year 1, the taxable income is equal to the before-tax cashflow of −$200 minus the depreciation charge of $4,000, yielding −$4,200. The income tax on this is −$2,100. The negative sign indicates a tax refund rather than a payment. Negative numbers in other rows means negative cashflows and losses.
4. The data in the last column are the after-tax cashflows. For each row, the entry equals before-tax cashflow minus income tax. For example, for year 1, it is −$200 − (−$2,100) = $1,900; its positive sign indicates a cash inflow (due to the $2,100 tax refund).

We next determine the EUAC of the after-tax cashflows, as follows. Note that the −ve signs of these cashflows are being treated as +ve since EUAC represents cost. If you have difficulty comprehending the following equation[23], help yourself by sketching an appropriate cashflow diagram.

---

[21] The posting of $3,000 capital loss in row 0 means that the associated tax saving is being realized at the time investment is made (year 0). If the saving is delayed due to accounting or tax cycle, then this entry should be made in the following year (year 1), and calculations modified accordingly.

[22] The capital loss would reduce income tax for the year. The forfeiture of this opportunity if the defender is not replaced is equivalent to an increase in taxable income by that amount. The equivalency is based on the fact that by not realizing the capital loss now the company is not reducing its taxable income by $3,000, and thus its tax by $1,500. In other words, not replacing the defender results in $1,500 more payable tax.

[23] To tailor the cashflow pattern, consider the $1,900 cashflow divided into two components: $2,000 and −$100. This makes the $100 cash outflow form a uniform series. In the absence of tailoring, evaluation would have been longer, as follows:

$$EUAC_{defender} = \{6,500 - 1,900(P/A,15\%,2)\}(A/P,15\%,5)$$

$$+ 100(F/A,15\%,3)(A/F,15\%,5)$$

$$= (6,500 - 1,900 \times 1.626) \times 0.2983 + 100 \times 3.472 \times 0.1483$$

$$= 1,017 + 51$$

$$= \$1,068$$

$$EUAC_{defender} = \{6,500 - 2,000(P/A,15\%,2)\}(A/P,15\%,5) + 100$$

$$= (6,500 - 2,000 \times 1.626) \times 0.2983 + 100$$

$$= (6,500 - 3,252) \times 0.2983 + 100$$

$$= 969 + 100$$

$$= \$1,069$$

Challenger (deluxe model):

$$Initial\,Cost = \$25,000$$

$$Useful\,Life = 5\,years$$

$$Estimated\,Salvage = \$0$$

$$Annual\,Maintenance = \$300$$

$$Annual\,Savings\,(Benefit) = \$10,000$$

$$Annual\,Depreciation\,(\$25,000 - 0/5) = \$5,000$$

$$Net\,Annual\,Benefit = \$10,000 - \$300$$

$$Savings - Maintenance\,Cost\,\$9,700$$

A table similar to that for the defender is prepared for the challenger, as follows. The explanation given earlier on the entries in the defender's table should be helpful in comprehending the entries in this table too.

Year	Before-Tax Cashflow	Depr. Loss	Taxable Income	Tax @ 50%	After-Tax Cashflow
0	–$25,000				–$25,000
1	$ 9,700	$5,000	$4,700	$2,350	$ 7,350
2	$ 9,700	$5,000	$4,700	$2,350	$ 7,350
3	$ 9,700	$5,000	$4,700	$2,350	$ 7,350
4	$ 9,700	$5,000	$4,700	$2,350	$ 7,350
5	$ 9,700	$5,000	$4,700	$2,350	$ 7,350

The EUAC of the after-tax cashflows for the challenger is

$$EUAC_{challenger} = 25,000(A/P,15\%,5) - 7,350$$

$$= 25,000 \times 0.2983 - 7,350$$

$$= \$108$$

With the $EUAC_{challenger}$ being lower, the challenger is attractive, and hence the deluxe model should replace the existing welding robot.

If the decision of the replacement analysis is to keep the defender whose economic life is greater than 1 year, the defender should be reanalyzed closer to the end of its economic life. However, if the salvage and maintenance cost data are less reliable or have changed since the last analysis, or a stronger challenger has become available in the market, then the defender should be reanalyzed earlier, for example next year during capital budgeting.

12.4.2 Unequal Lives

When the defender's remaining economic life is different from the challenger's economic life, the comparison becomes slightly more complex. We have discussed the analysis of two alternatives of unequal lives in earlier chapters in Part II. In replacement analysis, the defender and the challenger are the two alternatives. We determine a common analysis period based on the least common multiple (LCM) of their lives and follow the procedures outlined in Part II for two alternatives.

12.5 MARGINAL-COST-BASED ANALYSIS

The EUAC-based approach discussed in the previous section is not appropriate in all cases. We now discuss another approach—based on marginal cost. We begin with the question: When to use one approach instead of the other?

As mentioned earlier, an asset involves two costs: one relating to the recovery of the capital invested in it and the other to its operation and maintenance. The capital recovery cost decreases with the age of the asset. The operating cost, on the other hand, increases with age as maintenance becomes more expensive. In general, the total of the two costs initially decreases with age and then increases, as shown in Figure 12.7. The total-cost curve usually displays a minimum, the coordinates of which are $EUAC_{min}$ and economic life. Example 12.3 and Figure 12.5 have illustrated this earlier.

The total-cost curve in Figure 12.7 displays two distinct zones. In zone A, when the equipment is relatively new, the total cost decreases, primarily due to faster recovery of the capital cost arising from rapid decrease in salvage value. In zone B, when the equipment is relatively old, the total cost increases due to a sharper increase in the operating cost. In this zone, the salvage value remains almost constant[24], and thus capital recovery cost is

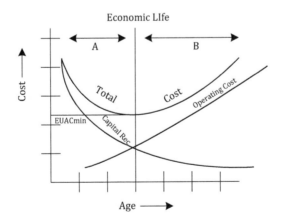

Figure 12.7 Total cost of an asset as it ages.

---

[24] When the asset is old (latter part of zone B) its salvage value depends mostly on the value of material it is made of. The salvage value therefore remains almost constant, resulting in capital recovery cost that is insignificant in comparison to the operating cost.

insignificant. Note that the operating cost rather than being linear, as in Figure 12.7, may be nonlinear and increase exponentially.

The EUAC-based replacement analysis discussed in Section 12.4 is applicable when the replacement decision is being made within zone A and the early part of zone B. If the replacement analysis encompasses the latter part of zone B, in which the total cost continues to increase with the asset age, with no minimum, the marginal-cost-based approach is applicable.

The marginal cost of an asset is the year-by-year total cost of keeping it operational. It differs from the EUAC. It relates only to the following year, while the EUAC can relate to any time duration. The marginal cost too is the total cost (comprising capital recovery cost and operating cost, but only for the next year). Thus, the time period to which marginal cost refers is always 1 year.

In many replacement analyses, the marginal cost of the defender makes better sense, since it looks at the next year's cost only. Example 12.7 illustrates how marginal costs are determined, while Example 12.8 illustrates marginal-cost-based replacement analysis.

Example 12.7: Marginal Cost

A vision-based robotic system is being considered for acquisition at a cost of $30,000. Its annual maintenance cost for the first 5 years is estimated to be $1,000. Thereafter it will increase each year by $1,000. The cost of insurance and utilities for the first year is expected to be $2,000, increasing thereafter by $750 each year. For 15% MARR, determine the asset's marginal cost over its 10-year useful life if its estimated market values are:

Year	Market Value	Year	Market Value
0	$30,000 (first cost)	6	$8,500
1	$24,000	7	$7,500
2	$19,000	8	$6,750
3	$15,000	9	$6,250
4	$12,000	10	$6,000
5	$10,000		

SOLUTION:

Let us first gather the various costs of operating the asset, and post them in the previous table, as follows.

Year	Market Value	Maint. Cost	Utility Cost
0	$30,000		
1	$24,000	$1,000	$2,000
2	$19,000	$1,000	$2,750
3	$15,000	$1,000	$3,500
4	$12,000	$1,000	$4,250
5	$10,000	$1,000	$5,000
6	$ 8,500	$2,000	$5,750
7	$ 7,500	$3,000	$6,500
8	$ 6,750	$4,000	$7,250
9	$ 6,250	$5,000	$8,000
10	$ 6,000	$6,000	$8,750

Next, we determine the marginal cost for each year as:

Marginal cost = Total cost for the year

$$= \text{Capital recovery cost} + \text{Operating cost}$$

For year 1:

Capital recovery cost = Loss in market value + Interest lost

$$= (30,000 - 24,000) + 15\% \text{ of } \$30,000$$

$$= 6,000 + 4,500$$

$$= \$10,500$$

Operating cost = Maintenance and utility costs

$$= \$1,000 + \$2,000$$

$$= \$3,000$$

Marginal cost = $\$10,500 + \$3,000$

$$= \$13,500$$

For year 2:

Capital recovery cost = Loss in market value + Interest lost

$$= (24,000 - 19,000) + 15\% \text{ of } \$24,000$$

$$= 5,000 + 3,600$$

$$= \$8,600$$

Operating cost = Maintenance and utility costs

$$= \$1,000 + \$2,750$$

$$= \$3,750$$

Marginal cost = $\$8,600 + \$3,750$

$$= \$12,350$$

For year 3:

Capital recovery cost = Loss in market value + Interest lost

$$= (19,000 - 15,000) + 15\% \text{ of } \$19,000$$

$$= 4,000 + 2,850$$

$$= \$6,850$$

Operating cost = Maintenance and utility costs

$$= \$1,000 + \$3,500$$

$$= \$4,500$$

Marginal cost = $\$6,850 + \$4,500$

$$= \$11,350$$

The marginal cost for the other years can be determined the same way, and the results summarized as follows.

Year	Loss in Market Value	Interest Foregone	Operating Cost	Marginal Cost
1	$6,000	$4,500	$1,000 + $2,000	$13,500
2	$5,000	$3,600	$1,000 + $2,750	$12,350
3	$4,000	$2.850	$1,000 + $3,500	$11,350
4	$3,000	$2,250	$1,000 + $4,250	$10,500
5	$2,000	$1,800	$1,000 + $5,000	$ 9,800
6	$1,500	$1,500	$2,000 + $5,750	$10,750
7	$1,000	$1,275	$3,000 + $6,500	$11,775
8	$ 50	$1,125	$4,000 + $7,250	$13,125
9	$ 500	$1,013	$5,000 + $8,000	$14,513
10	$ 250	$ 938	$6,000 + $8,750	$15,938

In this example, the marginal cost decreases initially, reaching a minimum of $9,800 at year 5, and then increases. Therefore, the EUAC-based approach of Section 12.4 is appropriate for replacement analysis of this vision-based robotic system.

However, if a defender's marginal cost increases each year, with no minimum, then the rule of replacement decision is: *Keep the defender if its marginal cost is lower than the challenger's EUAC; if not, replace it with the challenger.* Example 12.8 illustrates the procedure.

Example 12.8: Marginal Cost Based Analysis

The following data pertain to a defender:

$$\text{Purchase Price } (3 \text{ yrs ago}) = \$50,000$$

$$\text{Current Market Value} = \$10,000$$

$$\text{Decrease in Market Value next 6 yrs} \quad \$750/\text{year}$$

$$\text{Current Operating Cost} = \$7,500$$

$$\text{Operating Cost Increase by: } \$1,000/\text{year}$$

Considering the system of Example 12.7 to be its challenger, and assuming 15% MARR, should the defender be replaced?

SOLUTION:

We first analyze the defender by determining its marginal costs following the procedure of Example 12.7. The results are summarized in the following table. With their marginal costs known, the defender and the challenger are compared later.

Year	Loss in Market Value	Interest Foregone	Operating Cost	Marginal Cost
1	$750	0.15 × $10,000 = $1,500	$ 7,500	$ 9,750
2	$750	0.15 × $ 9,250 = $1,388	$ 8,500	$10,638
3	$750	0.15 × $ 8,500 = $1,275	$ 9,500	$11,525
4	$750	0.15 × $ 7,750 = $1,163	$10,500	$12,413
5	$750	0.15 × $ 7,000 = $1,050	$11,500	$13,300
6	$750	0.15 × $ 6,250 = $ 938	$12,500	$14,188

Since the defender's marginal cost increases during the 6-year analysis period, we compare its marginal cost, year-by-year, with the corresponding EUAC of the challenger to decide when to replace it.

However, what we have in the last column of the final table in Example 12.7 are the marginal costs, not EUACs, of the challenger. From these marginal cost data, we determine the EUAC corresponding to the year, and then compare it with the defender's marginal cost, as follows. The defender is replaced when its marginal cost exceeds the challenger's EUAC.

Year 1:
The analysis period for determining the challenger's EUAC is 1 year. From Example 12.7, its marginal cost for year 1 is $13,500. Since there is no other cashflow, the EUAC equals the marginal cost. With the defender's $9,750 marginal cost for year 1 (see the previous table) being less than the challenger's $13,500 EUAC, the defender is not replaced.

Year 2:
The analysis period for determining the challenger's EUAC is 2 years. The marginal costs of the challenger (Example 12.7) for years 1 and 2 are $13,500 and $12,350 respectively. These costs at periods 1 and 2 are converted (sketch a cashflow diagram, if helpful) to the EUAC by first determining their P-value and then converting the P-value into A-value as

$$EUAC_2 = \left\{13,500(P/F,15\%,1) + 12,350(P/F,15\%,2)\right\}(A/P,15\%,2)$$

$$= (13,500 \times 0.8696 + 12,350 \times 0.7561) \times 0.6151$$

$$= (11,740 + 9,338) \times 0.6151$$

$$= \$12,965$$

Since the defender's $10,638 marginal cost for year 2 is less than the above $EUAC_2$ of the challenger, the defender is not replaced.

Year 3:
The marginal costs of the challenger (Example 12.7) for years 1, 2, and 3 are $13,500, $12,350, and $11,350 respectively. These costs at periods 1, 2, and 3 are converted to their EUAC by first determining their P-value and then converting the P-value into A-value as

$$EUAC_3 = \left\{13,500(P/F,15\%,1) + 12,350(P/F,15\%,2)\right.$$

$$\left. + 11,350(P/F,15\%,3)\right\}(A/P,15\%,3)$$

$$= (13,500 \times 0.8696 + 12,350 \times 0.7561 + 11,350 \times 0.6575) \times 0.4380$$

$$= (11,738 + 9,338 + 7,463) \times 0.4380$$

$$= \$12,500$$

Since the defender's marginal cost for year 3 ($11,525 from the previous table) is less than the above $EUAC_3$ of the challenger ($12,500), the defender is not replaced.

The calculations for the other years are carried out the same way. A table is used to summarize the results and make the replacement decision, as follows.

Year	Defender's Marginal Cost	Challenger's EUAC	Decision to Replace
1	$ 9,750	$13,500	No
2	$10,638	$12,965	No
3	$11,525	$12,500	No
4	$12,413	$12,100	Yes

Since the defender's marginal cost at year 4 ($12,413) exceeds $EUAC_4$ of the challenger ($12,100), the defender is replaced at year 4. Thus, the defender is kept operational for 4 years, at the end of which it is replaced by its challenger (of Example 12.7).

12.6 MAPI METHOD

To facilitate economic analyses of engineering projects, some companies develop their own manuals or monographs. For example, AT&T's manual[25] has been in use for a long time. The primary purposes of developing own manuals are to streamline, simplify, and guide the decision-making process throughout the company on the method, calculations, and other associated matters such as sources of data and their reliability. One major topic within such manuals invariably is replacement analysis.

Replacement analysis is often cumbersome, as seen in the previous sections. Companies try to make such analyses easier for their decision makers by developing how-to procedures. A special feature of such step-by-step procedures[26] is the simplicity derived from a stream-lined analysis with minimum mathematical efforts. This is achieved by using tables and charts, rather than equations. In the case of replacement analysis, a popular method has been the one developed by George Terborgh for the Machinery and Allied Products Institute (MAPI), a trade organization. In this section, we discuss the MAPI method and illustrate its use through an example.

The MAPI method is fully described in the MAPI publication *Business Investment Management*, obtainable from the Institute's office in Washington, D.C. The method relies on the use of worksheets or forms that help organize the given data. The sheets are filled in, leading to the determination of after-tax ROR based on incremental approach discussed in Chapter 9. The filling-in of the worksheets requires referring to one of the two charts provided. Thus, the use of the MAPI method requires knowing which chart[27] to refer and how to fill in the forms for the given data. It is this simplicity, primarily due to the absence of equations that has made the MAPI method popular with practicing engineers. Note, however, that the charts and forms are based on the fundamental principles discussed throughout the text.

---

[25] AT&T, *Engineering Economy,* third edition, McGraw-Hill, 1977.

[26] If your company practices any special procedure or uses a manual for replacement analysis, or for that matter any economic decision-making, that procedure is usually preferred to other methods. However, such procedures must not violate the basic principles of engineering economics discussed throughout the text.

[27] The MAPI method makes several assumptions on which the charts and forms are based. The user must be aware of these assumptions. One should develop another set of charts and forms if the assumptions differ from those in the MAPI method. The discussions in Section 12.6, and the associated problems at the end of this chapter, are limited to those pertaining to MAPI assumptions.

12.6.1 Concept of Capital Consumption

According to the MAPI method, investment in and use of any resource is basically *capital consumption*. The cost incurred in procuring equipment and keeping it operational can be conceptualized as the cost of buying a service. A resource's service life starts with its installation and is 100% at the beginning. As it is used, the percentage of remaining service life decreases, becoming zero by the end of useful life. This concept is illustrated in Figure 12.8. The horizontal axis represents the percentage of service life expired, which is zero in the beginning (when installed) and 100% at the end of useful life. The vertical axis represents the projected before-tax earnings as a percentage of the initial investment, which is 100% in the beginning and zero at the end. The straight-line variation assumes that the earnings decrease linearly as the service-life-expired[28] increases.

The utility of the equipment, or the service it provides, is derived from the consumption of the capital invested. The capital consumption for each period decreases in such a way that the cumulative capital consumption by the end of the useful life equals the initial investment.

In the MAPI method, we determine annual capital consumption, which is the difference between unrecovered investment at the beginning of the year and the equipment value at the end of the year. The determination is based on the following assumed allocations of the before-tax earnings:

1. Pay income tax at the 50% rate on the difference between the earnings and the capital cost. The capital cost is the sum of the asset's depreciation and the interest paid on the debt portion of the investment.
2. Pay 3% interest on one-fourth of the unrecovered investment which is assumed borrowed.
3. Provide a 10% after-tax return on the three-fourths of the unrecovered investment which is assumed equity capital, i.e., raised within the company (not borrowed).

Based on these[29] allocations, the MAPI method relates before-tax earnings with the initial investment. For a given investment and other relevant data, a set of before-tax earnings for each year of use yields capital consumptions in such a way that the cumulative capital consumption equals the investment by the end of the service life.

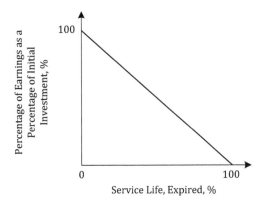

Figure 12.8   Concept of capital consumption.

---

[28] Service-life-expired is the opposite of remaining life. The former increases as the resource ages, while the latter decreases. In the beginning, service-life-expired is 0%, while the remaining life is 100%. By the end of the useful life, they change to 100% and 0% respectively.

[29] For allocations other than these, another set of charts and forms can, and should, be developed and used.

Table 12.1 Computation of Yearly Capital Consumptions (Initial Investment = $1,000, 5-Year Service Life, Straight-Line Depreciation, No Salvage Value)

	Year				
	1	*2*	*3*	*4*	*5*
1. Before-Tax Earnings	$ 475	$380	$285	$190	$ 95
2. Straight-Line Depr.	$ 200	$200	$200	$200	$200
3. Interest Payment	$ 8	$ 6	$ 6	$ 2	$ 1
4. Taxable Earnings	$ 267	$174	$ 81	−$ 12	−$106
5. Income Tax @ 50%	$ 134	$ 87	$ 41	−$ 6	−$ 53
6. After-Tax Earnings	$ 341	$293	$244	$196	$148
7. Unrecovered Investment	$1,000	$741	$510	$308	$137
8. Equity Capital Return @10%	$ 75	$ 56	$ 38	$ 23	$ 10
9. Capital Consumption	$ 259	$231	$202	$171	$137

For the purpose of illustration, let us consider an investment of $1,000, 5-year useful life, straight-line depreciation, and zero salvage value. The computations for the yearly capital consumptions are summarized in Table 12.1, which is followed by an explanation about the entries.

Explanation:

1. The before-tax earnings for year 1 (row 1) of $475 were calculated based on MAPI assumptions. Their values for other years are also given in this row. These values render the sum[30] of capital consumptions for the 5 years of service (row 9) equal to the $1,000 initial investment (note that the salvage value is zero). The data in row 1 are based on the linear relationship in Figure 12.8, as illustrated in Figure 12.9. For example, the before-tax earnings from Figure 12.9 for year 3 is 60% of that for year 1, i.e., 0.60 × $475 = $285.

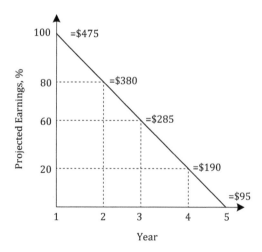

Figure 12.9 Consumption rate of $1,000 of capital.

---

[30] 259 + 231 + 202 + 171 + 137 = 1,000.

2. In row 2, the straight-line depreciation for zero salvage value and a 5-year useful life is obtained from (P-S)/N = ($1,000 – 0)/5 = $200.

3. Row 3 data are obtained by charging 3% interest on 25% of the unrecovered investment. For example, for year 1, the unrecovered investment in row 7 is $1,000. The 3% interest on 25% of this sum = 0.03 × (0.25 × $1,000) = 0.03 × $250 ≈ $8.

The computation of interest for the other years is slightly complex, since it requires the value of unrecovered investment for the year. Let us consider year 3. The unrecovered investment for this year is equal to the initial investment of $1,000 minus the sum of capital consumptions for years 1 and 2. From row 9, these consumptions are $259 and $231. Thus, the unrecovered investment at the end of year 2 is $1,000 – ($259 + $231) = $510. This $510 becomes the unrecovered investment for year 3, whose 25% equals $127.50. Thus, the interest for year 3 = 3% of $127.50 ≈ $4.

4. The data in the fourth row (taxable earnings) are simply before-tax earnings minus the sum of interest and depreciation charge. For example, for the second year, it is $380 – ($200 + $6) = $174. Note that the taxable earnings for years 4 and 5 are negative.

5. Row 5 contains income tax obligations which are 50% of the taxable earnings. For example, for the third year, it is 50% of $81 ≈ $41.

6. The data in the sixth row are the differences between before-tax earnings and the corresponding income tax (row 1 minus row 5). For year 2, for example, it is $380 – $87 = $293. For year 4, it is $190 – (–$6) = $196.

7. The unrecovered investment for the first year is simply the initial investment. For other years, it is the value of investment remaining after cumulative capital consumption up to the previous year. For example, for year 2, it is $1,000 minus the capital consumption for year 1 which (in row 9) is $259. Thus, the entry in row 7 under year 2 = $1,000 – $259 = $741. For year 4, the unrecovered investment[31] likewise is $1,000 minus the cumulative capital consumption up to year 3 = $1,000 – ($259 + $231 + $202) = $308.

8. The data in the eighth row represent 10% return on the equity capital, which in the MAPI method is assumed to be 75% of the unrecovered investment. For the third year, for example, the equity capital being 75% of the unrecovered investment of $510 is $382.50. Thus, the return for the third year is 10% of $382.50 ≈ $38.

9. The data in the last row are the amounts of capital consumed each year. For any year, it is equal to after-tax earnings (row 6) minus the interest paid (row 3) minus the return on equity capital (row 8). Thus, for year 4, it is $196 – $2 – $23 = $171. Note that the total of capital consumptions (sum of the data in row 9) equals the initial investment of $1,000, a basic premise of the MAPI method. Such an equality check ascertains the accuracy of the computations.

---

[31] The unrecovered investment at the beginning of year 4 can be evaluated alternatively, in fact easily, as being

= Unrecovered investment at the beginning of year 3

–Capital consumption for year 3

= $510 – $202

= $308

12.6.2 Retention Value

During the course of determining the year-by-year capital consumption, as illustrated previously, equipment's retention values get computed. The retention value is simply the value of the asset at the year-end. Thus,

$$\begin{matrix} \text{Retention value} \\ \text{at Year-End} \end{matrix} = \begin{matrix} \text{Retention value} \\ \text{at BOY} \end{matrix} - \begin{matrix} \text{Yearly Capital} \\ \text{consumption} \end{matrix}$$

Let us reconsider the illustration of the previous subsection, in which the initial cost is $1,000. For the first year, capital consumption is $259; hence, the retention value[32] at the end of first year = $1,000 – $259 = $741. The retention value is sometimes expressed as percentage of the initial cost. For year 1, it is $741/$1,000 = 0.741 = 74.1%. For year 2, the capital consumption is $231. Thus,

Retention value at the end of year 2

= Retention value at the beginning of year 2

–Capital consumption for year 2

= Retention value at theend of year 1

–Capital consumption for year 2

= $741 – $231

= $510

= 51% of the $1,000-initial cost

The retention values for the other years[33] are determined the same way, yielding[34]:

Year	1	2	3	4	5
Retention Value, %	74.1	51.0	30.8	13.7	0

The retention values are plotted in a chart, as Figure 12.10, where the y-axis denotes the percentage while the x-axis the service life. Such a chart does away with the need to compute retention values for a given problem; the value is simply read off the chart. The curves in Figure 12.10 correspond to a set of salvage ratios. The salvage ratio is the ratio of asset's salvage value to the initial cost, and is expressed in percentage. For the illustrative example under discussion, the salvage value is zero, yielding a salvage ratio of zero. For a service life of 5 years, the retention value can be read off the zero salvage-ratio curve as 74%, approximately the same as computed previously. The instruction for using the chart is given in the figure caption. Note that this chart is valid for a *one-year comparison period* under straight-line depreciation.

For comparison periods longer than one year, another chart is used. Such a chart is shown in Figure 12.11, where the x-axis represents comparison period as a percentage of service life. Again, the instruction for using the chart is given in the figure caption. For a 5-year piece of

---

[32] Same as unrecovered investment for the following year (row 7 under year 2) in Table 12.1. Thus, the retention value at the end of any year is the unrecovered investment for the following year.

[33] Note that the percentage values relate to the data in row 7 of Table 12.1, being equal to the data for the year divided by $1,000—the investment.

[34] The retention value decreases to zero at the end of useful life, as can be noted above.

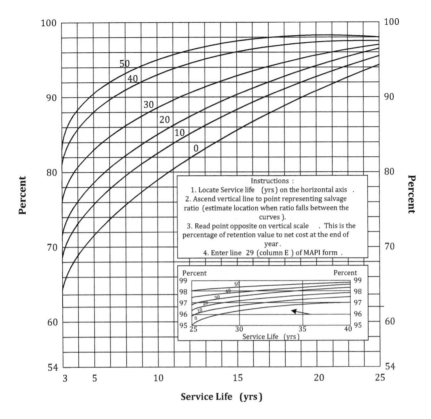

Figure 12.10 Retention values – service life (1-year comparison).

equipment, the comparison period for the first year is obviously 20% of the service life. For the other years, it is 40% for year 2, 60% for year 3, 80% for year 4, and 100% for year 5. The retention-value percentages computed in the illustrative example previously are easily read off the zero salvage ratio curve of Figure 12.11, corresponding to the comparison period percentages on the x-axis. For example, for year 1 (20% on the x-axis, the retention value on the zero salvage ratio curve is 74.1%, while for year 3 (60% on the x-axis) it is 30.8%. Note that while Figure 12.11 can be used for 1-year comparison periods, Figure 12.10 gives higher resolution in reading off the retention values, and hence is more accurate for such periods.

MAPI Chart No. 3A

(ONE-YEAR COMPARISON PERIOD AND STRAIGHT-LINE DEPRECIATION)

12.6.3 Summary Forms

Along with the charts of retention values, a worksheet called a *MAPI Summary Form* is used in replacement analysis by the MAPI method. Such a form is shown in Figure 12.12. The summary form contains 40 items grouped in two parts. Part I, comprising items 1 through 25, enables the computation of the challenger's operating advantage, while Part II comprising items 26 through 40 details the investments and returns. Each part has three sections: A, B, and C. The before-tax and after-tax rates of return get calculated at items 34 and 40. A step-by-step procedure of filling-in the summary form leads to the evaluation of a project's after-tax ROR; the objective of the MAPI method.

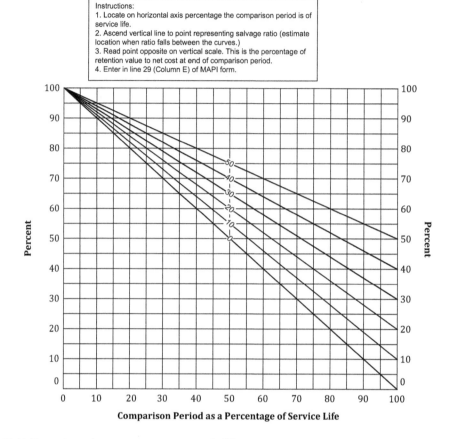

Figure 12.11 Retention values – comparison periods (%).

MAPI Chart No. 3B

(LONGER THAN ONE-YEAR COMPARISON PERIODS AND STRAIGHT-LINE TAX DEPRECIATION)

An explanation of the summary form (Figure 12.12) may be helpful. The analysis begins by filling in the relevant information, such as project title and comparison period[35], at the top of the form.

Under Part I, OPERATING ADVANTAGE, we enter in section A the financial impact of the project on revenues; for example, item 1 allows for such an impact due to changes in product quality, which may increase or decrease the revenue. The effects on operating costs are considered in section B (items 4 through 22). In the non-MAPI methods, several of these costs might get left out unless the analysis is comprehensive and the analyst is alert. The existence of a list, as here, prompts the analyst to consider all the possible costs and revenues. Section C considers the combined effect of all the revenues and costs, summarizing the investment's annual operating advantage at item 25.

Under Part II, INVESTMENT AND RETURN, the initial net investment is worked out in section A, while the terminal investment in B. The retention value percentage from one of the charts (Figure 12.10 or Figure 12.11) is entered in section B as item 29E. In section

---

[35] The comparison period is usually a year, for which Figure 12.10 is applicable. For longer comparison periods, Figure 12.11 is used.

Project no._____

MAPI SUMMARY FORM
(AVERAGE SHORTCUT)

Project _____

Alternative _____

Comparison Period (yrs)   (P) _____

Assumed Operating Rate on Project (hours per year   _____

I. OPERATING ADVANTAGE

(Next-Year for a 1-Year Comparison Period,* Annual Averages for longer Periods)

A. EFFECT OF PROJECT ON REVENUE

	Increase	Decrease	
1 From Change in Quality of Products	$	$	1
2 From Change in Volume of Output			2
3 Total	$ X	$ Y	3

B. EFFECT ON OPERATING COSTS

	Increase	Decrease	
4 Direct Labor	$	$	4
5 Indirect Labor			5
6 Fringe Benefits			6
7 Maintenance			7
8 Tooling			8
9 Materials and Supplies			9
10 Inspection			10
11 Assembly			11
12 Scrap and Rework			12
13 Down Time			13
14 Power			14
15 Floor Space			15
16 Property Taxes and Insurance			16
17 Subcontracting			17
18 Inventory			18
19 Safety			19
20 Flexibility			20
21 Other			21
22 Total	$ Y	$ X	22

C. COMBINED EFFECT

23 Net Increase in Revenue (3x - 3y)	$	23
24 Net Decrease in Operating Costs (22x - 22y)	$	24
25 Annual Operating Advantage (23 + 24)	$	25

* Next year means the first year of proj. operation. For proj.with a significant break-in period, use performance after break in

Figure 12.12 MAPI summary form.

C, the before-tax return is worked out at item 34, while after-tax return at item 40. Note that the variable P in item 32 is the comparison period used, whose value is one in the case of 1-year comparative analysis.

With the charts and the summary form explained, a step-by-step procedure of analyzing projects by the MAPI method is illustrated in Example 12.9.

Example 12.9: MAPI Method

A pick-and-place robot whose economic life is 1 year is being considered for replacement by its deluxe model. The operating advantages of the deluxe model have been posted in Part I of the MAPI summary form as Figure 12.14. The deluxe model costs $40,000 and qualifies for a 25% tax credit.

    Assume straight-line depreciation
    10 years as the estimated service life
    10% of the net costs as salvage value
    $1,000 as next-year increase in depreciation and interest deductions
    $5,000 as defender's disposal value
    $4,000 as challenger's salvage value

Based on the MAPI method, should the defender be replaced if MARR is 10%?

II. INVESTMENT AND RETURN
A. INITIAL INVESTMENT

26 Installed Cost of Project $ _____
    Minus Initial Tax Benefit of $ _____ (Net Cost) $ _____ 26
27 Investment in Alternative
    Capital Additions Minus Initial Tax Benefit $ _____
    Plus: Disposal Value of Assets Retired
      By Project* $ _____ $ _____ 27
28 <u>Initial Net Investment</u> (26-27) $ _____ 28

B. TERMINAL INVESTMENT

29 Retention Value of Project at End of Comparison Period
    (Estimate for Assets, if any, that cannot be depreciated or expensed, for others, estimate of use MAPI charts.)

Item or Group	Installed Cost, (-) Initial Tax Benefit A	Service Life (yrs) B	Disposal Value, End of Life (% of net cost) C	MAPI Chart Number D	Chart % E	Retention Value (A x E/100) F
	$					

Estimated from Charts (total of column F) $ _____
    Plus: Otherwise Estimated $ _____ $ _____ 29
30 Disposal Value of Alternative at End of Period* $ _____ 30
31 <u>Terminal Net Investment</u> (29-30) $ _____ 31

C. RETURN

32 Average Net Capital Consumption (28-31/P) $ _____ 32
33 <u>Terminal Net Investment</u> (29-30) $ _____ 33
34 <u>Before-Tax Return</u> (25-32/33 x 100)   % _____ 34
35 Increase in Depreciation and Interest Deductions $ _____ 35
36 Taxable Operating Advantage (25-35) $ _____ 36
37 Increase in Income Tax (36 x Tax Rate) $ _____ 37
38 After-Tax Operating Advantage (25-27) $ _____ 38
39 Available for Return on Investment (38-32) $ _____ 39
40 <u>After-Tax Return</u> (39/33 x 100)   % _____ 40

* After Terminal Adjustments

Figure 12.13 MAPI summary form (continued).

SOLUTION:

MAPI analysis involves a careful filling-in of the summary form with the given data, as presented in Figures 12.15 and 16, along with the associated simple computations. The procedure for Part I comprises:

    Fill-in the top portion of the summary form.
    Do the total of items 1 and 2 at item 3.
    Do the total of items 4 through 21 at item 22.
    Compute the values at items 23, 24, and then 25.

For the given data, the annual operating advantage, as computed at item 25, is $4,100.

    Next, Part II of the form is filled in with the appropriate data. The $40,000 installed cost and the 25% tax credit yield $30,000 as the net cost in Section A. With $5,000 as defender's disposal value, the initial net investment is found at item 28 to be $25,000. In Section B, the 88.8% data at item 29E has been obtained from Figure 12.10 corresponding to a 10-year service life and 10% salvage ratio ($4,000/$40,000 = 0.1 = 10%). The terminal net investment as computed at item 31 is $22,640. Finally, the after-tax return as computed at item 40 is 12.2%. Since it is higher than the 10% MARR, the deluxe model should replace the defender.

    This example has illustrated the simplicity and comprehensiveness of the MAPI method for replacement analysis.

PROJECT NO._____   SHEET 1

MAPI SUMMARY FORM
(AVERAGING SHORTCUT)

PROJECT_____

ALTERNATIVE_____

COMPARISON PERIOD (YEARS)   (P)___1 year___

ASSUMED OPERATING RATE OF PROJECT (HOURS PER YEAR)   _____

I. OPERATING ADVANTAGE
(NEXT-YEAR FOR A 1-YEAR COMPARISON PERIOD,* ANNUAL AVERAGES FOR LONGER PERIODS)

A. EFFECT OF PROJECT ON REVENUE

	INCREASE	DECREASE	
1 FROM CHANGE IN QUALITY OF PRODUCTS	$ 6000	$	1
2 FROM CHANGE IN VOLUME OF OUTPUT	1500		2
3 TOTAL	$ X	$ Y	3

B. EFFECT ON OPERATING COSTS

	INCREASE	DECREASE	
4 DIRECT LABOR	$	$ 200	4
5 INDIRECT LABOR	300		5
6 FRINGE BENEFITS	50		6
7 MAINTENANCE		500	7
8 TOOLING		200	8
9 MATERIALS AND SUPPLIES		1900	9
10 INSPECTION		100	10
11 ASSEMBLY			11
12 SCRAP AND REWORK		500	12
13 DOWN TIME			13
14 POWER		50	14
15 FLOOR SPACE			15
16 PROPERTY TAXES AND INSURANCE	200		16
17 SUBCONTRACTING			17
18 INVENTORY			18
19 SAFETY		100	19
20 FLEXIBILITY		100	20
21 OTHER			21
22 TOTAL	$ Y	$ X	22

C. COMBINED EFFECT

23 NET INCREASE IN REVENUE (3X − 3Y)	$_____	23
24 NET DECREASE IN OPERATING COSTS (22X − 22Y)	$_____	24
25 ANNUAL OPERATING ADVANTAGE (23 + 24)	$_____	25

* Next year means the first year of project operation. For projects with a significant break-in period, use performance after break-in.

Figure 12.14 Data for Example 12.9.

12.7 RETIREMENT

At times, engineers and technologists make the decision to retire a resource. Retirement does not involve replacement. It is simply a decision of whether to continue operating the resource or not, as illustrated in Example 12.10.

Example 12.10: Retirement

A private company is considering whether to retire an irrigation canal it built 50 years ago at a cost of $2 million. The original estimated life of the canal was 75 years. The construction of an interstate highway has reduced the water discharge in the canal. Revenue collection from irrigation fees is estimated to be $120,000 per year. The cost to maintain the canal is $80,000 per year, which is likely to increase each year by $5,000. Assuming zero salvage value for the canal, should it be retired (abandoned!) if the applicable annual compound interest rate is 6%?

SOLUTION:

The initial construction cost of $2 million being sunk cost is irrelevant. Since a decision criterion is not specified, we can choose to use any of the six discussed in the text. Let us use the PW-criterion under which we abandon the canal if its PW is negative.

PROJECT NO._____   SHEET 1

MAPI SUMMARY FORM
(AVERAGING SHORTCUT)

PROJECT_____*Pick-and-Place   Robot-Deluxe Model*_____
ALTERNATIVE_____*Continue with the Current Model*_____
COMPARISON PERIOD (YEARS)   (P)____*1 year*____
ASSUMED OPERATING RATE OF PROJECT (HOURS PER YEAR)   _____

I. OPERATING ADVANTAGE
(NEXT-YEAR FOR A 1-YEAR COMPARISON PERIOD,* ANNUAL AVERAGES FOR LONGER PERIODS)

A.   EFFECT OF PROJECT ON REVENUE

		INCREASE	DECREASE	
1	FROM CHANGE IN QUALITY OF PRODUCTS	$ 600	$	1
2	FROM CHANGE IN VOLUME OF OUTPUT	1500		2
3	TOTAL	$ 7500 X	$ Y	3

B.   EFFECT ON OPERATING COSTS

		INCREASE	DECREASE	
4	DIRECT LABOR	$	$ 200	4
5	INDIRECT LABOR	300		5
6	FRINGE BENEFITS	50		6
7	MAINTENANCE		500	7
8	TOOLING		200	8
9	MATERIALS AND SUPPLIES		1500	9
10	INSPECTION		100	10
11	ASSEMBLY			11
12	SCRAP AND REWORK		500	12
13	DOWN TIME			13
14	POWER		50	14
15	FLOOR SPACE			15
16	PROPERTY TAXES AND INSURANCE	200		16
17	SUBCONTRACTING			17
18	INVENTORY			18
19	SAFETY		100	19
20	FLEXIBILITY		100	20
21	OTHER			21
22	TOTAL	$ 550 Y	$ 3150 X	22

C.   COMBINED EFFECT

23	NET INCREASE IN REVENUE (3X – 3Y)	$ 7500	23
24	NET DECREASE IN OPERATING COSTS (22X – 22Y)	$ 2600	24
25	ANNUAL OPERATING ADVANTAGE (23 + 24)	$ 10100	25

* Next year means the first year of project operation. For projects with a significant break-in period, use performance after break-in.

Figure 12.15 Solution of Example 12.9.

Noting that the canal's remaining life is 25 years, n = 25. As given, i = 6%. For the given cashflow data (sketch a diagram if necessary),

$$PW = PW_{benefits} - PW_{costs}$$

$$= 120,000(P/A,6\%,25) - \{80,000(P/A,6\%,25) + 5,000(P/G,6\%,25)\}$$

$$= (120,000 - 80,000)(P/A,6\%,25) - 5,000(P/G,6\%,25)$$

$$= 40,000 \times 12.783 - 5,000 \times 115.9732$$

$$= 511,320 - 579,866$$

$$= -\$68,546$$

Since its present worth is negative, the canal should be retired.

II. INVESTMENT AND RETURN

A. INITIAL INVESTMENT

26 INSTALLED COST OF PROJECT $ *40,000*
   MINUS INITIAL TAX BENEFIT OF *25% credit* $ *10,000* (Net Cost) $ *30,000* 26
27 INVESTMENT IN ALTERNATIVE
   CAPITAL ADDITIONS MINUS INITIAL TAX BENEFIT $ *0*
   PLUS: DISPOSAL VALUE OF ASSETS RETIRED
   BY PROJECT* $ *5,000* $ *5,000* 27
28 INITIAL NET INVESTMENT (26 – 27) $ *25,000* 28

B. TERMINAL INVESTMENT

29 RETENTION VALUE OF PROJECT AT END OF COMPARISON PERIOD
   (ESTIMATE FOR ASSETS, IF ANY, THAT CANNOT BE DEPRECIATED OR EXPENSED; FOR OTHERS, ESTIMATE OR USE MAPI CHARTS.)

Item or Group	Installed Cost, Minus Initial Tax Benefit (Net Cost) A	Service Life (Years) B	Disposal Value, End of Life (Percent of Net Cost) C	MAPI Chart Number D	Chart Percentage E	Retention Value ($\frac{A \times E}{100}$) F
Deluxe model robot	$ *30,000*	*10*	*10%*	*3A (Fig. 14.10)*	*88.8*	$ *26,640*

ESTIMATED FROM CHARTS (TOTAL OF COLUMN F) $ *26,640*
PLUS: OTHERWISE ESTIMATED $ *—* $ *26,640* 29
30 DISPOSAL VALUE OF ALTERNATIVE AT END OF PERIOD* $ *4,000* 30
31 TERMINAL NET INVESTMENT (29 – 30) $ *22,640* 31

C. RETURN

32 AVERAGE NET CAPITAL CONSUMPTION $\left(\frac{28-31}{P}\right)$ $ *2,300* 32

33 AVERAGE NET INVESTMENT $\left(\frac{28+31}{2}\right)$ $ *23,820* 33

34 BEFORE-TAX RETURN $\left(\frac{25-32}{33} \times 100\right)$ % *32.5* 34

35 INCREASE IN DEPRECIATION AND INTEREST DEDUCTIONS $ *1,000* 35
36 TAXABLE OPERATING ADVANTAGE (25 – 35) $ *9,100* 36
37 INCREASE IN INCOME TAX (36 × TAX RATE) *50%* $ *4,550* 37
38 AFTER-TAX OPERATING ADVANTAGE (25 – 37) $ *5,550* 38
39 AVAILABLE FOR RETURN ON INVESTMENT (38 – 32) $ *3,190* 39

40 AFTER-TAX RETURN $\left(\frac{39}{33} \times 100\right)$ % *13.4%* 40

Figure 12.16 Solution of Example 12.9 (continued).

12.8 SUMMARY

Replacement analysis is a type of specialized engineering decision-making in which existing equipment, called the defender, is analyzed for possible replacement by its challenger. It is based on the basic principles of engineering economics discussed in the earlier chapters. The analyst should ensure that sunk costs are not included in the analysis. The defender is analyzed for its economic worth—usually its market value—and remaining economic life. Likewise, the challenger is analyzed for its economic life, and costs and benefits. An asset's economic life corresponds to the time period at which its equivalent uniform annual cost (EUAC) is minimum. In the EUAC-based replacement analysis, the defender and the challenger are compared for their EUACs; the one with lower EUAC is preferred. Problems involving unequal economic lives of the defender and challenger are relatively more complex. For assets whose total cost increases each year, with no minimum, replacement analysis is based on marginal cost which is the year-by-year total cost of keeping the asset operational. In the marginal-cost-based analyses too, the defender and the challenger are compared, and the defender is replaced if its marginal cost exceeds the challenger's EUAC.

Companies sometimes use their own "home-grown" procedure to simplify replacement analysis. Company-specific procedures, such as the MAPI method, are based on specific assumptions and the associated charts and forms, rather than on equations. Based on the concept of capital consumption, the MAPI method is relatively simple, and yet comprehensive. Retirement analysis in which the asset is retired or abandoned, rather than replaced, is a special type of replacement analysis, and is usually simpler.

DISCUSSION QUESTIONS

12.1   Why is market value the best indicator of a defender's economic worth? Illustrate through an example from everyday life, such as selling an old car.
12.2   How useful is the concept of defender and challenger in a replacement analysis? Does it make the analysis any simpler?
12.3   Why has replacement analysis been given prominence in the text by devoting a complete chapter on it?
12.4   List the major assumptions of the MAPI method.
12.5   Compare the MAPI method with EUAC and marginal-cost-based replacement methods.
12.6   Under what circumstance will you opt for marginal-cost-based replacement analysis?
12.7   Why are sunk costs considered irrelevant in replacement analysis?
12.8   Explain what specific information is presented about an asset evaluated through the MAPI method.
12.9   Discuss the similarities and differences, if any, in the concept of capital consumption and depreciation expense.
12.10   Compare and contrast between book and market values.
12.11   List and describe the steps in replacement analysis.
12.12   Describe a scenario where a company may find it necessary to develop their own method of replacement analysis.

MULTIPLE-CHOICE QUESTIONS

12.13   A defender's economic worth is usually its
  a. book value
  b. sunk cost
  c. market value
  d. trade-in value
12.14   In a replacement analysis, existing equipment is the
  a. challenger
  b. defender
  c. either a or b, depending on whether its remaining life is longer than the useful life of the new one
  d. either a or b, depending on whether its cost is higher than that of the new one
12.15   Trade-in value is usually _____ the market value.
  a. equal to
  b. smaller than
  c. greater than
  d. half of
12.16   In a replacement analysis, the defender is considered an investment worth its
  a. original cost minus the cumulative depreciation

    b. market value

    c. trade-in value

    d. book value

12.17 In general, there are two choices in a replacement analysis. One is to replace the defender now, and the other is to keep it:

    a. for its remaining useful life

    b. for another year

    c. until a better challenger is available

    d. none of the above

12.18 Economic life corresponds to

    a. minimum EUAC

    b. maximum EUAC

    c. minimum PW

    d. maximum PW

12.19 Depreciation affects the

    a. first cost

    b. salvage value

    c. market value

    d. book value

12.20 The defender is kept in use until it yields how much EUAC?

    a. minimum

    b. maximum

    c. useful economic life

    d. none of the above

12.21 When a defender and challenger have unequal lives, a common analysis period is based on _____.

    a. resulting marginal profit

    b. useful lives

    c. LCM

    d. all of the above

12.22 Which of the following cost is not associated with the cost of an asset?

    a. capital recovery

    b. maintenance

    c. tax, titles, and insurance

    d. both b and c

12.23 The total cost of an asset generally tends to _____ throughout the life of the asset.

    a. increase then decrease

    b. decrease then increase

    c. remain constant

    d. decrease exponentially

12.24 Which of the following in not considered an initial cost?

    a. shipping

    b. installation

    c. purchase of equipment to set up

    d. insurance

12.25 The rule of the replacement decision is:

    a. accept the challenger if the EUAC is lower than the defender's EUAC

    b. replace assets every 5 years

    c. keep the defender if its marginal cost is lower than the challenger's EUAC

    d. keep the defender if its marginal cost is higher than the challenger's EUAC

12.26 The end result of the MAPI summary form is:
  a. evaluation of a project's after-tax ROR
  b. evaluation of a project's before-tax ROR
  c. determination of the economic life of a project
  d. none of the above

NUMERICAL PROBLEMS

12.27 You are considering the replacement of a 20-year-old stamping machine, for which you visit the local dealer where a salesperson says:

The new machine you are interested in has been very popular. Unfortunately, its list price has gone up to $29,500 only last month. But I think I would be able to convince the manager, who graduated from the same college as you, to sell to you at the old price of $28,000, thus saving you $1,500. For your old machine our trade-in offer is $8,000, which is $2,000 better than its market value you have checked yourself to be $6,000. Thus, the new machine will cost you only $20,000. If you don't buy from us, you will be paying $23,500 (= $29,500 – $6,000). Thus, buying from us saves you $3,500—a real good bargain.

From the viewpoint of replacement analysis:
What is the defender's economic worth?
What is the challenger's first cost?

12.28 Best Bread had purchased a personal computer system for $20,000 3 years ago. At that time the system was estimated to have a useful life of 5 years and a salvage value of $10,000. The company uses MACRS for depreciation. It invested another $5,000 this year in software upgrades. The current operating cost is $3,000 per year. The anticipated salvage value at the end of its useful life is now reduced to $4,000. The system is being analyzed for possible replacement, which has determined that it can be sold for $9,000. What value is relevant in the analysis? For an income tax rate of 50%, what is the net proceeds from the disposal of the computer system?

12.29 Since Mira cannot afford to replace her old car now, worth $500, she decides to keep it for one more year. Based on the car's repair history, she estimates its repair cost for the next year to be $650. She will not be able to sell the car next year for more than $200. How much will it cost her to keep the car operational for one more year if i = 10% per year?

12.30 A forklift truck is being analyzed as a challenger. It costs $30,000 to buy. Its salvage value and maintenance costs are as follows:

Year	Maint. Cost	Salvage Value
1	$ 250	$25,000
2	$ 750	$20,000
3	$1,500	$15,000
4	$2,500	$10,000
5	$4,000	$ 5,000

Assuming MARR of 8%, what is the truck's economic life?

12.31 An old lathe with market value of $5,000 is undergoing its replacement analysis. Determine its economic life if $i = 8\%$ per year and its estimated salvage values and maintenance costs are as follows.

Remaining Life (*n*, yrs)	Salvage Value, **S** @ EOY *n*	Maint. Cost
0	$5,000	
1	$4,000	$ 100
2	$3,500	$ 200
3	$3,000	$ 350
4	$2,500	$ 550
5	$2,000	$ 800
6	$1,500	$1,100
7	$1,500	$1,450
8	$1,500	$1,900

12.32 A consulting firm maintains a small aircraft for its use. The aircraft was bought 15 years ago for $300,000. The accounting department feels that it is costing too much to keep this aircraft operational. The firm is considering replacing it with an identical aircraft. The maintenance and cost estimating departments have gathered the following data.

Year	Maint. Cost	Salvage Value
1	$ 5,000	$30,000
2	$ 7,000	$20,000
3	$ 9,000	$12,000
4	$11,000	$ 5,000
5	$13,000	$0

Besides the maintenance there are other costs as well, which together total $15,000 per year. Assuming that the aircraft has already been fully depreciated and its market value is $60,000, what is its economic life if MARR is 20%?

12.33 A defender with a market value of $1,000 is being analyzed for replacement by a challenger whose first cost is $1,500. Both have a service life of 3 years and no salvage value. During the 3 years the defender's operating costs are estimated to be $350, $400, and $500. These data for the challenger are $100, $200, and $300 respectively. For a MARR of 15%, should the defender be replaced?

12.34 In Problem 12.21, assume that a similar challenger is available from the distributor who offers a $1,250 trade-in for the defender. The challenger, however, costs $1,800. Should the distributor's offer be accepted?

12.35 A CNC lathe purchased 5 years ago for $120,000 is being considered for replacement by a turning cell. The market value of the CNC lathe is $25,000, but its value will decrease during the next 5 years as per the following table, which also shows the operating costs. If MARR is 15%, when should the CNC be replaced?

Year	Salvage Value	Operating Cost
1	$23,000	$3,500
2	$20,000	$4,000
3	$16,000	$4,750
4	$ 9,000	$5,750
5	$0	$7,000

12.36 The replacement analysis of a concrete mixer yielded the following results:

Year (n)	Defender's EUAC if kept n yrs	Challenger's EUAC if kept n yrs
1	$3,675	$6,500
2	$4,500	$5,875
3	$5,230	$5,125
4	$5,985	$4,580
5	$6,450	$6,150

What are the economic lives of the defender and the challenger? When should the defender be replaced?

12.37 A stamping machine was installed 5 years ago at a cost of $47,500. Its annual operating cost was then estimated to be $2,500, increasing each year by $1,000. Its market value was estimated to decrease each year by 10%. The current projection is that the machine can be used for the next 5 years. However, the supplier of the machine offers an improved version for $60,000. This cost includes free maintenance for the next 5 years; its EUAC will be minimum at the fifth year. Ignoring other operating costs, and assuming MARR of 12%, when should the defender be replaced?

12.38 A company is considering replacing its forklift truck by an automated guided vehicle (AGV) for its material handling needs. The truck's economic life is 1 year, and it can be sold this year for $10,000.

The AGV costs $120,000 and qualifies for a 30% investment tax credit under the government's Manufacturing Modernization Mandate. It is estimated to be in service for 10 years and its end-of-life disposal value is estimated to be $10,000. The increase in next year's depreciation and interest charges is likely to be $5,000. The AGV will be depreciated by the straight-line method. It offers the following annual operating advantages and disadvantages:

Direct labor down by:	$25,000
Maintenance down by:	$ 300
Benefit from improved flexibility:	$ 250
Benefit from reduced downtime:	$ 150

What is the after-tax ROR from investment in the AGV as per the MAPI method, if the company's tax obligations are 50%?

12.39 A U.S. company built a chemical plant in Ethiopia 25 years ago at a cost of $4 million. The original estimated life of the plant was 75 years. The net profit from the operations is estimated to be $250,000 per year. The annual operating cost is $125,000, which is likely to increase annually by $10,000. If the plant has zero salvage value, should it be abandoned? Assume an annually compounded interest rate of 10% per year.

12.40 A small yard maintenance company is considering replacing their primary riding mower. The mower was purchased 2 years ago for $8,000 and was expected to last 6 years. Salvage value at the end of its estimated useful life was expected to be $3,500. During the past 2 years, straight-line depreciation has been charged.

A new style of mower has come in the market for $5,000. A dealership is offering $3,000 on the original mower as a trade-in. The yard company manager has determined that the original mower can be sold in the yellow pages for around $4,000. Determine the following at the end of the mower's useful life:

a. market value
b. capital gain on mower transaction
c. book value

12.41 Now make an economic comparison between the old and new mowers from Problem 12.40. Determine whether the yard maintenance company should sell the old mower or keep the mower operational for one more year, during which the maintenance cost is expected to be $300. The annual benefit from the mower equals $1,500 and the salvage value at the end of year 3 is expected to be $3,500. Assume MARR of 8%.

12.42 Determine the annual cost of operating the primary riding mower in Problem 12.40 for the next 6 years. End of year salvage values and maintenance expense are:

Year	Salvage Value	Maintenance Cost
1	$7,000	$200
2	$6,500	$225
3	$6,000	$250
4	$5,000	$350
5	$4,500	$450
6	$3,500	$500

12.43 An outboard motor was installed on a work boat 4 years ago at a cost of $15,000. Its annual operating cost was estimated to be $800, increasing each year by 5%. The market value is estimated to decrease by 8.5% each year. Currently, managers think that the outboard will be able to provide service for another 3 years. Dealers have informed the managers of a new model of motor that is $18,000 and includes free maintenance for the entirety of service. The EUAC will be at a minimum in the fourth year. Ignoring operating cost, and assuming MARR of 10%, when should the defender be replaced?

12.44 A defender with a market value of $9,500 is being analyzed for replacement by a challenger whose first cost is $12,000. Both have a service life of 5 years and no salvage value. During the 5 years the defender's operating costs are estimated to be $400, $450, $500, $550, and $600. These data for the challenger are $200, $250, $300, $400, and $450 respectively. For a MARR of 9%, should the defender be replaced?

12.45 A computer manufacturing company is releasing a commercial operating systems backup program onto the market. The backup system will cost $30,000 and have free maintenance for the first 2 years of service. For the following salvage value and maintenance costs, determine the economic life of the system, assuming 5% interest.

Year	Salvage Value	Maintenance Cost
1	$27,000	$0
2	$25,500	$0
3	$24,500	$ 750
4	$22,500	$ 850
5	$21,000	$1,200
6	$19,500	$ 850
7	$18,000	$ 800
8	$17,000	$ 750
9	$15,500	$ 800
10	$13,000	$ 900

The maintenance cost in year 5 is due to an overhaul of the system, which lowers the subsequent year's maintenance cost.

12.46 Look-Out-Below Painting Service purchased a new van 2 years ago for $23,000. At the time the van was estimated to have a useful life of 5 years and a salvage value of $7,500. The company uses straight-line depreciation for the 5-year useful life. This year the company invested another $3,500 in a new engine. The current operating costs are $1,000 per year. The van is now estimated to have a salvage value of $6,000. The owners of the service are considering replacing the van and have discovered that, currently, the van can be sold for $14,760. Determine the capital loss or gain if the van were sold on the open market.

12.47 An industrial planking machine purchased 4 years ago for $50,000 is being considered for replacement by a new model. The market value of the planking machine is $30,000, but its value will decrease during the next 4 years as per the following table, which also shows the operating costs. If MARR is 12%, when should the machine be replaced?

Year	Salvage Value	Operating Cost
1	$25,000	$3,000
2	$23,000	$3,500
3	$18,000	$4,000
4	$15,000	$5,750

12.48 A company is considering replacing its straddle retrieval system with a straddle reach system for its pallet retrieval needs. The straddle's economic life is 1 year, and it can be sold this year for $25,000.

The straddle reach system costs $40,000 and qualifies for a 25% investment tax credit under the government's Warehousing Modernization Mandate. It is estimated to be in service for 12 years and its end-of-life disposal value is estimated to be $8,000. The increase in next year's depreciation and interest charges is likely to be $4,500. The straddle reach will be depreciated by the straight-line method. It offers the following annual operating advantages and disadvantages:

Direct labor down by:	$15,500
Maintenance down by:	$ 1,000
Benefit from improved flexibility:	$ 4,000
Benefit from reduced downtime:	$ 3,500

What is the after-tax ROR from investment in the straddle reach system as per the MAPI method, if company's tax obligations are 50%?

12.49 Since Arnold cannot afford to replace his personal computer now, worth $600, he decides to keep it for one more year. Based on the computer's repair history, he estimates its repair cost for the next year to be $150. He will not be able to sell the computer next year for more than $400. How much will it cost him to keep the car operational for one more year if $i = 7.5\%$ per year?

12.50 Rhonda is considering replacing a 10-year-old golf cart. The salesperson tells her that a new off-road model of golf cart will cost $10,500. Rhonda has done her research and knows that these prices are the best in three states. In the surrounding area, these off-road golf carts sell for $13,000. The dealership is offering a trade-in value of $3,500 for Rhonda's old golf-cart, which is equal to the book value. The defender's initial cost was $12,000. Rhonda likes the deal but wants to evaluate her options.

From the viewpoint of replacement analysis:
what is the defender's economic worth?
what is the challenger's first cost?

FE EXAM PREP QUESTIONS

12.51 Depreciation affects the
a. first cost
b. salvage value
c. market value
d. book value

12.52 Trade-in value is usually _____ the market value.
a. equal to
b. smaller than
c. greater than
d. half of

12.53 Assets should be replaced at the end of their economic life when:
a. new technology is available
b. serviceability and limited economic useful life are assumed
c. the required service life is short
d. none of the above

12.54 Valid reason(s) for replacing an existing asset are:
a. it does not meet needed output capacity
b. new technology is available to replace old
c. maintenance cost has increased significantly
d. all of the above

12.55 The economic life of an asset is defined as:
a. year where operating costs exceed depreciation expense
b. year of minimized annual operating expenses and capital costs
c. maximum years of available service
d. age of maximum depreciation expense

12.56 A bank of solar panels purchased 4 years ago for $75,000 is being considered for replacement by a more efficient model. The market value of the solar panels is $50,000, but its value will decrease during the next 5 years as per the following table, which also shows the operating costs. Assume MARR is 10%.

Year	Maint. Cost	Salvage Value
1	$5,300	$45,000
2	$3,500	$40,000
3	$4,000	$30,000
4	$4,500	$10,000
5	$4,500	$ 7,500

What is the economic life of the solar panels?
a. 10 years
b. 20 years
c. 8.5 years
d. none of the above

12.57 Should the machine be replaced in Problem 12.56?
a. yes
b. no
c. no, retain for one more year
d. none of the above

12.58 In a replacement analysis, alternative equipment is the
    a. challenger
    b. defender
    c. either a or b, depending on whether its remaining life is longer than the useful life of the new one
    d. either a or b, depending on whether its cost is higher than that of the new one

12.59 Salvage value is usually _____ market value?
    a. less than
    b. equal to
    c. less than or equal to
    d. none of the above

12.60 More technologically advanced assets are generally assumed to:
    a. have lower costs and greater salvage cost
    b. have increased output capacity
    c. have higher operating costs
    d. all of the above

Chapter 13

Break-Even Analysis

LEARNING OUTCOMES

- What is break-even analysis?
- Understand the relevance of break-even analysis in business decisions.
- What is break-even point?
- How to calculate break-even point.

Break-even analysis is key to any good business decision. Break-even analysis helps engineers and businesses to understand if a project or business venture is worth the risk. The break-even analysis has been updated and extended for the use in more business situations. It has been a widely used method for cost, profitability, and volume. It not just a mathematical calculation; rather it is used in a variety of situations for decision-making in business.

There is a measure of risk in most decisions. Risk can be divided into two classes: known and unknown risk. Known risk can be prepared for. Unknown risk can be detrimental to a business or engineering project. Uncertainty can plague a business decision as well. Uncertainty can include: cash flow, economic uncertainty, variation of products or procedures, and variation of planned period. The break-even analysis is an inexpensive tool used to evaluate the payback period of a project. Key terms associated with the break-even analysis are: fixed cost, variable cost, unit price. As the same with certain risk, fixed cost can be accounted for easily; however, due to the nature of variable cost, the project must consider the unknowns of variable cost. The analysis period is the time or span in which a project will be assessed. The analysis period can be determined in an equal or unequal life span of the time in which a project or business decision will be undertaken.

The widely used break-even model is a basic liner model. The uncertainty has not been openly treated, except in sales volume. The most widely used break-even model is the basic linear model. Both revenue and total costs are assumed to be linear in this model, i.e., the price and the variable cost per unit are constants. Jaedicke & Robichek (1964) have explicitly introduced uncertainty in the linear model. The model has been extended several times, but the utility of the model has not been reduced (Fig. 13.1).

13.1 USES OF BREAK-EVEN ANALYSIS

Break-even analysis has been used for various decisions such as:

- This analysis helps in very crucial decision-making in the business that is to be made to buy a particular item.

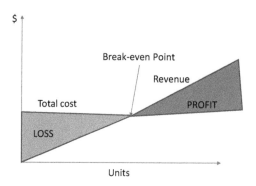

Figure 13.1 Break-even analysis.

- It helps in deciding whether the machine be replaced or not replaced.
- It helps in deciding the feasibility of launching a new product in the market.
- It helps in deciding the profitability of the project.
- It helps in estimating the profitability of a new product or project.
- It helps in making choices.
- It helps in estimating the impact of change in price and costs.

13.2 BASIC BREAK-EVEN MODEL

$$\frac{\text{Fixed costs}}{\text{Sales price per unit} - \text{variable cost per unit}} = \text{Break even in dollars to be sold BE}_s$$

Contribution margin per unit = Sales price per unit-variable cost per unit
The above formula gives the total number of units that the company should sell to get enough profits to cover all its expenses. By multiplying the price of each unit by the output from the formula we can calculate sales dollars.

$$\text{Break Even Point in Dollars} = \text{Sales price per unit} * \text{Break Even Point in Units}$$

The above formula gives the total dollar amount we need to achieve to have zero profit and zero loss. Through this we can calculate the number of products that needs to be sold to achieve a specific profitability level. Dividing desired profit in dollars by contribution margin per unit gives the number of units to be sold to get profits without considering the fixed costs. Then the break-even point number of units must be added.

$$\text{No. of units to produce desired profit} = \frac{\text{Desired profit in \$}}{\text{Contribution margin per unit}}$$

$$+ \text{Break even no. of units}$$

Example:
   Tim is the managerial accountant in charge of a large toy factory's production line and supply chain. He isn't sure the current year's couch models are going to turn a profit and what to measure the number of units they will have to produce and sell in order to

cover their expenses and make $500,000 in profit. The production statistics are given as follows:

Total fixed costs = $500,000
Variable costs per unit = $300
Sales price per unit = $500
Desired profits = $200,000

Break-even point per unit must be calculated first.

$$\text{Break-even point} = \frac{\$500,000}{\$500 - \$300} = 2500\,\text{units}$$

The toy factory must sell 2,500 units to compensate its variable and fixed costs.
Total sales dollars can be calculated by multiplying the number of units and total sales price of each unit.
Total sales dollars = 2500 units × $500 per unit = $1,250,000
The company must sell 2,500 units or toys equivalent to $1,250,000 before profits are realized.

Total no. of toys to be sold to achieve $200,000 profit goal

$$= \frac{\$200,000}{\$500 - \$300} + 2500\,\text{units} = 3,500\,\text{units}$$

The difference between the number of units required to meet a profit goal and the required units that must be sold to cover the expenses is called margin of safety which is one of the most important concepts. In this example 2,500 units had to be sold to meet the company's expenses and 3,500 units had to be produced to meet the profit limit. The margin of safety here is 1,000 units.

13.3 LINEAR BREAK-EVEN MODEL

In the linear break-even model, the revenue and costs are proportional to output. All costs and selling price are constant, and all units produced are sold.
The linear break-even model is as follows:

$$P = (sV - cV - F - Db - I) - (sV - cV - F - D - I)T$$

Notations
P = after-tax profit per unit of time
V = volume of sales per unit of time
s = selling price per unit
c = variable cost per unit including raw materials, direct labor, direct supplies, etc.
F = fixed costs per unit of time – property taxes, rent, insurance, etc. ***Fixed cost per unit time excludes book depreciation and debt interest expenses
Db = book depreciation per unit of time
I = debt interest expense per unit of time
D = tax depreciation per unit of time

T = tax rate
sV = gross income (revenue) per unit of time
cV = variable costs per unit of time

From this model, we wish to determine at what point for which the profit and the income taxes are zero. The two main break-even points of interest are:

- the volume of sales at which profit is zero
- unit sales price at which profit is zero for a given sales volume

For the first point of interest, the formula for the volume of production at which profit is zero is:

$$V_b = \frac{F + I + D_b}{S - c}$$

For the second point of interest, the formula for the unit sales prices for which profit is zero for a given sales volume is:

$$s_b = c + \frac{F + I + D_b}{V}$$

The profit before tax can be calculated as follows:

$$P_b = sV - cV - F - D_b - I$$

where,
P_b = before-tax profit per unit of time
V = volume of sales per unit of time
s = selling price per unit
c = variable cost per unit – raw material, direct labor, direct supplies, etc.
F = fixed costs per unit of time – property taxes, rent, insurance, etc. ***Fixed cost per unit time excludes book depreciation and debt interest expenses
D_b = book depreciation per unit of time
I = debt interest expense per unit of time
sV = gross income (revenue) per unit of time
cV = variable costs per unit of time

If one or more components in the profit equation are not linear, then a non-linear break-even model will be formed (Fig. 13.2).

13.4 NON-LINEAR BREAK-EVEN MODEL

There is no linear pattern in costs and revenues. The average fixed cost decreases with the increase in outputs.
Average fixed cost = Fixed cost/number of units
The average variable cost first decreases with the increase in output, and then increases with the increase in output (Fig. 13.3).

$$\text{Average variable cost} = \frac{\text{variable cost}}{\text{number of units}}$$

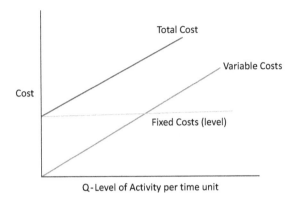

Figure 13.2 Basic linear cost relationship.

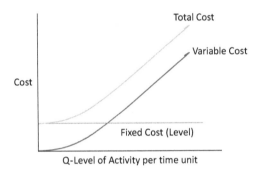

Figure 13.3 Basic non-linear cost relationship.

$$\text{Average total cost} = \text{average fixed cost} + \text{average variable cost}$$

$$= \frac{\text{Total cost}}{\text{number of units}}$$

13.5 THE BREAK-EVEN POINT

The break-even point, notated QBE, is the point where the revenue and total cost relationships intersect. Since variable costs have a direct impact on total profit, we can see from Fig. 13.4 that the break-even point may change if variable costs are lowered.

13.6 SOLVING FOR A BREAK-EVEN VALUE

The following methods can be applied when solving for an unknown parameter:

- Direct Solution—manually if only one interest factor is involved in the setup
- Trial and Error—manually if multiple factors are present in the formulation
- Spreadsheet model where the Excel financial functions are part of the modeling process

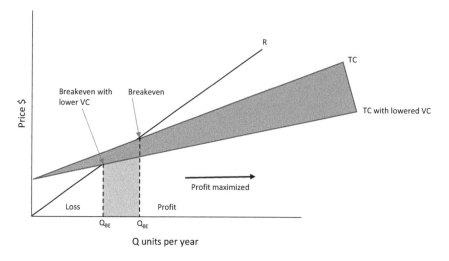

Figure 13.4 Effects of lowering variable costs on break-even point.

13.6.1 Cost–Revenue Model Approach

Within this approach, the relationship between cost, revenue, and volume is examined. The functions are defined, and linear or non-linear cost or revenue relationships are assumed. We want to find the parameter that will either minimize costs or maximize profits.

13.6.1.1 Fixed Cost

Fixed costs are those that will be incurred regardless of production or activity levels (Table 13.1). This is the expense of a company's general operating capacity and must be generally expected to remain constant. Before altering fixed costs downward, all activity must be ceased. To buffer these costs, efficiencies of operations must be improved. These costs include:

- Costs of buildings
- Insurance
- Fixed overhead
- Depreciation of buildings
- Machinery and equipment
- Salaries of administrative personnel

13.6.1.2 Variable Cost

Variable costs are expected to change accordingly with the level of activity within the company (Table 13.2). Variable costs tend to increase or decrease in direct correlation with the

Table 13.1 Examples of Fixed Cost

	One-Time	Accumulative
High	Equipment	Rent
Low	Licenses	Interest

Table 13.2 Examples of Variable Cost

	Demand	Supply
Production	Shipping Charges	Delivery Charges
Profitability	Seasonal Workers' Wages	Cost of Goods

volume of production within the company. They are impacted by efficiency of operation, improved designs, quality, safety, and higher sales volumes. Examples of these costs include:

• Direct labor
• Materials
• Indirect costs
• Marketing
• Advertising
• Warranties

Direct Variable Cost

Direct variable costs are those costs that can be inferable to the production of a specific product or service and allocated to a specific cost center. Some of the examples of direct variable cost are raw materials and wages for the workers working on the production line.

Indirect Variable Cost

Indirect variable costs cannot be directly inferable from production, but they do alter with output. Some examples of indirect cost variables are depreciation, maintenance, and certain labor costs.

Semi-Variable Cost

Semi-variable costs are those which remain constant through a wide range of business volumes but changes when the volume goes out of range. For example, in a call center the wages for the operator is fixed when the calls range from 0–100 calls per day. If the calls volume increases between 101 and 200, a second operator is required. Within the call volume range 0–100, the cost for the operator is fixed, but across all the ranges 0–200 the cost for the operator is semi-variable.

13.6.1.3 Revenue

Revenue is the total amount a business gets by selling its goods to customers during a specific period. Revenue is the gross income which means the discounts and deductions have already been adjusted. Profit and loss can be calculated through revenue later by deducting other types of costs. Revenue can be calculated by multiplying the number of units sold by the price at which the unit is sold.

13.6.1.4 Contribution Margin

Contribution margin is calculated by subtracting variable costs from revenue. It shows how the fixed costs will be covered through the company's revenue. Contribution margin can be expressed in the form of a percentage of net sales or per unit basis or for the total amount.

13.6.1.5 Total Cost

Total costs are equal to the sum of fixed costs and variable costs. Total profit is equal to the difference of generated revenue and total costs. Revenue is equal to the volume of sales times the unit sales price. Variable cost is equal to the volume of sales times the unit variable cost.

$$TC = FC + VC$$
$$Profit = Revenue - Total\,Cost$$
$$Revenue = Volume\,of\,Sales \times Unit\,Sales\,Price$$
$$VC = Volume\,of\,Sales \times Unit\,Variable\,Cost$$
$$P = R - TC$$
$$P = R - (FC + VC)$$

Example 13.1

Sample Data	
Variable Cost	$ 35
Fixed Cost	$80,000
Unit Price	$ 250

The construction manager reported a fixed cost for administration, operation, and warehouse facilities of $80,000. Variable cost was $35.00 per unit. Each item in the resource center has a $250.00 unit price. The point (sales and units) above which the resource center will realize a profit is calculated according to the following methods.

VARIABLE COST RATIO: (VCR)

$$VCR = \frac{Variable\,Cost}{Unit\,Price} = \frac{\$35}{\$150} = 0.233$$

The relationship between variable cost and unit price is called the *variable cost ratio* (VRC).

BREAK-EVEN DOLLAR SALES POINT: (BES)

$$BES = \frac{Fixed\,Cost}{1 - VCR} = \frac{\$80,000}{1 - 0.233} = 104,302.47$$

The amount that must be made to balance fixed and variable costs given the unit price is called the *break-even dollar sales point* (BES).

BREAK-EVEN UNIT SALES POINT: (BEU)

$$BEU = \frac{BES}{Unit\,Price} = \frac{104,302.47}{250} = 417.21(418)\,units$$

The number of items that must be sold in order to reach the break-even point is known as the *break-even unit* (BEU).

Example 13.2

RPM dba Dominos borrowed $100,000 at 9.8% APR. Their cost is the following:

- $30,000 rent
- $60,000 salary
- Each shift requires 3 labor hours and $1.20 of material for production
- Depreciation cost is $15,000 per year
- Hourly labor rate is $9.75/hr
- Selling price of each product is $11.00
- Income tax is 30%

Given factors:

$$F = \$30,000 + \$60,000 = \$90,000$$

$$Db = \frac{\$15,000}{yr}$$

$$I = \$100,000 * 0.098 = 9800$$

$$s = \$41.00 / \text{in units} / \text{hr}$$

$$c = \$9.75 * 3 + 1.20 = \$30.45$$

Calculation of the break-even point (BEP):

$$Vb = \frac{F + I + Db}{S - c} = \frac{\$90,000 + \$15,000 + 9800}{\$41 - 30.45} = 10881.52 \text{ units}$$

IF 20,000 units were sold, the profit would be:

$$P = R - (FC + VC) = \$41.00 * 20,000 - (90,000 + 30.45 * 20,000)$$

$$= \$121,000.00$$

profit
If the market demand was 100,000 units, the sale price for the company to break even would be:

$$sb = c + \frac{F + I + Db}{v} = \$30.45 + \frac{90,000 + 9800 + 15000}{100,000 \text{ units}} = 30.85$$

Example 13.3

Company ABC borrowed $200,000 at 12% APR. Their costs are the following:

- $20,000 rent
- $35,000 management salary
- Each product requires 2 labor hours and $5 of material for production
- Depreciation cost is $25,000 per year
- Hourly labor rate is $10/hr
- Selling price of each product is $30
- Income tax is 37%

(a) Calculate the break-even point.
(b) Calculate the profit if 40,000 units are sold.
(c) If the market demand is 30,000 units, what will the unit sales price of the product be for the company to break even?

Given factors:

F = $20,000 + $35,000 = $55,000
Db = $25,000/yr
I = $200,000 * 0.12 = $24,000
s = $30/unit
c = $10 *2 + 5 = $25

(a) Calculate the BEP.

$$V_b = \frac{F+I+D_b}{S-c}$$

$$= \frac{\left(\$55,000+\$25,000+\$24,000\right)}{\left(\$30-\$25\right)}$$

$$= \$20,800 \text{ units}$$

(b) What is the profit if 40,000 units are sold?

$$P = R-\left(FC+VC\right)$$

$$= \$30*40,000-\left(\$55,000+\$25*40,000\right)$$

$$= \$145,000$$

(c) If the market demand is 30,000 units, what will the unit sales price of the product be for the company to break even?

$$s_b = c+\frac{F+I+D_b}{V}$$

$$= \$25+\frac{\left(\$55,000+\$25,000+\$24,000\right)}{30,000}$$

$$= \$28.47$$

Example 13.4

The unit price is $80, and the number of units sold is 150. The cost per unit is $20 and the fixed cost is $1,500. In Excel, find the income, variable cost, and break-even which is profit made due to selling the 150 units.

SOLUTION:

First fill the Excel spreadsheet with the known values in the manner given below.

	A	B	C
1	Unit Price	$80	
2	Units sold	150	
3	Income		
4	Cost/unit	$20	
5	Variable costs		
6	Fixed costs	$1,500	
7	Profit		
8			

To find income type "=B1*B2"
To find variable cost type "=B4*B2"
To find profit (break-even) type "=B3-B5-B6"

	A	B	C
1	Unit Price	$80	
2	Units sold	150	
3	Income	$12,000	
4	Cost/unit	$20	
5	Variable costs	$3,000	
6	Fixed costs	$1,500	
7	Profit	$7,500	
8			

Example 13.5

The fixed cost is $24,000, variable cost is $8,000, and unit price is $11,000. The volume of production is between 5 and 15. Find the break-even point.

	A	B	C	D	E	F	G
1	Fixed costs	$24,000					
2	Variable costs	$8,000.00					
3	Unit price	$11,000					
4							
5	Volume of production	Fixed costs	Variable costs	Total costs	Income	Marginal income	Net profit
6	5	$24,000	$40,000.00	$64,000.00	$55,000	$15,000.00	($9,000.00)
7	6	$24,000	$48,000.00	$72,000.00	$66,000	$18,000.00	($6,000.00)
8	7	$24,000	$56,000.00	$80,000.00	$77,000	$21,000.00	($3,000.00)
9	8	$24,000	$64,000.00	$88,000.00	$88,000	$24,000.00	$0.00
10	9	$24,000	$72,000.00	$96,000.00	$99,000	$27,000.00	$3,000.00
11	10	$24,000	$80,000.00	$104,000.00	$110,000	$30,000.00	$6,000.00
12	11	$24,000	$88,000.00	$112,000.00	$121,000	$33,000.00	$9,000.00
13	12	$24,000	$96,000.00	$120,000.00	$132,000	$36,000.00	$12,000.00
14	13	$24,000	$104,000.00	$128,000.00	$143,000	$39,000.00	$15,000.00
15	14	$24,000	$112,000.00	$136,000.00	$154,000	$42,000.00	$18,000.00
16	15	$24,000	$120,000.00	$144,000.00	$165,000	$45,000.00	$21,000.00

In Excel:

Variable costs = Volume of production * $8,000
Total costs = Fixed cost + variable cost
Income = Volume of production * unit price
Marginal income = Income-variable cost
Net profit = Income-variable cost-fixed cost

The volume of production is 8 units since the net profit becomes positive from the 9th output. The revenue is $3,000.

To get the break-even analysis chart select total cost, income, and net profit data in Excel and choose the line graph.

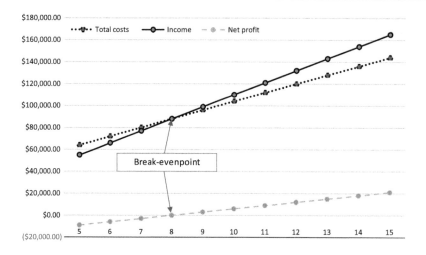

Example 13.6

The number of units sold is given as 50. The cost per unit is $15 and fixed cost is $1,200. Find the break-even point and represent it in a graph.

First enter the given data which is the number of units sold, cost per unit, and fixed cost in the manner given as follows.

	A	B
1	Units sold	60
2	Cost per unit	$15
3	Fixed costs	$1,200

Then name four columns as unit price, revenue, costs, and profit. Fill in the unit price column from 5–55 as shown.

	A	B	C	D	E
1	Units sold	60			
2	Cost per unit	$15			
3	Fixed costs	$1,200			
4					
5	Unit price	Revenue	Costs	Profit	
6	5	$300	$2,100	-$1,800	
7	10	$600	$2,100	-$1,500	
8	15	$900	$2,100	-$1,200	
9	20	$1,200	$2,100	-$900	
10	25	$1,500	$2,100	-$600	
11	30	$1,800	$2,100	-$300	
12	35	$2,100	$2,100	$0	
13	40	$2,400	$2,100	$300	
14	45	$2,700	$2,100	$600	
15	50	$3,000	$2,100	$900	
16	55	$3,300	$2,100	$1,200	
17					

To calculate revenue type "=A6*B1." Change the values according to the row numbers.
To calculate cost type "=B1*B2+B3."
To calculate profit type "=B6-E6." Change the values according to the row numbers.

To get the graph select the revenue, cost, and profit columns and select the line graph and adjust the values.

The break-even point is at unit price 35 since after that the profit value becomes positive. The revenue is $2,400.

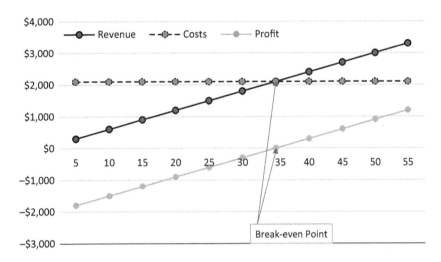

13.7 LIMITATIONS OF BREAK-EVEN ANALYSIS

There are certain limitations in using break-even analysis such as:

- Break-even analysis is useful for a set range of volume which should not be too distant from expected values.
- The break-even analysis is based on estimated projection of values.
- It is assumed that there is always a linear relationship between costs and revenues, which may not always be true for all business situations.

13.8 APPLICATIONS OF BREAK-EVEN ANALYSIS

- Cost calculation—To avoid losses, the number of products to be sold can be known using break-even analysis.
- Budgeting and setting targets—Since the number of products to be sold to break-even is known budgeting becomes easy.
- Motivational tool—Since break-even analysis shows profits at various points of time it helps to motivate staff especially the sales department.

13.9 SUMMARY

Accurately forecasting the cost and sales, producing a break-even analysis, can be done by mathematical equations. Businesses use break-even analysis when trying to propose the total sales or revenue needed to equal total expenses. At the break-even point, profit has not been made; however, no losses have occurred. Break-even analysis is critical for any business because the break-even point allows the business to see the lower limit of profit. Understanding the break-even point allows the business to set profit margins according to capital spent. Break-even analysis is necessary for any business.

DISCUSSION QUESTIONS

13.1 What is the significance of using break-even analysis?

13.2 What is the break-even point? What does it reflect?

13.3 How do you calculate the break-even point?

13.4 How do you plan profit through the break-even point?

13.5 What are the limitations of using break-even analysis?

13.6 What is the difference between fixed cost and variable cost?

13.7 What will be the effect on break-even quantity of
    a. a decrease in unit price
    b. a decrease in average variable cost
    c. a decrease in fixed cost

13.8 How profit would be accounted in a break-even formula
    a. when profit is set as requirement for a year
    b. when profit is set as amount per unit

MULTIPLE-CHOICE QUESTIONS

13.9 The selling price per unit is $100, and contribution per unit is *75. Calculate total variable costs for 150 units?
    a. $15,000
    b. $11,250
    c. Not possible to calculate
    d. $3,750

13.10 Select the limitations of using break-even analysis?
    a. Does not take credit crunch into account.
    b. There may be a change in selling price.
    c. The product may become more fashionable.
    d. The demand is affected by new entrants in the market.

13.11 Total contribution =
    a. variable costs less fixed costs
    b. total sales less total variable costs
    c. total selling price less fixed costs
    d. total sales fixed costs

13.12 At the break-even point
    a. total expenses > total revenue
    b. total expenses = total revenue
    c. total expenses < total revenue
    d. any of the above

13.13 The profit mainly depends on
    a. production output
    b. revenue
    c. production cost
    d. all of the above

13.14 An example of a fixed expense would be 15% sales commission.

True False

13.15 Fixed expenses may include property taxes and rent.

True False

13.16 Variable expenses change with the change in volume.

True False

13.17 Manager's monthly salary is an example of variable expense.

True False

13.18 A retailer's cost of goods is variable expense.

True False

13.19 The break-even point there is
   a.   loss
   b.   profit
   c.   no profit or loss
   d.   none of these

13.20 The following assumption is not applicable in break-even analysis
   a.   All variable costs are fixed.
   b.   All fixed costs are fixed.
   c.   The prices of input factors are constant.
   d.   Volume of production and sales are equal.

13.21 The margin of safety is
   a.   actual sales + sales at break-even point
   b.   actual sales – sales at break-even point
   c.   actual sales / sales at break-even point
   d.   actual sales × sales at break-even point

13.22 At the break-even point, the contribution margin equals to
   a.   variable cost
   b.   administrative costs
   c.   sales revenues
   d.   fixed costs

13.23 Boat has sales of $200,000, a contribution margin of 20%, and a margin of safety is $80,000. What is Boat's fixed cost?
   a. $80,000
   b. $24,000
   c. $16,000
   d. $96,000

13.24 Abby Corp. has a fixed cost of $100,000 and break-even sales of $800,000. What would be the projected profit at $1,200,000?
   a. $150,000
   b. $400,000

c. $ 50,000
d. $200,000

13.25 An automobile company produces tires exclusively for trucks. The annual fixed costs are $50,000, and variables costs are $150 per unit. The company can sell tires for $200.
  a. What quantity of tires must the company sell to break-even?
  b. Determine break-even revenue.
  c. The company sold 4,000 units last year. Calculate the profit.
  d. The fixed cost next year is expected to rise to $55,000. Determine break-even quantity.

13.26 Two companies Abar Inc. and Behar Inc. are competitors. They manufacture and sell lawn mowers. Abar is an older company and requires a variable cost of $150 per lawn mower; its fixed costs are $100,000 per year. Behar is more automated and has lower variable cost per unit; it is $120. Its fixed costs are $200,000. Being competitors, both companies sell their products at a price of $200 per unit.
  a. Determine the break-even for each company.
  b. At which quantity would the two companies have equal profits?
  c. If 5,000 units per year are required to be sold, which company will make more profits?

13.27 The company produces and sells sea salt at the price of $9. Its fixed costs are $100,000 per year and the unit variable cost per bag is $3.
  a. Calculate the break-even quantity.
  b. What will the profit, if company sets a sales target of 200,000 bags?

NUMERICAL PROBLEMS

13.28 The selling price per unit is $75.00, fixed cost is $500,000, and variable cost is $42.00 per unit. The maximum capacity of Demon inc. is 40,000 units and the company anticipates selling 25,000 units. Design a break-even chart with break-even point and margin of safety at present.

13.29 The Hadon Company manufactures bricks for the construction industry. The production capacity for the year is 200,000 bricks. The selling price per brick is $2.5, variable cost is $1.1 per brick, and the fixed cost is $55,000 per annum. Determine the break-even point in terms of revenue and output.

13.30 ABAC Inc. sells TV tables for the price of $55 per unit. The variable cost is $435 per unit and fixed cost is $200,000 per period. Construct the break-even chart and determine the sales value to reach a profit of $30,000.

13.31 Calculate breakeven for a barber shop.

Fixed Cost	
Owner's salary	$25,000
Laundry rental	$ 6,000
Fixed salary	$ 8,000
Electricity bills	$ 1,000
Telephone bills	$ 400
Magazine subscriptions	$ 180
Maintenance	$ 550
Security	$ 450
Total fixed cost	$74,220

Variable cost per client		
Consumables	$	10
Barber's fee	$	10
Total	$	20

13.32 Calculate the break-even for a ceiling manufacturing business.

Fixed Cost	
Rental	$15,000
Owner's salary	$15,000
Salaries to employees	$35,000
Electricity bills	$ 8,000
Telephone bills	$ 1,000
Security	$ 1,000
Transportation cost	$ 8,000
Total Cost	$83,000

The cost of sales of manufacturing the ceiling fan is $2,564.
   The price of the ceiling fan is $3,500.

13.33 Calculate break-even point in units and dollars from the following information:

Price per unit	$	25
Variable cost per unit	$	12
Total fixed cost	$8,000	

13.34 Betton Inc. manufactures TV tables. The fixed cost is $600,000 and each table sells for $150. The variable cost is $75 per unit. The maximum capacity of the factory is 35,000 units and the company anticipates selling 18,000 units each period. Construct the break-even chart, showing the break-even point and the margin of safety.

13.35 Down Hill Inc manufactures soft toys for the American market. The cost is as follows:

	In dollars
Machinery	5,000
Wages per soft toy	8
Packaging per soft toy	5
Rent	4,000
Materials per soft toy	7
Advertising and selling	5,000

13.36 The average selling price of a soft toy is $15 and the capacity of the company is 5,000 soft toys per year.
   Calculate:
   a.   break-even in units of output
   b.   break-even level of sales revenue

13.37 The following is the cost analysis for six months of Salless Limited.

Price per car	$4,000
Material cost per car	$2,500
Labor cost per car	$1,000
Variable coast	$500 per car
Overhead	$500,000 for 6 months of production

a. Calculate the break-even quantity.
b. Determine the margin of safety for 6 months.
c. Calculate profit for this period.

13.38 A television costs $100 per unit to produce and is sold at $150 per unit. The fixed cost is $55,000. What quantity must be sold to break even?

13.39 A television costs $100 per unit to produce and is sold at $150 per unit. The fixed cost is $55,000. What quantity must be sold to break even. If the fixed cost is increased by 22%, then what percent of an increase is needed in sales revenue to reach breakeven?

13.40 The variable cost of making and selling a chair is 65% of sales revenue. The fixed cost is $25,000 per period. What will be the break-even point in terms of sales revenue?

13.41 Dagar Enterprises is contemplating modernization of its plant. The company now sells its tube lights for $15 each; the variable cost per unit is $6, and fixed costs are $650,000 per year. Calculate the break-even quantity.
a. After modernization the company will have the fixed costs of $550,000 per year, but variable costs will increase by $3. Calculate the new break-even point.
b. In the new plant the company would decrease the price to $12 to improve its competitive position in the market. At which quantity will the profit of the old and new plant be equal? And how much would the new profit be?

13.42 The Ahara book company sells paper bags at an average price of $8. It fixed costs are $300,000 per year and the unit variable cost of each bag is $3.
a. Calculate the company's break-even quantity.
b. The company sales target for the year is 200,000 units. What will be its profit?

13.43 The following is an illustration of the application of break-even analysis to a restaurant. The numbers are based on a survey. It is a comparison of the restaurant offering a full menu and the restaurant with a selective menu.

	Restaurant offering full menu	Restaurant offering selective menu
Fixed cost	$345,500	$240,500
Variable cost	$249,400	$350,900
Revenue	$850,000	$711,100
Profit	$ 55,500	$ 70,500

a. Which restaurant has the higher profit margin?
b. What are the break-even point points for both the restaurants?

13.44 Hamar Computers is a small firm that sells computer programs for microcomputers. The following data have been collected on development and production costs for the program and documentation.

Fixed Costs	
Advertising	$20,000
Development cost	$20,000
Manual preparation	$ 5,000
Total	$45,000

Variable costs per units

Transportation	$ 2.00
Printing	$ 4.00
Blank disc	$ 5.00
Total	$11.00

a. Determine the break-even number of programs.
b. The company has a target profit of $100,000. Determine the unit and sales volume (in dollars) required. Determine the unit and sales volume (in dollars) required to meet the goal.
c. Market price drops by 25%; determine the new break- even point in units and dollars.

Chapter 14

Risk Analysis

LEARNING OUTCOMES

- What is risk?
- How to develop a risk management plan.
- How to conduct project risk analysis.
- What are the risk responses?
- How to develop risk responses.
- How to monitor and control risk.

When the terms "risk" and "uncertainty" come to mind, some may think of these terms as synonymous; there are, however, some prevailing differences. Both risk and uncertainty present the problem of not knowing what future conditions will be, however, risk offers estimates of probabilities for possible outcomes, whereas uncertainty does not. Some would argue that everything in life involves risk. In the business world, the risk is prevalent in nearly every project. When the decision is made to undertake a project, many factors must be analyzed to determine the nature and the amount of risk that could possibly result. In the production world, factors that must be considered, including a market for the product, the growth of the product in the market, how the product life cycle will affect its growth and success, prices of the product, the competition in the market, and the capital investments that must be made for the product to make it to the mark.

The listed examples are known risks, regardless of to what extent the risk will be present. These risks are to be expected and kept in mind when the project is being taken into consideration. However, it must be clearly understood that risks are like forecasts. The best estimation for risk will never be accurate, only approximated. Also, there is no absolute method available to evaluate the probability of the risk occurring. The risk can be known or unknown. Unknown risks can produce devastating results for a company and sometimes be the determining factor in whether the company remains viable. Examples of unknown risks include natural disasters, terrorist attacks, wars, epidemics, sudden price fluctuations, and logistical delays. Risk analysis is defined as "the assignment of probabilities to the various outcomes of an investment project" (Park & Tippet, 2005). The factors that are examined to perform risk analysis have a considerable measure of uncertainty. In other words, there is a degree to which one is unsure a factor may or may not affect a project at hand.

RISK CATEGORIES

Risk categories can be broad including the sources of risks that the organization has experienced. Some of the categories could be:

- External risks: government related, regulatory, environmental, market related. Internal risks: service related, customer satisfaction related, cost related, quality related.
- Technical risks: Any change in technology related.
- Unforeseeable risks: Some risks of 9–10% can be unforeseeable risks.

14.1 HOW TO DEVELOP A RISK MANAGEMENT PLAN

There are some steps in the risk management planning process. These are described as follows.

14.1.1 Establish Project Background

Identify, assess, and document potential risks. This involves mapping the following:

- The social scope of risk management (what are our stakeholders facing).
- The identifications and objectives of stakeholders (Do we want to ensure minimal financial impact, programmatic impact, etc.?)
- What resources are available to us to help mitigate the effects of the risks?
- What structures do we have in place to cope with the scenarios that could present themselves?

14.1.2 Risk Identification

This step is most effective when done very early in the project. Having a brainstorming session with a team member to list out several potential risk items is a good beginning. Include all potential risks, including the risks that are already known and assumed, such as scope creep. Include threats that may arise from human threats, operational issues, procedural impacts, financial threats, and natural events. We can talk to the industry experts who may have experience in our project type to get a different perspective. Identify not only the threats but also any opportunities that may impact our project. Opportunities may assist us in bringing the project in on schedule, perhaps with better deliverables or make it more profitable.

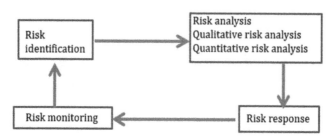

The process of identifying the risk that may negatively affect the project: To identify risk, the project team should review schedule, program scope, cost estimation, key performance parameters, performance, challenges, expectations vs. current plan, external and internal dependencies, implementation challenges, integration, interoperability, supportability, supply-chain vulnerabilities, ability to handle threats, cost deviations, test event expectations, safety, security, and more. To get an insight into areas of consideration of risk, historical data from similar projects, stakeholder interviews, and risk lists can be checked. It is a continuous process, as risk may also appear during project progression. Previously identified risks may also disappear during project progression. All information of risks is then documented in the risk register. The risk register is amended with project progression and risk analysis.

14.1.3 Assessment

Once risks have been identified, they must be assessed for potential severity of loss and probability of occurrence. These quantities can be either simple to measure for example the value of a lost building, or impossible to know such as the probability of an unlikely event occurring. Therefore, in the assessment process, it is critical for making the best guesses possible to properly prioritize (Tables 14.1 and 14.2).

This will help you prioritize the risks as well as allocate resources appropriately.

14.1.4 Potential Risk Treatments

Once the risks have been identified, it is important to outline how we will manage the risk.
Possible Solutions:

- Avoidance (elimination): We can avoid a risky activity, i.e., changing the travel routes, avoiding areas deemed unsafe, etc.
- Reduction (mitigation): Involves methods/solutions that reduce the severity of the loss e.g., equipping staff with health and safety kits, keeping emergency numbers, fire equipment, backing up files, etc.
- Retention: Involves accepting the loss when it occurs.
- Transfer: We can transfer to another party to accept the risk. This can be typically done through insurance, outsourcing services, etc.

14.1.5 Create a Risk Management Plan

Select appropriate controls or countermeasures to treat each risk. The mitigation needs to be approved by the appropriate level of management (Table 14.3).

Table 14.1 Sample Risk Matrix for Grading Risks

Likelihood of occurrence	Level of severity		
	Low	Medium	High
Low			
Medium			
High			

Table 14.2 Sample Risk Prioritization Method

Description of Risk	Occurrence	Severity	Rank (Prioritization)	Status
Loss of property	Low	Medium	2	New possibility
Loss of staff members	Medium	High	1	Increasing

Table 14.3 Sample Risk Management Table

Description of Risk	Occurrence	Severity	Rank	Status	Action	Who	Cost
Loss of property	Low	Medium	2	New possibility	Insure all property		
Loss of staff members	Medium	High	I	Increasing	Periodical security alerts issued to staff		

14.1.6 Implementation

Follow all the planned methods for mitigating the effect of the risks.

14.1.7 Evaluate and Review

Initial plans are never perfect or fully effective. Experience and change in circumstances will require changes in the plan and contribute information to allow different decisions to be made depending on the risk being faced.

14.2 HOW TO ANALYZE RISK

Project risk analysis is a process of analyzing and managing risk in a project. Effective risk analyses will increase the possibility of a project to be successful. It will also save time and cost of the project. The risk
   analysis includes analysis of the project environment, and of risk associated with the project. No two projects are the same, and hence, they have different project conditions and exposure to risk. Therefore, the approach and methods of risk analysis may be different for different projects. Risk analysis identifies risk associated with a project and then helps the project team in developing an effective risk management plan. It is a process through which the project team increases the chance of successful completion of a project, saves time, resources and cost.

14.2.1 Benefits of Risk Analysis

The following are the benefits of using project risk analysis:

- It develops a better understanding of a project.
- It helps in developing a better plan for successful project management.
- It provides a better understanding of the cost and time of any project.
- It provides an understanding of different risks associated with a project.
- Risk identification helps in developing effective risk responses.
- The understanding of risk improves decisions in selecting projects.
- A better understanding of risks helps in better project management, which further increases the project benefits.

14.2.2 Planning for Risk Analysis

Many companies undertake ambitious and very complex projects which must be executed successfully often in a risky and uncertain environment. The project team determines the risk management process. This process includes the analysis of all elements in the project. Those elements may be projected scope, priorities, and deliverables. The risk may be of high or low impact, but every project has some risk and uncertainties. The uncertainty is also an exposure to risk. The risks and uncertainties can be lack of confidence in technology, the improper structure of finances, non-availability of right people, non-availability of right resource, etc.

Risk management needs the right knowledge and planning. Since it is essential for the success of a project, the project manager must be careful in developing and implementing a risk management plan. It identifies the frequency, probability, and the impact of risk. The risk management plan will further help project managers in developing resources and strategies to overcome risk. The risk management encourages discussion among team members on the issues of scalability of a project, the frequency of risk, risk management plan, and estimation of risk management activities, and maintenance of risk register, etc. There are two steps in the planning of the risk analysis process. These are described as follows.

14.2.3 Identify Threats

The first step to risk analysis is to identify the existing and possible threats we might face. These can come from many different sources. For instance, these could be:

- Human—illness, death, injury, or other loss of a key individual.
- Operational—disruption to supplies and operations, loss of access to essential assets, or failures in distribution.
- Reputational—loss of customer or employee confidence, or damage to market reputation.
- Procedural—failures of accountability, internal systems, or controls, or from fraud.
- Project—going over budget, taking too long on key tasks, or experiencing issues with product or service quality.
- Financial—business failure, stock market fluctuations, interest rate changes, or non-availability of funding.
- Technical—advances in technology, or from technical failure.
- Natural—weather, natural disasters, or disease.
- Political—changes in tax, public opinion, government policy, or foreign influence.
- Structural—dangerous chemicals, poor lighting, falling boxes, or any situation where staff, products, or technology can be harmed.

We can use several different approaches to carry out a thorough analysis. Such as:

- Run through a list such as the one just mentioned to see if any of these threats are relevant.
- Think about the systems, processes, or structures that we use, and analyze risks to any part of these. What vulnerabilities can we spot within them?

We can ask others who might have different perspectives. If we are leading a team, ask for input from our people, and consult others in our organization, or those who have run similar projects. Tools such as SWOT analysis and Failure Mode and Effects Analysis (FMEA) can also help us to find new threats, while scenario analysis helps us to explore possible future threats.

Having a brainstorming session with team members to list out several potential risk items is a good start. Include all potential risks, including the risks that are already known and assumed, such as scope creep. Include threats that may arise from human threats, operational issues, procedural impacts, financial threats, and natural events. We can talk to the industry experts who may have experience in our project type to get a different perspective.

Identifying risks is not only the threats, but also any opportunities that may impact our project. Opportunities may assist us in bringing the project in on schedule, perhaps with better deliverables or make it more profitable. Communication at this stage is crucial. Including communication of risk as part of all meetings is effective to illustrate the importance of risk management, share the risk potentials, and provide a platform for discussion.

14.2.4 Estimate Risk

Once we have identified the threats we might face, we need to calculate out both the likelihood of these threats being realized and their possible impact. One way of doing this is to make our best estimate of the probability of the event occurring, and then to multiply this by the amount it will cost us to set things right if it happens. This gives us a value for the risk.

$$\text{Risk value} = \text{Probability of Event} \times \text{Cost of Event}$$

As a simple example, imagine that we have identified a risk that our rent may increase substantially. We may think that there is an 80% chance of this happening within the next year because our landlord has recently increased rents for other businesses. If this happens, it will cost our business an extra $500,000 over the next year. So, the risk value of the rent increase is:

$$0.8(\text{Probability of Event}) \times \$500,000(\text{Cost of Event}) = \$400,000(\text{Risk Value})$$

We can also use a Risk Impact/Probability chart to assess risk. This will help us to identify which risks we need to focus on.

14.2.5 Qualitative Risk Analysis

The quantification of risk in a project is sometimes not possible. The qualitative risk analysis method includes judgment and intuitions. It uses a pre-defined rating scale. The risks are scored based on their impact on the project, likelihood of occurrence. But, the qualitative risk analysis is not effective for high-impact risks, and it is also not possible to estimate the complete need of resources and time. The qualitative risk analysis uses past project experiences, identified risks, budget, probability and impact of risks, etc. A few of qualitative risk analysis methods are:

14.2.5.1 Probability and Impact Assessment

This method examines the probability of risk occurrence and its impact on the project. It also classifies whether the risk is an opportunity or a threat to a project. The probability and impact are only used for risks, not for projects. Experts are invited to discuss on the probability of risk occurrence, and its impact on the project. All information is then documented in the risk management plan. Probability and impact assessment are generally performed by interviewing stakeholders, investigating the ongoing work and documenting the results of everything.

14.2.5.2 Probability and Impact Matrix

As a responsible manager, one should be aware of any possible risks. But this does not mean each risk has to be addressed immediately. If every risk is addressed immediately it will be too expensive both in time and resources. Instead, the risks should be prioritized. Probability and impact matrix help to prioritize risks. A probability and impact matrix are prepared after the estimation of probability and impact of risk. The values range from 0 to 1. The risk which has high probability and high impact is more important. The company may classify possible risks in different shades (as shown in Table 14.4). The red color is classified as high risk, yellow is a moderate risk, and the green is low risk. These values further help in developing risk responses.

14.2.5.3 Risk and Data Quality Assessment

This method examines risk based on its understanding and available quality of data. The quality and accuracy of data are necessary for trustworthy qualitative risk analysis. It examines both the understanding of risk, and the reliability and availability of data.

14.2.5.4 Risk Categorization

Risk categorization is a process of identifying and evaluating the probability and impact of risks. There are several methods for risk categorization. The risk probability can be categorized as extreme, medium, low, and minimal; the same classification can also be used for categorizing risk impact. The risk categorization helps the project manager in deciding which risk needs more attention. A table of risk categories can be developed as (Table 14.5):

- **Estimate or prioritize risks:** Once the risks are identified, the next step is to assess the likelihood of the threat being realized. Some risks will have a much higher impact. One approach is to make a best estimate of the probability and multiply this by the amount it will cost to set things right, if it happens. This will provide an impact value

Table 14.4 Risk Matrix (Chittoor, 2013)

Impact	1	2	3	4	5
Probability (In Percent)	Negligible	Minor	Moderate	Significant	Severe
81–100	Low Risk	Moderate Risk	High Risk	Extreme Risk	Extreme Risk
61–80	Minimum Risk	Low Risk	Moderate Risk	High Risk	Extreme Risk
41–60	Minimum Risk	Low Risk	Moderate Risk	High Risk	High Risk
21–40	Minimum Risk	Low Risk	Low Risk	Moderate Risk	High Risk
0–20	Minimum Risk	Minimum Risk	Low Risk	Moderate Risk	High Risk

Table 14.5 Risk Categories

	Impact		
Probability	High	Medium	Low
High	Extreme	high	Medium
Medium	High	Medium	Low
Low	medium	low	Minimal

associated with the risk. Another approach is to assign each risk a numerical rating, such as a scale from 1 to 5. By scaling we can find any potentially large events that can cause huge losses or gains. These will be the number one priority. We should make sure that our priorities are used consistently and focus on the biggest risks first and the lesser priority risks as applicable.

- **Manage the risk:** Plan out and implement a response for each risk. Typically, we will have four options: (1) Transfer the risk (subcontracting scope or adding contractual clauses), (2) Risk avoidance (eliminating the source of the risk, such as changing a vendor), (3) Risk minimization (influencing the impact), and (4) Risk acceptance. We can create a contingency plan for the largest risks. This would encompass all actions taken if a risk were to occur.
- **Create a risk register:** This will enable us to view progress and stay on top of each risk. A good risk register, or log, will include a risk description, ownership, and the analysis of cause and effect. This register will also include the associated tasks. A good risk register is a valuable tool in communication project status. It should be easily maintained and updated. By remaining current and up to date, the risk register will be viewed as a relevant and useful tool throughout the project lifecycle.

Once a solid risk management process is established, it forms the basis for crisis prevention and cost effectiveness. Risk management involves adapting the use of existing resources, contingency planning, and resource allotment. This process does not need to be complicated. By implementing a project risk management process at the beginning of each project, the team can prepare for whatever may occur and maximize the project results.

14.2.5.5 Risk Urgency Assessment

This method classifies risk based on time. The risk is urgent if it has the possibility of appearing in the near future. These are called urgent risks and are treated urgently. Time is the only indicator used in this method. A few risks, those that are expected to have a high impact on a project, may not be urgent in this assessment due to their possibility of occurrence later.

- Low impact/low probability—These risks are considered low low-level risks and can be ignored.
- Low impact/high probability—These risks are of moderate importance. If the project encounters these types of risks, they can cope with them and move on. But the likelihood of these risks occurring should be reduced.
- High impact/low probability—These risks are considered of high importance. The impact of this risk should be reduced at any cost if they occur, and a contingency plan should be developed in case they occur.
- High impact/high probability—These risks are of critical importance. These risks should be any project's top priority and close attention must be paid to them.

14.2.5.6 Delphi Technique

The Delphi technique is used to forecast the likelihood and outcome of future events. It is a risk brainstorming session but is different to the traditional risk brainstorming. In this technique a group of expert's opinions are used to identify, analyze, and evaluate risks. They individually give their estimations and assumptions to a facilitator who in turn reviews the data and issues a summary report. This is a continuous process.

Example: Assume you are the project manager and you want to know if a product will be successful in the market. You decide to use the Delphi technique and invite the top 100 people within your organization to participate. The questionnaire is constructed in the following way and each participant is asked to rate the following options to achieve the goal:

- Improve development team productivity.
- Provide tiered product pricing.
- Increase the size of the sales team.
- Respond rapidly to customer feedback.
- Other (this option allows the participants to add their own idea).

Since there are five questions, you ask people to assign points between 5–1 to each option. The following table shows a snapshot of how the participants might answer the questionnaire (Table 14.6).

The process of receiving feedback and sending out questionnaires is repeated until an agreed number of rounds has been completed or the standard deviation is low. Low standard deviations will tell us that we have a low variance for each item in the list. In the example, we can see that there is a low variance for the "Launch in China" option, telling us that most people agree as to how well this option will help us achieve our goal. Once the low variance is achieved we can then pick the items with highest mean values to focus our effort on them. The highest mean value items on the list will be the most crucial items that the organization should focus on to ensure success.

14.2.5.7 SWIFT Analysis

In a "Structured What If Technique" (SWIFT), a team investigates how the changes to a design or plan affect the project in a workshop environment. Mostly this technique is used to evaluate the viability of opportunity risks. This method is mostly used in healthcare, radioactive waste management, offshore installations, control of major accident hazards plants, and medical devices.

14.2.5.8 Bow-Tie Analysis

This method helps to identify risk mitigations. In this method the first step is to start looking at the risk events, then it is projected in two directions: on the left side all the potential causes are listed and to the right the potential consequences of the event are listed. This makes it easier to identify and apply mitigation techniques to each of the cases separately

Table 14.6 Delphi Technique

Goal	Person 1	Person 2	Person 3	Person 4	Mean	Standard Deviation
Improve development team productivity	1	5	3	1	2.5	1.9149
Provide tiered product pricing	3	3	1	2	2.25	0.9574
Increase the size of the sales team	4	2	4	3	3.25	0.9574
Respond rapidly to customer feedback	2	1	2	5	2.5	1.7321
Other (launch in China)	5	4	5	4	4.5	0.5774

and effectively which mitigates the probability of risk occurrence and the impact of the risk in case it occurs.

14.2.6 Quantitative Risk Analysis

The quantitative risk analysis quantitatively examines each risk. The possibility of risk occurrence and its effect is numerically examined and estimated. The quantitative risk analysis quantifies all probabilities of risk occurrence, probabilities of project completion, and the probability of getting project benefits. It helps project managers in taking various decisions and in developing a contingency plan. It also determines which risk needs the most attention. It tries to predict the possibility of project success in uncertainty. The quantitative risk analysis needs plenty of data and information before any analysis. This data can be measurable project objectives such as a risk management plan, schedules, budget, project cost details, list of identified risk, risk register, description of risk probability and impact, the scope of the project, and the nature of the organization, etc. Different techniques are used to collect this data.

14.2.6.1 When to Perform Quantitative Risk Analysis

- Projects that require a Contingency Reserve for the schedule and spending plan.
- Large, complex projects that require Go/No Go choices (the Go/No Go choice may happen on various occasions in a project).
- Projects where upper administration needs more insight concerning the likelihood of finishing the task on schedule and budget plan.

14.2.6.2 Sensitivity Analysis

Due to uncertainty and variation in parameter estimates, there is always some degree of risk in taking up any project. Sensitivity analysis is used to determine the effect of variation. Sensitivity generally means relative magnitude of change in the measurement of merit caused by one or more changes in estimated study factor values. Sensitivity analysis helps in examining how much effect the variation in project elements will have on project objectives. It also determines the most impactful risk, meaning the risk which has the most potential of affecting a project. The sensitivity analysis can be of two types:

One-Way Sensitivity Analysis

It is the simplest form of this analysis. This model creates changes in only one value in the model and then examines the impact of that change on the result. For example, it might be shown that changing the effectiveness of intervention by 20%, the cost-effectiveness ratio falls by 30%. This is called one-way sensitivity analysis because only one parameter is changed at once. This analysis can be done using different approaches which are useful for different purposes. Assume that an expert might want to test which parameters have the best impact on a model's outcomes. For this situation, every parameter in the model (or, at any rate, every one of the key parameters) could be changed by an explicit sum. Say, for instance, that all parameters were to be expanded and diminished by 30% of their original parameters. For every parameter change, the expert may record the rate effect on the model's result, which can be demonstrated graphically as a tornado chart. A tornado diagram is helpful in exhibiting the effect that a settled change in every parameter has on the primary results; it isn't valuable in representing to the certainty that an expert may have in the model's input.

Multi-Way Sensitivity Analysis

This model makes changes in more than one value, and after that inspects the connection between the change in values, and its effect on the outcome. It should be noted that in any case, the representation and interpretation of multi-way sensitivity analysis turn out to be complex and difficult as the number of parameters included increases. One technique that is used to survey the certainty around all parameters is to attempt extreme sensitivity analysis, by differing most of the parameters in a model to their "best" and "worst" case. The best- and worst-case scenario values ought to be chosen from the point of view of the intervention that is being evaluated.

Advantages of Financial Sensitivity Analysis

There are many important reasons to perform sensitivity analysis:

- Sensitivity analysis adds reliability to a financial model by testing the model over a wide arrangement of conceivable outcomes.
- Financial sensitivity analysis enables the expert to be adaptable with the limits within which to test the reactivity of the dependent variable to the independent variable. For instance, the model to think about the impact of a 5-point change in interest rate on bond costs would be not quite the same as the financial model that would be utilized to review the impact of a 20-point change in interest rates on bond prices.
- Sensitivity analysis helps to make better decisions. Decision makers utilize the model to see how responsive the yield is to changes in specific factors. This relationship can help an expert in determining unmistakable ends and be instrumental in settling on ideal choices.

14.2.6.3 Expected Monetary Value Analysis (EMV)

Expected monetary value (EMV) is a statistical procedure in risk management that is utilized to evaluate the risks, which in turn, helps the expert to ascertain the contingency reserve. This method examines and estimates the average value of the outcome. It helps in calculating the amount required to manage all identified risks and in selecting the choice which involves less money to manage the risks.

EMV = probability of risk occurrence × impact

Example 1:

You have identified a risk with a 30% chance of occurring. If this risk occurs, it may cost you 500 USD. Calculate the expected monetary value (EMV) for this risk event.

SOLUTION:

The probability of risk = 30%

Impact of risk = 500 USD

Expected monetary value (EMV) = probability * impact

= 0.3 * 500

= 150

The expected monetary value (EMV) of the risk event is −150 USD.

Example 2:

Assume the probability of risk occurrence in a project is 60%. The risk may have a positive impact of $100,000, with a probability of 60%, and the negative impact of $75,000 with a probability of 40%. What will be the expected monetary value of risk?

SOLUTION:

Total probability of positive impact $= 0.6 \times 0.6 = .36$

EMV for positive impact $= .36 \times \$100,000 = \$36000$

Total probability for negative impact $= 0.6 \times 0.4 = .24$

EMV for negative impact $= .24 \times \$75,000 = \$18000$

The EMV for risk is $= \$36000 - \$18000 = \$18000$

Benefits of Expected Monetary Value (EMV) Analysis

- It gives an average result of all recognized uncertain occasions.
- It helps to select the best choice with a reinforcement of objective data.
- It helps to calculate contingency reserve.
- It helps with a make or purchase choice amid the procurement planning process.
- It helps in decision tree analysis. Decision tree analysis is a graphical diagrammatic technique which helps to comprehend the issue and solution effortlessly.
- This system does not require any expensive assets, just the specialists' opinion.

Drawbacks of Expected Monetary Value (EMV) Analysis

- This method is generally not utilized in small and small-medium estimated projects.
- This procedure includes expert opinions to conclude the probability and impact of the risk. In this manner, the individual inclination may influence the outcome.
- This technique functions impressively when we have numerous risks since it helps spread the effect of the risks.
- Sometimes we may miss the incorporation of positive dangers, which may influence the result.
- The dependability of this analysis depends on the information given as input. Hence, the information quality evaluation ought to be performed thoroughly.

14.2.6.4 Decision Tree Analysis

Decision tree analysis is developed in a tree structure. The decision tree reflects the description of decisions and the consequence of selecting that decision. It describes the possible cost and benefit of selecting an alternative decision. It is a diagrammatic representation of path and events. It helps project managers in selecting the most feasible decision that has more certainty and returns. The decision tree analysis uses pictures for representing path and events.

A box represents choices to be made

A circle represents probable outcomes

A line connects probable outcomes and choices

Key Points about Decision Tree Analysis:

- DTA considers future uncertain events. The events are named inside square shapes, from which option lines are drawn.
- When a decision tree is drawn, it has decision points and numerous chance points. Each point has different symbols.
- Because this arrangement results in an outline that takes after a tree stretching from left to right, a decision tree is a suitable name. To investigate a decision tree, move from left to right, beginning from the decision node. This is the place the branching begins. Each branch can prompt a chance node.

From the branch node, there can be further branching. At last, a branch will end with the end-of-branch symbol.

- The likelihood estimate will regularly be referenced on the node or a branch, while the cost impact is toward the end.
- To compute, move from right to left on the tree. The cost value can be on the end of the branch or on the node.
- The best choice is the alternative that gives the most elevated positive value or least negative value, depending upon the situation.

Steps to Use Decision Tree Analysis in Project Risk Management

- Document a decision in a decision tree.
- Assign a probability of occurrence for the risk pertaining to that decision.
- Assign monetary value of the impact of the risk when it occurs.
- Compute the expected monetary value for each decision path.

Example 1:

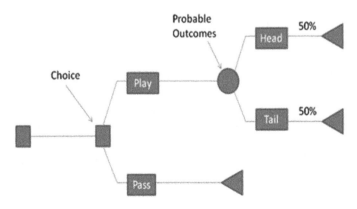

Example 2:

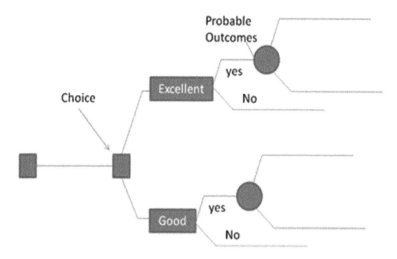

SIMULATION:

The simulation is used to identify uncertainties, and to estimate its possible impact on the project. The Monte Carlo technique is used for simulation.

14.2.6.5 Monte Carlo Analysis

The Monte Carlo analysis involves developing a mathematical model, and then running a simulation to identify the impact of project risk. It uses random sampling for developing a range of possible outcomes in different scenarios. It is an effective technique of forecasting outcome. This method is important in project management because probable total cost of a project can be calculated, and it helps to find a range or a potential date of completion for the project. This process also helps with better communication between project managers and senior managers. A Monte Carlo analysis:

- Requires a computer-based program
- Evaluates the overall risk in the project
- Determines the probability of completing the project on any specific day, or for any specific cost
- Determines the probability of any activity being on the critical path
- Path convergence is considered
- Cost and schedule impacts can be assessed
- Results in a probability distribution

14.3 RISK RESPONSES

Risk response is the process of developing strategic options, and determining actions, to enhance opportunities and reduce threats to the project's objectives. A project team member is assigned to take responsibility for each risk response. This process ensures that a team member is monitoring the responses, although the team member may delegate implementation of a response to someone else.

14.3.1 Planning for Risk Responses

The risk response reduces the probability and possibility of the negative impact of risk on the project. It includes selection of options and actions. It is committed towards increasing the possibility of getting positive returns. The effectiveness of risk response will decide the impact of risks on a project. It makes the project team responsible for making choices/decisions. They can develop risk responses by using:

- Identified risks
- Risk priority list
- Risk management plan
- Project cost and schedule
- Outcome of quantitative and qualitative risk analysis
- Causes and outcome of risk
- Possibility of risk occurrence

14.3.2 How to Develop Risk Responses

The risk responses are developed to lower the impact of risk. There are several methods and techniques for developing risk responses. First, the project team identifies risk categories and then develops an appropriate risk response.

14.3.2.1 Risk Avoidance

Risk avoidance is basically avoiding risk. To eliminate risks, change scope of the project, extend the schedule to eliminate a risk to timely project completion, change project objectives, clarify requirements to eliminate ambiguities and misunderstandings, gain expertise to remove technical risks.

14.3.2.2 Transference of Risk

This involves moving the impact of the risk to the third party. Though the risk is still there, the third party now manages it. Direct methods such as insurance, warranties, or performance bonds can be used. Indirect methods include unit price contracts, legal options. Subcontracting a phase of the work can also transfer risk but ensuring that the party can handle the risk better than you is important.

14.3.2.3 Risk Mitigation

The cost and impact of risk is either reduced or eliminated in risk mitigation. The risk may still appear, but the cost and impact will be lower. A mitigation plan is prepared in advance and is implemented when a risk crosses the threshold. The project team can simplify and develop a prototype or simulation for risk mitigation.

14.3.2.4 Risk Acceptance

The project teams are advised to accept risk. They should clearly identify the probability and impact of risks. The risk acceptance helps in developing a feasible contingency plan. It can further help in making better decisions during the project. It prepares them well to respond better to any risk. The risk acceptance may be passive or active. The passive acceptance prepares them well in understanding the intensity of risk, while active acceptance encourages preparing a better contingency plan. This sets an acceptability point in a project.

14.3.2.5 Risk Register

The risk register includes a list of risks, a list of potential responses, root causes of risks, and updated risk categories. Risk register is developed prior to the development of a risk response plan. It is a list of important risk events that could affect the project's critical success factors. Risks are prioritized based on probability and impact (Table 14.7).

$$RISK = Probability \times Impact$$

The probability and impact score of small projects can be between 1 and 10 and the risk score is a simple multiplication.

Example: In this risk register example, Risk #2 is the most important risk, followed by Risk #1 and lastly Risk #3. Risk response plan for Risk #2:
Determine trigger condition:

- What defines bad weather?
- Who decides it is bad enough?

"Trigger Condition: The site foreman drives the haul road each morning and decides whether the haul road is safe for haul trucks."
Decide which risk response type to use: avoid, transfer, mitigate, or accept.
Mitigate by re-assigning the haul trucks to move other materials down the highway. Develop the response plan: It must be:

- Cost effective
- Scaled to the magnitude of the risk
- Acceptable to the stakeholders
- Achievable

"The project manager will mitigate the impact of the lack of production by re-assigning the haul trucks to haul other materials that need to be moved but can be moved down the paved highway."
This response falls into the category of risk mitigation, not avoidance, because there is no change to the project's scope, schedule, or objectives. Avoidance would be creating a condition whereby the haul road does not have to be used anymore.

14.3.2.6 Risk Communication

Communication during a crisis can be more important than the response itself. Therefore, because the strength of the response to an unexpected event is often judged on communication,

Table 14.7 Example of Risk Register

ID	Description	Probability	Impact	Risk Score
1	Haul truck breaks down	10%	$50,000	$ 5,000
2	Weather makes haul road impassible	20%	$50,000	$10,000
3	Truck driver calls in sick	2%	$20,000	$ 400

it is important that the risk register, and response plans, be communicated to the applicable stakeholders. Because of this, the risk registers and response plans should be communicated to the appropriate stakeholders in advance during project planning. Then, when an unexpected event occurs the stakeholders will not only be more supportive of the response, but the final judgment will be much more favorable. The project manager will be off to a running start.

14.3.2.7 Responding to Risks

Following identification and analysis of project risks, the PRMT acts to improve the condition in favor of project success. Ultimately, it is not possible to eliminate all threats or take advantage of all opportunities—but they will be documented to provide awareness that they exist and have been identified. Successful risk response will change the risk profile through the project life cycle, and risk exposure will diminish. Risk response involves:

- The PRMT should determine which risks warrant a response and identify which strategy is best for each risk.
- Assigning an action to the risk owner to identify options for reducing the probability or impacts of each risk. The risk owner takes the lead and can involve experts available to the project.
- Evaluating each option for a potential reduction in the risk and cost of implementing the option. Selecting the best option for the project.
- Requesting additional contingency, if needed.
- Assigning an action to the risk owner to execute the selected response action. The risk owner is the lead and may assign specific tasks to other resources to have the response implemented and documented.

If the PRMT judges that a risk should be accepted, it may assign an action to the risk owner to prepare a contingency plan if deemed necessary.

Risk Strategies:

For Threats	For Opportunities
Avoid: Risk can be avoided by removing the cause of the risk or executing the project in a different way while still aiming to achieve project objectives. Not all risks can be avoided or eliminated, and for others, this approach may be too expensive or time-consuming. However, this should be the first strategy considered.	**Exploit:** The aim is to ensure that the opportunity is realized. This strategy seeks to eliminate the uncertainty associated with a particular upside risk by making the opportunity definitely happen. Exploit is an aggressive response strategy, best reserved for those "golden opportunities" having high probability and impacts.
Transfer: Transferring risk involves finding another party who is willing to take responsibility for its management, and who will bear the liability of the risk should it occur. The aim is to ensure that the risk is owned and managed by the third party best able to deal with it effectively. Risk transfer usually involves payment of a premium, and the cost-effectiveness of this must be considered when deciding whether to adopt a transfer strategy.	**Share:** Allocate risk ownership of an opportunity to another party who is best able to maximize its probability of occurrence and increase the potential benefits if it does occur. Transferring threats and sharing opportunities are similar in that a third party is used. Those to whom threats are transferred take on the liability and those to whom opportunities are allocated should be allowed to share in the potential benefits.

Mitigate: Risk mitigation reduces the probability and/or impact of an adverse risk event to an acceptable threshold. Taking early action to reduce the probability and/or impact of a risk is often more effective than trying to repair the damage after the risk has occurred. Risk mitigation may require resources or time and thus presents a tradeoff between doing nothing versus the cost of mitigating the risk.

Enhance: This response aims to modify the "size" of positive risk. The opportunity is enhanced by increasing its probability and/or impact, thereby maximizing benefits realized for the project. If the probability can be increased to 100%, this is effectively an exploit response.

Acceptance: This strategy is adopted when it is not possible or practical to respond to the risk by the other strategies, or a response is not warranted by the importance of the risk. When the project manager and the project team decide to accept a risk, they are agreeing to address the risk when it occurs.

Contingency: Making plans to handle the risk if it occurs. For example, back-out procedures that can restore a system if a launch fails.

Examples of Risk Responses:

	Risk Statement	Risk Response
Design	Inaccuracies or incomplete information in the survey file could lead to rework of the design.	**Mitigate:** Work with Surveys to verify that the survey file is accurate and complete. Perform additional surveys as needed.
	A design change that is outside of the parameters contemplated in the Environmental Document triggers a supplemental EIS which causes a delay due to the public comment period.	**Avoid:** Monitor design changes against ED to avoid reassessment of ED unless the opportunity outweighs the threat.
Environmental	Potential lawsuits may challenge the environmental report, delaying the start of construction or threatening loss of funding.	**Mitigate:** Address concerns of stakeholders and the public during the environmental process. Schedule additional public outreach.
	Nesting birds, protected from harassment under the Migratory Bird Treaty Act, may delay construction during the nesting season.	**Mitigate:** Schedule contract work to avoid the nesting season or remove nesting habitat before starting work.
R/W	Due to the complex nature of the staging, additional right of way or construction easements may be required to complete the work as contemplated, resulting in additional cost to the project.	**Mitigate:** Re-sequence the work to enable ROW Certification.
	Due to a large number of parcels and businesses, the condemnation process may have to be used to acquire R/W, which could delay the start of construction by up to 1 year, increasing construction costs and extending the time for COS.	**Mitigate:** Work with right-of-way and project management to prioritize work and secure additional right-of-way resources to reduce impact.
Construction	Hazardous materials encountered during construction will require an on-site storage area and potential additional costs of disposal.	**Accept:** Ensure storage space will be available.
	Unanticipated buried man-made objects uncovered during construction require removal and disposal resulting in additional costs.	**Accept:** Include a Supplemental Work item to cover this risk.

14.3.2.8 A Risk Perspective Can Enhance Decisions

When considering risk mitigation methodology, it is important to recognize the impacts of the decision. The impact of responding to a risk may make sense in the short term (e.g., saves design costs, allows the team to meet the schedule), but the impact of the risk needs to be taken.

For example, the impact of just a few unknown conditions can affect the construction schedule to the point where an environmental work window requires the project to be suspended. It is important to recognize how much of an impact when deciding.

14.3.2.9 Entering Risk Responses into the Risk Register

The risk response action for each risk is entered the "Response Actions" column of the risk register. Risk responses are options and actions that enhance opportunities or reduce threats. The PMRT, PRM, PM, or project team decide upon the response action to risks listed in the risk register. The response action is then assigned to one person, the person responsible for executing and monitoring the risk response that is chosen. Planned risk responses must be appropriate to the significance of the risk, cost-effective in meeting the challenge, realistic within the project context, and agreed upon by all parties involved and owned by a single person.

14.4 HOW TO MONITOR AND CONTROL RISK

Risk monitoring and control is required to:

- Ensure the execution of the risk plans and evaluate their effectiveness in reducing risk.
- Keep track of the identified risks, including the watch list.
- Monitor trigger conditions for contingencies.
- Monitor residual risks and identify new risks arising during project execution.
- Update the organizational process assets.

14.4.1 Purpose of Risk Monitoring

- Risk responses have been implemented as planned.
- Risk response actions are as effective as expected or if new responses should be developed.
- Project assumptions are still valid.
- Risk exposure has changed from its prior state, with analysis of trends.
- A risk trigger has occurred.
- Proper policies and procedures are followed.
- New risks have occurred that were not previously identified.

14.4.2 Inputs to Risk Monitoring and Control

- Risk management plan.
- Risk register contains outputs of the other processes, for example, identified risks and owners, risk responses, triggers, and warning signs.
- Approved change requests. Approved changes include modifications such as scope, schedule, the method of work, or contract terms. This may often require new risk analysis to consider the impact on the existing plan and identifying new risks and corresponding responses.
- Work performance information.

Project status and performance reports are necessary for risk monitoring and control of risks.

14.4.3 Tools and Techniques for Risk Monitoring and Control

There are various tools and techniques for risk monitoring and control as follows:

- **Risk reassessment:** It includes project risk reviews at all team meetings, major reviews at major milestones, risk ratings and prioritization may change during the life of the project. Changes may require additional qualitative or quantitative risk analysis.
- **Risk audits:** Examine and document the effectiveness of the risk response planning in controlling risk and the effectiveness of the risk owner.
- **Variance and trend analysis:** This method used for monitoring overall project cost and schedule performance against a baseline plan. Significant deviations indicate that updated risk identification and analysis should be performed.
- **Reserve analysis as execution progresses:** Some risk events may happen with a positive or negative impact on cost or schedule contingency reserves. Reserve analysis compares available reserves with the amount of risk remaining at that time and determines whether reserves are enough.
- **Status meetings:** Risk management can be addressed regularly by including the subject in project meetings.

14.4.4 Outputs from Risk Monitoring and Control

Based on the outputs from risk monitoring and control, projects can be corrected through different actions. Implementing contingency plans or workarounds frequently results in a requirement to change the project plan to respond to risks. The result is an issuance of a change request that is managed by overall change control. Following are two major types of correction.

- **Corrective action:** Corrective action consists of performing the contingency plan or workaround. Workarounds are previously unplanned responses for emerging risks. Workarounds must be properly documented and incorporated into the project plan and risk response plan.
- **Preventive action:** These actions are used to direct the project towards compliance with the project management plan.

14.5 SUMMARY

Although the chapter provides enough information and process of risk estimation and development of risk responses, there are still some situations where the complex nature of a project, technological changes, and other elements can make the process challenging. Scholars are still developing the best method for risk identification and management.

DISCUSSION QUESTIONS

14.1 What is the significance of using risk analysis?
14.2 What are the different techniques of quantitative risk analysis?
14.3 What are the different techniques of qualitative risk analysis?
14.4 What is the significance of planning for risk responses?

14.5 What is risk avoidance?
14.6 What is risk mitigation?
14.7 What is sensitivity analysis?

MULTIPLE-CHOICE QUESTIONS

14.8 Name the document which records all risks in hierarchical order.
    a. risk list
    b. risk breakdown structure
    c. risk management
14.9 State the statement which is true.
    a. Risks are always negative.
    b. Risks details are documented in the risk register.
    c. Prediction of probability of risk occurrence is not determined.
14.10 Which document should be considered to identify risk response?
    a. register
    b. risk management plan
    c. risk response plan
14.11 At which step is the risk priority on probability and impact set?
    a. risk management
    b. quantitative risk analysis
    c.Nqualitative risk analysis
14.12 Which step come after the risk management plan?
    a. qualitative risk analysis
    b. risk responses
    c. risk identification
14.13 In which method is the questionnaire used to collect experts' input for risk identification?
    a. brain storming
    b. interview method
    c. Delphi technique
14.14 Which one out of following is not a part of risk management?
    a. risk avoidance
    b. risk identification
    c. risk management plan
14.15 The elimination of threat in risk response is called
    a. avoidance
    b. transfer
    c. mitigation
14.16 Risk avoidance should be done
    a. when it is transferable
    b. when risk cannot be avoided
    c. when there is high probability of occurrence and impact
14.17 Which of the following are external risk areas?
    a. regulatory
    b. cost
    c. schedules

14.18 Which one of the following is not a factor in risk assessment?
  a. risk probability
  b. risk impact
  c. insurance

14.19 If the probability of risk occurrence is 55% and the impact is $20,000, what will be the expected monetary value?
  a. $ 11,000
  b. $1,100,000
  c. $ 13,000
  d. $1,800,000

14.20 Compute risk probability if the risk impact is $25,000, and the expected monetary value is $15,000.
  a. 7.5%
  b. 0.6%
  c. 0.8%
  d. 1.5%

14.21 The risk is identified in which step of the risk management process?
  a. quantitative risk analysis
  b. qualitative risk analysis
  c. risk monitoring and control

14.22 Risk tolerance helps in
  a. estimation of the project
  b. determination of the action plan
  c. project schedule
  d. risk ranking

14.23 Insurance is an example of
  a. risk avoidance
  b. risk transfer
  c. risk mitigation

NUMERICAL PROBLEMS

14.24 Three alternatives are being evaluated for the protection of electric circuits, with the following required investments and possibilities of future:

Alternative	Capital Investment	Probability of Loss in any Year
A	$ 75,000	0.30
B	$ 90,000	0.10
C	$100,000	0.05

If a loss does occur, it will cost $95,000 with a probability of 0.85 and $130,000 with a probability of 0.25. The probabilities of loss in any year are independent of the probabilities associated with the resultant cost of a loss if one does occur. Each alternative has a useful life of 7 years and no estimated market value at that time. The MARR is 10% per year, and annual maintenance expenses are expected to be 8% of the capital investment. It is desired to determine which alternative is best based on expected total annual cost.

14.25 A drainage channel in a community where flash floods are experienced has capacity sufficient to carry 700 cubic feet per second. Engineering studies produce the following data regarding the probability that a given water flow in any one year will be exceeded and the cost of enlarging the channel:

Water Flow (ft³/sec)	Probability of a Greater Flow Occurring in Any One Year	Capital Investment to Enlarge Channel to Carry This Flow
700	0.25	
1,000	0.15	$25,000
1,300	0.10	$35,000
1,600	0.07	$50,000
1,900	0.06	$65,000

Records indicate that the average property damage amounts to $20,000 when serious overflow occurs. It is believed that this would be the average damage whenever the storm flow is greater than the capacity of the channel. Reconstruction of the channel would be financed by 40-year bonds bearing 8% interest per year. It is thus computed that the capital recovery amount for debt repayment (principal of the bond plus interest) would be 8.39% of the capital investment, because (A/P,8%,40) = 0.0839. It is desired to determine the most economical channel size (water-flow capacity).

	A	B	C	D	E	F
1						
2	Interest rate	8%				
3	Years	40				
4	(A/P,8%,40)	0.0839				
5	Average property damage amount			$20,000		
6						
7	Water flow	Probability	Capital investment	Capital recovery amount	Expected annual proprty damage	Total expected annual cost
8	700	0.25			$5,000	$5,000
9	1,000	0.15	$25,000	$2,097.50	$3,000	$5,098
10	1,300	0.1	$35,000	$2,936.50	$2,000	$4,936.50
11	1,600	0.07	$50,000	$4,195.00	$1,400	$5,595.00
12	1,900	0.06	$65,000	$5,453.50	$1,200	$6,653.50

14.26 A flagship hotel in Cedar Falls must construct a retaining wall next to its parking lot due to the widening of the city's main thoroughfare located in front of the hotel. The amount of rainfall experienced in a short time may cause damage in varying amounts, and the wall increases in cost to protect against larger and faster rainfalls. The probabilities of a specific amount of rainfall in a 30-minute period and wall cost estimates are as follows:

Rainfall Inches Per 30 Minutes	Probability of Greater Rainfall	First Cost of Wall, $
2.0	0.7	225,000
2.25	0.5	300,000
2.5	0.09	400,000
3.0	0.05	450,000
3.25	0.008	500,000

The wall will be financed through a 6% per year loan. The principal and interest will be repaid over a 10-year period. Records indicate that average damage of $50,000 has occurred with heavy rains, due to the relatively poor cohesive properties of the soil along the thoroughfare. A discount rate of 6% per year is applicable.

Find the amount of rainfall to protect against by choosing the retaining wall with the smallest AW value over the 10-year period.

14.27 Decision D4, which has three possible alternatives—x, y, or z—must be made in year 3 of a 6-year study period to maximize the expected value of present worth. Using a rate of return of 20% per year, the investment required in year 3 and the estimated cash flows for years 4 through 6, determine which decision should be made in year 3.

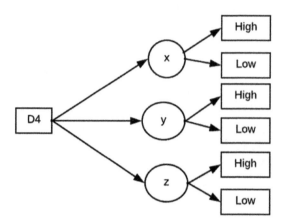

	Investment Required				
	Years				Outcome
High/Low	3	4	5	6	Probability
High(x)	200,000	500,000	500,000	500,000	0.7
Low(x)	200,000	400,000	300,000	200,000	0.3
High(y)	75,000	300,000	400,000	500,000	0.45
Low(y)	75,000	300,000	300,000	300,000	0.55
High(z)	350,000	190,000	170,000	150,000	0.7
Low (z)	350,000	−300,000	−300,000	−300,000	0.3

14.28 A company planning to borrow $15.5 million for a plant expansion is not sure what the interest rate will be when it applies for the loan. The rate could be as low as 10% per year or as high as 12% per year for a 7-year loan. The company will only move forward with the project if the annual worth of the expansion is below $6.5 million. The M&O cost is fixed at $4.2 million per year. The salvage could be $2 million if the interest rate is 10% or $2.5 million if it is 12% per year. Is the decision to move forward with the project sensitive to the interest rate and salvage value estimates?

Chapter 15

Other Considerations

LEARNING OBJECTIVES

- How to account for inflation
- Analysis of government projects
- Rationing of capital
- Selection of MARR
- Uncertainties in the data
- Computer as a tool
- An automation project case study

Throughout this text the basic principles of engineering economics have been emphasized. They have been illustrated through their applications in decision-making pertaining to investment in engineering resources. What you have learned so far is sufficient to analyze a large variety of industrial projects. Some projects, however, may require considerations beyond what we have covered so far. Some of these considerations are of an advanced nature; their comprehensive discussions are beyond the scope of this text. However, the important adjunct considerations are briefly presented in this last chapter to apprize you of their ramifications to economic analyses. Engineering economists and managers need to be aware of their possible effects on the decision, especially if the best alternative you are ready to select is only marginally better than the second best.

15.1 INFLATION

Inflation is a fact of life. It denotes a general increase in the prices of goods and services. Governments throughout the world try to keep inflation under control. Well-managed economies under stable governments, for example that of the U.S., usually succeed in maintaining a reasonable inflation rate. U.S. inflation in the 1990s was in the 2–3% range. In many countries, inflation is in double digits (10% and above). At times, it can get out of control, as in Europe during the Second World War, in the U.S. during the 1970s oil embargo, and in Russia in the 1990s following the demise of the Soviet Union. The term *hyperinflation* is used to describe extremely high inflation, usually out of control.

Prices may at times decrease rather than increase, resulting in *deflation*, as happened following the Second World War. Deflation results when the supply of goods or services is more than the demand. A deflationary economy may occur when the industrial infrastructure undergoes severe perturbation, usually due to political instability or collapse. In stable times, inflation is the norm.

15.1.1 Inflation Rate

Inflation is expressed as the *rate of increase* in the prices. The rate is measured by considering the prices of a "basket" of consumer goods and services. The use of the term basket is metaphorical, encompassing those items that fulfill the basic needs of an average consumer. If the cost of the items in this basket was $350 a year ago, but is $370 now, then the prices increased by a factor of 370/350 = 1.057, i.e., at an annual *inflation rate* of 5.7%.

Another yardstick for measuring inflation is the *consumer price index* (CPI). The prices prevailing in an arbitrary (reference) year are represented by an index of 100. The prices in the subsequent years are related to this index. In the U.S., an index of 100 was allocated most recently to the general prices of 1983. Since then the CPI has increased as per the following table. Based on the previous year's CPI, the annual increase for any year can be evaluated, as expressed in the last column. For example, for 1995 the increase was (CPI_{1995} − CPI_{1994})/CPI_{1994} = (152.4 − 148.2)/148.2 = 0.0283 = 2.83%. The percentage in the last column is thus a measure of the annual inflation rate.

Year	CPI	Annual Increase, %
1983	100	
....	....	
1990	130.7	
1991	136.2	4.21
1992	140.3	3.01
1993	144.5	2.99
1994	148.2	2.56
1995	152.4	2.83
1996	156.9	2.95
1997	160.3	2.16
1998	163.5	2.00
1999	167.8	2.00 (estimate, not actual)

When the value of CPI becomes large over time, the reference (or base) year is reset, i.e., advanced to another recent year, to which a new price index of 100 is allocated. This advancing of the base year keeps the CPI values in manageable three digits (before decimal). The CPI-based measurement of inflation rate is practiced by other countries as well, but the base year may vary.

The inflation rate in a country is the result of a complex interaction among several variables, such as money supply, government policies, currency exchange rates, quality of infrastructure, and the level of political stability.

15.1.2 Analysis

Inflation diminishes the purchasing power of future sums. Consider, for example, an investment last year of $100. At 8% interest rate, this has become $108 today—$8 more. During the year, however, the prices have risen too. That means what can be purchased this year with a sum is, due to inflation, less than what could have been bought last year with that sum. If the inflation rate f is assumed to be 2%, for example, then the inflation factor is 102/100 = 1.02. Thus, with reference to last year's purchasing power, the *worth* of today's $108 = $108/1.02 = $105.88. We denote $108 as *actual* dollars, while $105.88 as *inflation-modified* dollars. Referring to the cashflow diagram in Fig. 15.1, the vector $108 *has not*

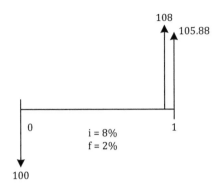

Figure 15.1 Inflation-modified cashflows.

accounted for inflation, while the $105.88 vector has. When the future dollars (cashflows) of a project are inflation-modified, they are called *current* dollars to signify that they represent the same purchasing power as of now. From an inflation viewpoint, they are the *real* dollars.

To account for inflation in engineering economic analyses, the cashflows are first modified. The inflation-modified cashflows are then analyzed the usual way, by any of the methods discussed in the text. Examples 15.1 through 15.3 illustrate the procedure.

Example 15.1: Analysis with Inflation

For 10% MARR, determine the present worth of the following cashflows by considering (a) no inflation, and (b) a 4% inflation.

Year	0	1	2	3	4
Cashflow	−$500	$300	$300	$300	$300

SOLUTION:

(a) The cashflows remain as given. The $500 investment yields a $300 annual benefit over 4 years (n = 4). Thus,

$$PW = PW_{benefits} - PW_{costs}$$

$$= 300\,(P\,/\,A, 10\%, 4) - 500$$

$$= 300 \times 3.170 - 500$$

$$= 951 - 500$$

$$= \$451$$

(b) To account for inflation, the cashflows are modified. Since the rate is 4%, this is done by dividing the future cashflows by 1.04 for each year into the future. Note that 4% inflation rate means an inflation factor of one plus fractional value of the inflation rate, i.e., 1 + 0.04 = 1.04. For an inflation rate f in percentage, the factor is 1+(f/100); where f is in fraction, it is (1 + f). Thus, the inflation-modified value of the:

First − year $300 cashflow = 300 / 1.04 = 288

Second − year $300 cashflow = $300 / (1.04 \times 1.04) = 277$

Proceeding the same way, the inflation-modified value of the:

Third − year $300 cashflow = $300 / (1.04 \times 1.04 \times 1.04) = 266$

Fourth − year $300 cashflow = $300 / (1.04 \times 1.04 \times 1.04 \times 1.04) = 256$

A more efficient way to modify the cashflows is to divide the cashflow-of-now by the inflation factor $(1 + f)^n$, where n is the year. For example, the third-year's modified cash-flow = $300/(1 + 0.04)^3 = 300/1.04^3 = 266$.

Another way is to use the previous year's value and divide it by $(1 + f)$. For example, the third-year's modified cashflow = (second-year's cashflow)/1.04 = 277/1.04 = 266.

The inflation-modified cashflow table becomes:

Year	0	1	2	3	4
Cashflow	−$500	$288	$277	$266	$256

Note that the consideration of inflation has disturbed the uniform-series pattern in the given cash inflows (benefits). Therefore, the functional notation (P/A,i,n) is no longer applicable. We have to handle each benefit separately using the functional notation (P/F,i,n). Thus, the present worth of the cashflows, taking inflation into account, is

$$PW = PW_{benefits} - PW_{costs}$$

$$= \{288(P/F,10\%,1) + 277(P/F,10\%,2) + 266(P/F,10\%,3)$$

$$+256(P/F,10\%,4)\} - 500$$

$$= (288 \times 0.9091 + 277 \times 0.8264 + 266 \times 0.7513 + 256 \times 0.6830) - 500$$

$$= (262 + 229 + 200 + 175) - 500$$

$$= 866 - 500$$

$$= \$366$$

Note that the inflation-modified PW ($366) is lower than that in part (a) where infla-tion was not considered ($451). By ignoring the inflation, the analyst might make an erroneous decision about a project, especially if the selected alternative is marginally better than the second best.

Example 15.2: Cashflow Tables

A hardness testing machine is being considered for acquisition at a cost of $10,000. Its useful life and salvage value are estimated to be 5 years and $3,000. The annual saving from in-house testing using this machine is likely to be $3,000. Prepare[1] the cashflow table if annual inflation during the machine's useful life is projected to be 4%.

---

[1] As a rule of thumb, if inflation is given in a problem, then it must be considered in the analysis even if not asked explicitly.

SOLUTION:

To prepare the cashflow table, we follow the procedure learned in Chapter 2. The skeleton of the table is as follows, where the first column is for the time period and the second column for the cashflows.

Year	0	1	2	3	4	5
Cashflow						

With the given cost and benefits posted, the cashflow table with no consideration of inflation becomes:

Year	0	1	2	3	4	5	Salvage Value
Cashflow	$10,000	$3,000	$3,000	$3,000	$3,000	$3,000	$3,000

Since inflation's effect on the cashflows is to be accounted for, we add two more columns for this purpose, and also combine the two cashflows for period 5, as follows. Under the third column we post the inflation factors, while the last column contains the inflation-modified cashflows.

Year	Cashflow	Inflation Factor	Modified Cashflow
0	$10,000		
1	$ 3,000		
2	$ 3,000		
3	$ 3,000		
4	$ 3,000		
5	$ 6,000		

Since annual inflation is 4%, the inflation factor[2] is $(1 + 0.04)^n$, where n is the year. On posting the given data, the cashflow table looks like:

Year	Cashflow	Inflation Factor	Modified Cashflow
0	−$10,000		−$10,000
1	$ 3,000	1.04^1	
2	$ 3,000	1.04^2	
3	$ 3,000	1.04^3	
4	$ 3,000	1.04^4	
5	$ 6,000	1.04^5	

Note that the $10,000 cashflow is of now, and hence not modified. Only the future cashflows are modified, by dividing them by their inflation factors. For example, for year 3, inflation-modified cashflow = third-year cashflow/$(1 + f)^3$ = $3,000/1.04^3$ = 2,667. The complete results are in the following table.

---

[2] The inflation factor equals $(1 + f)n$, where f is the inflation rate, and n the time period; n is usually expressed in years.

Year	Cashflow	Inflation Factor	Modified Cashflow
0	−$10,000		−$10,000
1	$ 3,000	1.04^1	$ 2,885
2	$ 3,000	1.04^2	$ 2,774
3	$ 3,000	1.04^3	$ 2,667
4	$ 3,000	1.04^4	$ 2,564
5	$ 6,000	1.04^5	$ 5,466

Example 15.3: Payback Considering Inflation

A small hydraulic machine is being planned for acquisition at a cost of $5,000. During its useful life of 7 years, it is expected to generate an annual profit of $3,400. Its annual maintenance cost is projected to be $300. The cost of space to house the machine and of the utilities is $900 per year. Determine the payback period for the machine if the annual inflation rate is expected to be 5%.

SOLUTION:

This is Example 5.1, except that inflation is to be considered here. To evaluate the payback period, we need to determine the duration in which the $5,000 investment will be recovered.

The recurring expenditure comprises the space and utility cost of $900 and the maintenance cost of $300, both annual. Thus, the total operating cost is $900 + $300 = $1,200 per year. The net benefit from the investment in the machine is obtained by subtracting this total from the profit. This yields a net benefit of $3,400 − $1,200 = $2,200 per year, which needs to be modified for inflation.

The modification for inflation disturbs the uniform-series pattern of benefits, as seen in Example 15.1. Consequently, the payback-period formula applicable for uniform cash inflows cannot be used. Instead, the approach explained in Example 5.2 for non-uniform benefits where the *cumulative* value of the net benefit is tracked is applicable. At the payback period the cumulative net benefit equals the $5,000 investment. The calculations can be streamlined by tabulating the data and the results, as follows.

Year	Net Benefit	Inflation Factor	Modified Net Benefit	Cumulative (MOD) Net Benefit
1	$2,200	1.05^1	$2,095	$2,095
2	$2,200	1.05^2	$1,995	$4,090
3	$2,200	1.05^3	$1,900	$5,990
4	$2,200	1.05^4		
5	$2,200	1.05^5		
6	$2,200	1.05^6		
7	$2,200	1.05^7		

The inflation factors in the third column should be obvious. The modified net benefit is obtained by dividing the net benefit by the corresponding inflation factor. For example, for year 2, it is $2,200/1.05^2 = 1,995$.

It is not essential to compute all the results for the last two columns of the table. To save time and effort, one should keep track of the cumulative value (last column) as calculations proceed. Once this value is equal to or greater than the investment ($5,000), the calculation ceases because the payback period becomes known by then.

The table shows that by the end of the third year, with cumulative net benefit of $5,990, the $5,000 investment is more than recovered. Hence the payback period is less than 3 years. The year-2 value of $4,090 suggests that it is greater than 2 years. Therefore, interpolate these two data, getting a payback period of 2.48 years, as shown as follows.

$$\text{Payback period} = 2 + (5,000 - 4,090) / (5,990 - 4,090)$$

$$= 2 + 910 / 1,900$$

$$= 2 + 0.48$$

$$= 2.48 \text{ years}$$

Should one interpolate to get the exact period for the payback? That depends on the accounting practice. If the bookkeeping follows year-end convention—usually so—then the payback period in this example is 3 years. However, if the revenues are realized round the year, and so accounted, then the exact answer of 2.48 years makes better sense.

In Example 5.1 where inflation was ignored, the payback period was determined to be 2.27 years. This example's payback period of 2.48 years is longer since inflation delays the recovery of investment by diminishing the value or purchasing power of future benefits.

The inflation rate f decreases the value of future sums by the factor $(1 + f)^n$. The interest rate i, on the other hand, increases a present sum by $(1 + i)^n$. Since the effect of f on future cashflows is also exponential, the compounding interest tables can be used to read off the values of inflation factors. The functional factor (P/F,i,n) can be used to account for the effect of inflation on future cashflows by considering it as the inflation factor (P/F,f,n). For example, under 6% inflation rate, a $350 future sum 5 years hence will in current dollars be 350(P/F,6%,5) = 350 × 0.7473 = $261.56.

In Chapters 1–12, we have been using interest rates with no consideration of inflation. Such a rate is called *market interest rate* since the market does not account for inflation while quoting the interest rates prevalent at the time. Inflation modifies the market interest rate as per

$$i_f = (i - f) / (1 + f) \tag{15.1}$$

where,
  i_f = "real" interest rate that compensates for inflation
  i = market interest rate, and
  f = inflation rate

Consider that a local bank offers its savers annually compounded 8% interest rate per year, i.e., $i = 8\%$. If inflation (f) is 2%, then from Equation (15.1), substituting 0.08 for i and 0.02 for f, we have

$$i_f = (0.08 - 0.02) / (1 + 0.02)$$

$$= 0.0588$$

$$= 5.88\%$$

In other words, savers *really* earn only 5.88%, not 8%. The inflation has "eaten" into the interest. Smart savers are aware of this fact, and while investing consider the *real* interest rate i_f that hedges against inflation.

If f is low[3], the 9.5% "real" interest rate approximately equals the 10% market interest rate. Compared to i, then from Equation (15.1) the real interest rate i_f approximately equals the market interest rate i. This is the situation we have been assuming in Chapters 1 through 12.

In Example 15.3, we saw that inflation extends the payback period of an investment. When ROR analysis is carried out to account for inflation, there is a similar effect: after-tax ROR is reduced. Thus, from the investor's viewpoint, inflation is bad since investment takes longer to recover, or the rate-of-return gets lower. From the borrower's viewpoint, however, inflation is good. The payments made in the future will be of reduced purchasing power. This is one reason why people borrow to finance their homes and other assets, especially in an inflationary economy.

If the project's analysis period is short, say 3 to 5 years, and inflation rate is low, then the effect of inflation on the cashflows can be ignored. This is what has been assumed in Chapters 1 through 12. Moreover, since the operational costs and benefits of all the alternatives of a project are likely to be affected by the same degree, inflation's impact on the cashflows may cancel out. Thus, in most cases, a decision without considering inflation is likely to be the same as that with inflation considered. However, if the best alternative is only marginally better than the second best, a reanalysis accounting for the inflation should be carried out to confirm the validity of the decision.

15.2 PUBLIC PROJECTS

Public projects are funded not for profit, but for the public good. They are usually government projects or those of nonprofit organizations. A number of organizations deal with public projects, such as school districts, charity hospitals, and the International Red Cross. The basic concept of costs and benefits applies to public projects as much as they apply to private projects. But the profit-oriented criteria such as ROR and payback period become irrelevant in public projects. Public projects are analyzed on the basis of benefit-cost ratio (or difference), as discussed in Chapter 8. For a project to be acceptable, its benefit-cost ratio should be greater than one. In the case of multi-alternative projects, the analyst must use the incremental method, accepting the higher-cost alternative only if it's incremental B/C ratio is greater than one.

Three major questions arise while analyzing government projects for funding considerations:

1. Who should pay for the project?
2. Who should derive the benefits?
3. What interest rate should be used in the analysis?

Based on commonsense, only those who will benefit from the project should pay. But how do you establish that, and implement it? If the school district is planning to build a new classroom, should the in-district senior citizens with no school-age children pay toward the construction? As another example, should the city restrict its park only to those who pay city taxes? How do you ensure that only those who paid for the facility use it? How do you police it? Should the city issue identity cards to its residents and require the card for admission into the park? Thus, the first two questions are really complex, with no easy answer.

---

[3] In the U.S., for example, during the 1990s, f was 2–3% while i was 8–12%. Approximating these as f = 2.5%, and i = 10%, from Equation (15.1),if = (0.10 – 0.025)/(1 + 0.025)= 0.0975/1.025= 0.095= 9.5%

Since they involve social and political considerations, the analysis of public projects may be more than economic.

The third question is equally perplexing. If the money to be spent on a government project has been raised through taxes, is the interest rate zero since it is not borrowed? Or, should the rate be what tax payers would have earned had they not been taxed? According to the U.S. federal government's Office of Management and Budget, it is economically unsound to take money from taxpayers who would have earned 12%, and invest it in government projects that yield 4%!

Another difficulty experienced in analyzing government projects is the treatment of the operating costs. Should the operating costs be paid out of the future revenues, or should they be budgeted in the beginning, and paid out later, as part of the investment cost? The benefit-cost ratio of a project differs depending on how the operating costs are treated. This has been illustrated in Chapter 8 through Example 8.2 and its appertaining discussions.

15.3 CAPITAL RATIONING

Most often than not companies have limited capital and several competing projects, and thus fail to fund all of them. In such situations capital is rationed, i.e., allocated to the most attractive projects. The rationing procedure involves evaluating the projects' RORs and ranking them in descending order. The allocation of the capital begins from the top of the list, going down until all the available capital has been assigned. Although Example 15.4 illustrates the procedure for ROR-criterion, capital can be rationed under any of the other five criteria.

Example 15.4: Capital Rationing

ABC Incorporated is considering the following projects for possible investment next year. Each project has a 5-year useful life and no salvage value. For 15% MARR, and a capital outlay of $25,000, which projects should be funded under ROR?

Project	1	2	3	4	5	6	7	8
First Cost	$5,000	$4,500	$2,500	$3,000	$5,000	$5,500	$6,000	$7,000
Annual Benefit	$1,319	$1,673	$ 836	$ 771	$1,492	$1,759	$1,963	$2,341

SOLUTION:

The solution procedure comprises the following steps. If the RORs are given, skip step 1.

Step 1: Evaluate the ROR of each project

All the projects have a 5-year useful life, i.e., n = 5. Calculate the ROR of each project as learned in Chapter 7. For example, for project 3, we have:

$$2,500 = 836(P/A, i, 5)$$

Thus,

$$(P/A, i, 5) = 2,500/836$$

$$= 2.9904$$

From the interest tables, for the above value of the factor, $i \approx 20\%$. Thus, the rate of return for the third project is 20%. Once all the RORs have been evaluated, they are tabulated, as follows.

Project	First Cost	Annual Benefit	Computed ROR, %
1	$5,000	$1,319	10
2	$4,500	$1,673	25
3	$2,500	$ 836	20
4	$3,000	$ 771	9
5	$5,000	$1,492	15
6	$5,500	$1,759	18
7	$6,000	$1,963	19
8	$7,000	$2,341	20

Step 2: Discard the unattractive projects

Check to see which of the projects have ROR less than the desired MARR, and discard them. In this case, for 15% MARR, projects 1 and 4 are discarded.

Step 3: Arrange the remaining projects

Arrange the remaining projects in descending order of ROR, as follows.

Project	First Cost	Annual Benefit	Computed ROR, %
2	$4,500	$1,673	25
3	$2,500	$ 836	20
8	$7,000	$2,341	20
7	$6,000	$1,963	19
6	$5,500	$1,759	18
5	$5,000	$1,492	15

Step 4: Allocate the available capital

Begin to allocate the available capital. Since project 2 tops the list due to its highest ROR, its required capital of $4,500 is allocated first. The next two projects, 3 and 8, have the same ROR, so they are allocated together, but 3 is preferred due to its lower capital need. The allocations can be tabulated in another column, as follows.

Project	First Cost	Annual Benefit	Computed ROR, %	Allocation
2	$4,500	$1,673	25	$4,500
3	$2,500	$ 836	20	$2,500
8	$7,000	$2,341	20	$7,000
7	$6,000	$1,963	19	
6	$5,500	$1,759	18	
5	$5,000	$1,492	15	

The allocation process is continued until all the available capital has been used up. To keep track of this constraint another column is created, as below, to record the cumulative allocation.

Project	First Cost	Annual Benefit	Computed ROR, %	Allocation	Cumulative Allocation
2	$4,500	$1,673	25	$4,500	$ 4,500
3	$2,500	$ 836	20	$2,500	$ 7,000
8	$7,000	$2,341	20	$7,000	$14,000
7	$6,000	$1,963	19		
6	$5,500	$1,759	18		
5	$5,000	$1,492	15		

As seen in the last column, we have thus far allocated $14,000 out of the available $25,000. So the allocation continues. With the next allocation, the cumulative allocation increases to $20,000.

Step 5: Is there any remaining capital?

After allocating to as many projects as possible, you may be left with some capital. What to do with the remaining capital? There are two choices depending on how much is left.

a. If the remaining capital is significantly lower than what is needed to fund the next best project, it is usually returned to the company capital pool for use elsewhere.
b. If the remaining capital is closer to what is needed to fund the next best project, and then try to acquire the required additional capital and fund the next project.

In this particular case, after allocating the best four projects at a total cost of $20,000, we are left with $5,000 (= $25,000 − $20,000). The next best project, namely 6, requires a capital of $5,500. If we had another $500 we could fund project 6 too. This fact should be discussed with the capital budgeting authority through the department manager. Negotiate with the budget officer, providing additional information on the project if needed, and highlighting its merits. The budgeting authority might release the additional $500 provided project 6 is among the other worthy ones within the company.

The best of the projects that could not be funded represents an investment opportunity lost due to a paucity of capital. In the example, in the given $25,000 capital and assuming no success with acquiring the additional $500, project 6 was the best unfunded project. The rate of return of the best unfunded project is called *opportunity cost*. In Example 15.4, therefore, the opportunity cost is 18%. Note that the unit of opportunity cost is %, not dollars. If the additional $500 could be acquired and project 6 funded, the opportunity cost would be 15%, the ROR of project 5.

15.4 MARR SELECTION

MARR is the minimum rate of return a project must be expected to earn before it can be approved for funding. It is fixed by the upper management on the basis of prevailing interest rates in the financial marketplace and other business considerations. Engineers or engineering technologists, especially in medium and larger companies, may not be involved in fixing the MARR. However, if working in small companies or in their own firms, they may face the task of deciding the value of MARR to be used in economic analyses. In this section we will discuss some basic considerations pertaining to this complex task.

The capital for investment may be available internally, from the profits generated within the company. While deciding to use its own profit as capital, the company must however consider the investment opportunities outside. The MARR to be used in the own-profit-as-capital case must be greater than, or at least equal to, the interest the capital can earn in the financial marketplace. For example, if the capital can be loaned to a financial institution or another company at 10%, then MARR must be 10% or more. How much more should depend on the relative risks in loaning the capital or investing it within the company on engineering projects.

Alternatively, the capital for investment may be borrowed from financial institutions. In such cases, the MARR should obviously be more than the interest rate for the borrowed capital. Again, how much more should depend on the risk in the engineering investment. Oftentimes, the required capital may have to be raised from more than one financial institution, whose interest rates may be different. In such a case, a weighted average interest rate is determined to account for the variations in interest rates of borrowed capital; MARR should be greater than this average interest rate.

After considering the opportunity cost discussed in section 15.3, the opportunity of loaning the capital to the financial marketplace, the cost or interest rate of borrowed capital, and any government investment incentive available, the value of MARR is decided upon. This value also takes into account the risk and uncertainty involved with the engineering investment, as well as the overall prevailing business climate, both within the company and without. As and when the financial market conditions change, the MARR is reviewed and readjusted if necessary. The decision about what MARR to use is thus complex, dynamic, and challenging.

15.5 DATA UNCERTAINTY

Another important consideration in economic analyses is to ensure that the data being used are reliable. Throughout the text we have assumed the data to be deterministic. If we state that the salvage value of machine A is $5,000, we are implying a 100% certainty[4] to this value. But how could we be so sure? We know that the salvage value relates to the future; we can be certain about its value only after it would be salvaged. But we need this data for analysis now—at the present time—when the investment decision is being made. The only data whose values we can be certain about are those of the past or pertaining to the present. For example, we can be sure of the cost of the machine under consideration for investment since the cost is based on a binding quotation. This cost as a cashflow is 100% certain because it is paid following the purchase.

Future data are nondeterministic; they can only be *estimated*. One cannot be 100% sure of their values. Further the data is in the future, the more the uncertainty in their values. For example, the useful life of a machine can only be estimated. The reliability of estimation is higher if the useful life, for example, is 3 years than if it is 12 years.

In engineering economics, as in other disciplines, data values should be estimated carefully, using statistical techniques wherever necessary. The basic approach is to analyze the past data, if available, and use the results to predict the future value. This process is called forecasting. It is also called *time-series analysis* because the past data form a series that occurred at different times.

---

[4] To avoid giving this impression, we have preferred to use throughout the text statements such as: The salvage value of machine A is *expected to be* $5,000.

A comprehensive economic analysis that takes into account the futuristic nature of the data and allows for statistical analyses in their estimation is beyond the scope of this text. However, some elementary considerations are presented in the next two subsections.

15.5.1 Optimistic-Pessimistic Approach

The statistical nature of the data can be accounted for by considering their extreme possible values along with the most likely value. Let us say that you are not sure of the useful life of a hydraulic machine being analyzed. You have determined its most likely value to be 6 years. But others in the company think it is likely to be 7 years, while some suggest it to be 5 years. Those who think it will last for 7 years are optimists by nature, while those who suggest 5 years are pessimists. Thus, there are three different values for the same data: 6 years as the most likely value, 7 years as the optimistic value, and 5 years as the pessimistic value. The question is: Which value is relevant and should be considered? The analyst may prefer to use the most likely value in making the decision, as done in Chapters 1 through 12. However, the optimists and pessimists consider the most likely value erroneous. Should the analyst ignore their values? These issues require statistical considerations in the analysis.

In the most elementary approach, the statistics is kept simple. The data values are incorporated in the analysis by using the mean value of the variable(s). The mean value is obtained by weighting the data, which may give equal or unequal importance to the most likely and pessimistic-optimistic values. With equal importance to all the values we obtain the arithmetic mean. One of the weighted mean is based on the assumption that the data satisfy a β-distribution. For such a case,

$$\text{Mean} = \frac{\text{Optimistic Value} + 4(\text{Most likely Value}) + \text{Pessimistic Value}}{6}$$

Example 15.5 illustrates the application of the most likely and optimist-pessimist data values in decision-making. Note how the solution has become lengthier even with simplistic statistics.

Example 15.5: Optimistic-Pessimistic Approach

A small hydraulic machine is being analyzed for acquisition. The engineering economist is sure only of its cost, which is $5,000. For the values of the other three data, he conducts some research. Based on input from an insurance company about the machine's useful life and salvage value, and from within the company about net annual benefit, he has determined the following most likely values, along with their optimistic and pessimistic values.

	Optimistic	Most Likely	Pessimistic
Net Annual Benefit	$3,000	$2,200	$1,000
Useful Life (yrs)	7	6	5
Salvage Value	$ 500	$ 300	$0

Should the machine be funded for investment if the payback period for approval is not to exceed 2 years, and MARR is 10%? Assume that the data follow β-distribution.

SOLUTION:

Under β-distribution, as just discussed, the most likely value is given fourfold weighting than the optimistic or pessimistic value. In other words, the most likely value is counted four times, giving a total of six data values. The mean value of the data is therefore obtained by adding these six data values and dividing the total by six, as follows:

$$\text{Mean} = \frac{\text{Optimistic Value} + 4(\text{Most likely Value}) + \text{Pessimistic Value}}{6}$$

For the net annual benefit, the mean value is:

$$\text{Benefit}_{\text{mean}} = \frac{\$3,000 + 4(\$2,200) + \$1,000}{6}$$

$$= \$2,133$$

The mean value of useful life is:

$$\text{Life}_{\text{mean}} = \frac{5 + 4 \times 6 + 7}{6}$$

$$= 6 \text{ years}$$

The mean salvage value is:

$$\text{Salvage}_{\text{mean}} = \frac{\$500 + 4 \times \$300 + 0}{6}$$

$$= \$283$$

The problem is now solved the usual way, as in other chapters, using the mean values of the parameters. So, A = $2,133, n = 6 years, and S = $283. Assuming the salvage value to be an element of the $5,000 investment which is reduced by recovery through salvaging,

$$\text{Investment} = \$5,000 - 283(P/F,10\%,6)$$

$$= \$5,000 - 283(P/F,10\%,6)$$

$$= \$5,000 - 283 \times 0.5645$$

$$= \$5,000 - \$160$$

$$= \$4,840$$

$$\text{Payback period} = \text{Investment} / \text{Annual benefit:}$$

$$= \$4,840 / \$2,133$$

$$= 2.27 \text{ years}$$

Since the payback period is greater than the desirable 2 years, the machine should not be funded for investment.

15.5.2 Statistical Analysis

In the more accurate statistical analyses, the past data are fully analyzed for their statistical characteristics. *Statistics* is concerned with the collection, analysis, and interpretation of quantitative data. Data may be of the attribute type or variable type. *Attribute data* can have only two values: conforming or non-conforming, pass or fail, go or no-go. *Variable data,* on the other hand, can have any value in steps of the measuring system's least count. Most engineering economics data are of the variable type. The useful life of equipment, expressed as 5 years, for example, is a variable data. The value of MARR is more like an attribute data, beyond which the project is funded and below which it is not.

After collecting the data, they are described through tables, charts, or graphs, and analyzed. The graphical analysis summarizes the data in the form of a *frequency distribution* which is modeled mathematically. The model is used to estimate the value of the parameter and to make predictions. For example, the useful life of equipment may be modeled by a *normal distribution*[5], also called the *bell-curve.* Such a distribution applies to a variety of industrial as well as natural processes. The height of students in a class follows normal distribution—so does the useful life of a machine.

Consider a machine's salvage value. How do we estimate it statistically? It is done by analyzing the salvage values of a large number of machines used in the past. The past data are collected and their frequencies plotted, as illustrated in Fig. 15.2. Note that the salvage values vary between $200 and $600, but they seem to cluster around $400. This $400 is the mean salvage value[6]. In statistical analyses we incorporate the data variation as well. We therefore quantify both the mean and the dispersion, based on the approximating normal distribution, or any other distribution that fits the data.

The normal distribution is mathematically defined as,

$$y(x) = \frac{1}{\sigma\sqrt{2\pi}} e^{-(x-\mu)/(2\sigma)}$$

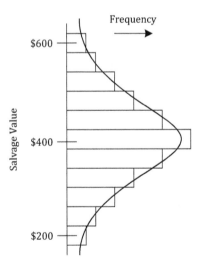

Figure 15.2   Concept of normal distribution.

---

[5] The data that follow this distribution display two characteristics: *central tendency* and *dispersion.* The central tendency is measured in terms of mean, median, or mode, while the dispersion is measured in terms of range or standard deviation.

[6] We have used such a mean value throughout the text as deterministic data having 100% certainty. In other words, we excluded any consideration of statistical variation, which may be acceptable in many industrial projects.

where

y(x) = ordinate of the curve corresponding to x,
μ = mean of all the x-values, and
σ = standard deviation.

The standard deviation is a measure of dispersion in the values of the variable x. Thus, by using μ and σ we are able to describe the salvage value in greater detail. We are not saying any more that the salvage value is $400, as done earlier in Chapters 1 through 12 where no statistical considerations were made. We are instead saying that the salvage value is likely to be $400, but this could be lower or higher as dictated by the value σ.

15.6 COMPUTERS AS A TOOL

Computers play an important role in engineering economics since they can perform calculations almost instantly. They are especially useful in ROR analyses where the calculations are repetitive. Recall how lengthy the trial and error approach to evaluating ROR becomes. However, computers must be programmed to carry out the calculations. The sequence of operations the hardware carries out under the control of a program is called software.

As a tool, the computer has had a phenomenal impact on engineering economics analyses. A plethora of software running on PCs helps engineers make faster and better economic decisions. The ROR-method, the most complex of the six, has benefitted enormously from the software-based analysis. While users may develop their own software by programming the relevant equations in a PC or programmable calculator, most prefer the commercial software that have been developed professionally, and are user-friendly and frequently upgraded. Besides offering a precise, accurate, and complete solution, the software offers what-if capabilities that lead to the optimum decision. Users should, however, be aware of the GIGO (garbage-in-garbage-out) nature of the computer environment. A few step-by-step sample calculations should be done on calculators to confirm the computer results.

While sophisticated software with advanced graphics capabilities are available for analyzing large engineering projects, spreadsheet analysis is usually sufficient for most needs. The spreadsheet is a collection of rows and columns in which equations can be programmed to act on the data contained in certain cells of the spreadsheet. The results are displayed in other cells. The spreadsheet-based analysis is relatively more user-friendly.

Any computer can be used for economic analysis. The personal computer (PC), however, is the most popular platform for engineering economists. A variety of PC-based software is available in the market. Buy the one that suits your needs. Talk to people who have used the software you intend to buy. Computer magazines and professional magazines such as *Engineering Economics* occasionally publish a comparative evaluation of the various software programs on the market.

15.7 AUTOMATION PROJECT—A CASE STUDY

Throughout the text we have emphasized the fundamental principles of engineering economics and their applications to decision-making. To achieve this, the problems were kept manageable. Most of the pertinent data were provided in the problem statement. In the real-world of industry, however, such data may not be readily available; they may have to be collected, rendering the decision-making task lengthy and difficult. Most projects may require efforts by several persons working as a team. As an engineer or engineering technologist, you may be a member of such a team. Other members may come from backgrounds different from yours, for example business, law, or computers.

In this section, we illustrate the economic analysis of a real-world automation project[7]. Since it involves the implementation of a robotic system, it has also been called robotic system, and new system or process. For simplicity of analysis, we apply the payback period criterion. As discussed earlier, the analysis will require two items of data, namely the cost of investment and the expected future benefits. These data may not be easily available and may have to be estimated. The engineer's task is to cost-justify the system based on payback criterion.

The first question the engineer faces is: What payback period is acceptable? His manager may offer a figure based on the company's investment policies. The next question is: Could a longer payback period be justified since the robotic system offers a non-tangible benefit of enhancing the company image as a high-tech firm?

The analysis must consider the likely benefits from the system in relation to the current operation. It has to account for the current costs and determine the likely costs with the robotic system. The difference in these costs represents the net benefits from the investment.

The pertinent data are gathered from company books and various departments, robot vendor and manufacturer, and other sources—both within and without. For example, the robot will cost $35,000 including shipping to the company plant. There are several other costs before the system becomes operational. These are:

15.7.1 Installation Cost

The installation cost includes the costs of moving the robot from where it was delivered to its location in the plant, connection for electricity, any foundational need recommended by the manufacturer, etc. These costs are gathered or calculated, adding up to $2,500.

15.7.2 Training Cost

The operator and other personnel associated with the robotic system will be trained. Determine the total hours of training required for each one and their hourly rates. In this case, an operator, his assistant, and the plant safety officer are to be trained. Note that the hourly rate for an employee is higher than the wage rate because of the fringe benefits, such as health care, vacation, retirement, social security taxes, and others. The operator is trained at a nearby community college under arrangement by the robot vendor. The hourly rate for the operator is $20, though his wage rate is only $15. He takes 25 hours of training at the college, costing the company $500 in lost production. His travel, boarding, and lodging expenses are $450. On his return from the college training he will train his assistant and the plant safety officer, once the robot is operational. The cost on assistant's and safety officer's training is $550. Thus, the total cost on training is $500 + $450 + $550 = $1,500.

15.7.3 End-of-Arm Tooling

A closer look at the variety of parts the robot will handle indicated a need to purchase some off-the-shelf tooling at a cost of $1,850. Other required tooling can be built in-house for which the estimated cost is $1,650. Thus, the total cost on the end-of-arm tooling is $3,500.

---

[7] Adapted from *Molding Systems*, Society of Manufacturing Engineers, September 1998, pp. 33–34.

15.7.4 Maintenance Equipment

The robotic system will require a $250 vibrometer for predictive maintenance.

15.7.5 Safety Guarding

According to the plant safety officer, a guarding is essential to comply with the Occupational Safety and Health Administration's (OSHA) requirements. The guarding will be fabricated in-house for which the material and labor costs are $500 and $1,000. Thus, the total cost on guarding is $1,500.

These costs being essential for rendering the robotic system operational are part of the investment. They can be summarized in a table, as follows. A summation of the costs shows that the total investment[8] in the system is $44,250.

Robot	$ 35,000
Installation	$ 2,500
Training	$ 1,500
End-of-Arm Tooling	$ 3,500
Maintenance Equipment	$ 250
Guarding	$ 1,500
Total	**$44,250**

In a similar way, we need to collect data on the benefits expected from the robotic system. The benefits are the savings from the system, which can be evaluated by comparing the cost of the current process with that of the new system. All types of cost items likely to be affected by the new system should be considered by listing them, and collecting the associated data. Such a task is usually referred to as *cost accounting*. For the present case study, we carry out this task in a table containing four columns with proper headings, as shown as follows. The first column is for the cost item, while columns two and three are for cost data on the current and new process (after automation). The last column will contain their difference as a benefit.

Cost Item	Current Process	After Automation	Benefit

The number and type of cost items gathered will vary from project to project. For the present case, the following items are applicable.

15.7.6 Value of Part

The part produced by the current process is valued at $0.50 which will remain unaffected by the automation. This value represents the revenue generated by the part. In many projects the value of the part may increase if the new system uses less material or extends the part life.

15.7.7 Effect on Output

The effect of automation on the output, if any, is determined next. To do this, compare the production rate under the current process with the new one. The current process produces

---

[8] We have called such data first cost or initial cost and invariably supplied them in the problem statements. Note how extensive the task can be in the real-world, as illustrated in this case study.

15,000 parts per week. The plant operates three 8-hour shifts each day, Monday through Friday, or 120 hours per week. Thus, the total time the process is operational

$$= 120\,hours\,/\,week \times 60\,minutes\,/\,hour \times 60\,seconds\,/\,minute = 432,000\,seconds\,/\,week$$

$$Average\,cycle\,time = Total\,time\,/\,Production\,rate$$

$$= (432,000\,seconds\,/\,week)\,/\,(15,000\,/\,week)$$

$$= 28.8\;seconds$$

It is estimated that the use of the robotic system will reduce[9] the cycle time by 2 seconds to 26.8 seconds. Thus, production will increase to (432,000 seconds/week)/ (26.8 seconds) ≈ 16,120 per week.

Thus, parts produced per year due to automation will increase from 780,000 (= 15,000/ week × 52 weeks/year) to 838,000 (= 16,120/week × 52 weeks/per year). The increased production[10] of 838,000 − 780,000 = 58,000 parts will generate an additional revenue of $29,000 since each part sells for $0.50.

Let us post these data in the table, as follows, before continuing with further data collection.

Cost Item	Current Process	After Automation	Benefit
Value of Part	$0.50	$0.50	
Annual Production	780,000	838,000	$29,120

15.7.8 Labor Cost

Currently one operator is used for each shift. Thus, each day three operators are required, whose annual cost is $121,680. It is estimated that the robotic system will decrease this cost by 25%. Thus, the saving will be 0.25 × $121,680 = $30,420. The labor cost after automation will be 0.75 × $121,680 = $91,120.

The labor cost per part is obtained by dividing the annual labor cost by the annual production. For the given data, the labor costs are:

$$Current\,process = \$121,000\,/\,780,000$$

$$= \$0.16$$

$$Automated\,process = \$91,260\,/\,838,000$$

$$= \$0.12$$

---

[9]   It is based on the actual cycle time of the current process obtained from time and motion study and the estimated cycle time with the robotic system. Note that if the actual cycle time data is unavailable, the engineering economist may have to seek help from industrial engineering staff who will conduct a time study on the current process. This illustrates how additional tasks may creep into the engineering economic analysis.

[10]   The new annual production could have been evaluated simply by multiplying the cycle time ratio with the current production as (28.8/26.8) × 780,000 ≈ 838,000.

These data can also be posted by updating the table as:

Cost Item	Current Process	After Automation	Benefit
Value of Part	$0.50	$0.50	
Annual Production	780,000	838,000	$29,120
Labor Cost	$121,000	$91,260	$30,420
Labor Per Part	$0.16	$0.12	

We continue to consider the other implications of the automation.

15.7.9 Quality Cost

The automation is expected to improve process quality, resulting in lower scraps. Under the current process the scrap is 3%, i.e., $0.03 \times 780,000 = 23,400$ parts per year. The new process is expected to reduce the scrap to 161 parts per week, i.e., $161 \times 52 = 8,370$ per year. The reduced scrap will result in an annual saving of $7,515, as follows:

$$\text{Parts saved annually through lower scarps} = 23,400 - 8,370$$

$$= 15,030$$

$$\text{Annual saving} = \$0.50 \times 15,030$$

$$= \$7,515$$

15.7.10 Maintenance Cost

The robotic-system-based new process, being sophisticated, costs more to maintain than the current one. Its annual maintenance cost will require an additional sum of $2,500. Note that it is a cost which must be paid out of the benefit from the new process. In the data table, it is entered with a negative sign to signify loss.

15.7.11 Insurance Cost

The robotic system will cost $150 less in insurance, which is a benefit.

15.7.12 Utilities Cost

The cost of power (electricity) and compressed air for the robotic system will be $400 more than that for the current system. This data too is a cost of the new system, and therefore carries a negative sign.

On posting this data, the complete table for the new system is as follows. By summing up the various benefits (last column), the net annual benefit from the automation system is determined to be $64,135.

Cost Item	Current Process	After Automation	Benefit
Value of Part	$0.50	$0.50	
Annual Production	780,000	838,000	$29,120
Labor Cost	$121,000	$91,260	$30,420

Labor Per Part	$0.16	$0.12
Quality of Part		$ 7,515
Maintenance		-$ 2,500
Insurance		$ 100
Utilities		-$ 400
	Total:	$64,135

The cost and benefit tables can now be consolidated, as follows.

	Costs		Benefits
Robot	$35,000	**Annual Production**	$29,000
Installation	$ 2,500	**Labor Cost**	$30,420
Training	$ 1,500	**Parts Quality**	$ 7,515
End-of-Arm Training	$ 3,500	**Maintenance**	-$ 2,500
Maintenance Equip	$ 250	**Insurance**	$ 100
Guarding	$ 1,500	**Utilities**	-$ 400
Total Cost:	$44,250	**Annual Benefits:**	$64,135

From the consolidated table,

Payback period = Total cost / Annual benefit

$$= \$44,250 / (\$64,135 \text{ per year})$$

$$= 0.69 \text{ year}$$

$$= 8.28 \text{ months}$$

Note that this calculation for the payback period was a simple task, once the two important data were known. Almost all the efforts of analyzing the automation project pertained to data collection. This is true of most engineering economics analyses in industry, where data gathering is usually more involved than the analysis itself. Throughout the text, we intentionally kept ourselves free of data gathering efforts so that we could focus on the analysis aspects of the problem. This was done by providing the data in the problem statements. In the real-world, data collection is a major part of the decision-making effort, as illustrated by this case study, and is usually time consuming.

15.8 SUMMARY

Realistic economic analyses of engineering projects may involve several considerations, of which the important ones have been discussed in this chapter. Inflation consideration can become pertinent in projects of long useful lives, especially if the inflation rate is high. Government projects have their own special features that must be considered in benefit-cost ratio analyses. Companies with limited capital may not be able to fund all the investment-worthy projects, so they ration the capital. The rationing is based usually on ROR. The fixing of MARR depends on the interest rates prevailing in the financial marketplace and the company's own business "health." Further difficulties in economic analyses arise from the uncertain nature of the data pertaining to the future. Statistical considerations enhance the accuracy and reliability of the decisions, but at the cost of analysis complexities. Economic analysis, especially the data collection, of real-world engineering projects is a lengthy process, as illustrated in this chapter through an automation project case study.

DISCUSSION QUESTIONS

15.1 Write a 200-word essay on inflation's effect on engineering economic analysis.

15.2 What is the inflation rate in your country? What has its trend been over the last 10 years?

15.3 Write a 200-word essay on economic analysis of not-for-profit engineering projects.

15.4 Why do companies ration the capital for engineering investments, and how do they do it?

15.5 Write a 200-word essay on statistical considerations in engineering economics analysis.

15.6 Explain the rationale behind selected a specific MARR on an investment.

15.7 Why is forecasting an accepted means of estimating future costs?

15.8 Describe three examples of costs that could be included in the initial price on an investment.

MULTIPLE-CHOICE QUESTIONS

15.9 Inflation is good for the
   a. lender
   b. borrower
   c. government
   d. a or b, depending on the prevailing interest rate

15.10 A personal computer costs $1,850 today. If the inflation rate f is 6% per year, its cost 2 years from now will be
   a. $1,875.34
   b. $1,978.89
   c. $2,009.66
   d. $2,078.66

15.11 The acronym CPI stands for
   a. current price information
   b. current price index
   c. consumer price index
   d. consumer product information

15.12 The notations μ and σ are associated with ____ distribution.
   a. exponential
   b. beta
   c. normal
   d. uniform

15.13 The notation σ is used to measure the _____ of a bell-shaped frequency distribution.
   a. central tendency
   b. dispersion
   c. height
   d. shape

15.14 Which of the following costs would not be included in the initial cost of an asset?
   a. installation

b. maintenance

c. training

d. none of the above

15.15 Labor cost per unit is equal to:

a. annual labor costs/annual production

b. annual production/annual labor costs

c. annual production/weekly working hours

d. none of the above

15.16 Mathematically modeled graphical analysis summarizes data in the form of:

a. MACRS percentage tables

b. a depreciation schedule

c. a frequency distribution

d. all of the above

15.17 Which type of distribution is commonly used for the useful life of a machine?

a. probability

b. scattered plot

c. exponential

d. normal

15.18 MARR should be greater than:

a. interest earned in the market

b. capital expenditures

c. labor cost

d. dividends paid on investments

15.19 Capital rationing is usually based on which of the following?

a. CPI

b. IRR

c. MARR

d. ROR

NUMERICAL PROBLEMS

15.20 The annual maintenance cost of a machine for the next 5 years is estimated to be $500. What is the present worth of the maintenance costs if inflation is (a) ignored, (b) expected to average 5% per year? Assume an annually compounded interest rate of 8% per year.

15.21 The maintenance cost of a machine is expected to be $500 this year. For the next 5 years it is likely to increase 5% per year due to machine wear and tear. What is the equivalent uniform annual maintenance cost if inflation is (a) ignored, (b) expected to average 5% per year? Assume an annually compounded interest rate of 7% per year.

15.22 If the inflation is 3% per year in Problem 15.21, what is the answer for part (b)?

15.23 Mary borrows $80,000 to buy a machining center. Beginning next month she will pay the lender $850 per month for the next 20 years. What ROR will the lender be enjoying if the annual inflation rate is 8%? (This is Problem 7.16, but with inflation.)

15.24 Two models of pollution-control equipment are under consideration to meet the legal air emission standard. The basic model costs $8,500 and will last for 5 years; the deluxe model costs $12,000 and will last for 10 years. If the annual inflation rate is 6%, which model should be purchased? Assume MARR = 12%. (Hint: Consider a 10-year analysis period, noting that the replacement cost of the basic model at the end of the fifth year will be higher due to inflation.)

15.25   Poonam earns $5,000 as an annual bonus and invests it for 10 years at an annual interest rate of 12%, compounded yearly. If the annual inflation is projected to be 3% per year during the next 5 years and 5% thereafter, how much will she get as (a) actual dollars, (b) current dollars? (Hint: In (b) the purchasing power is maintained.)

15.26   The optimistic, most likely, and pessimistic costs of a new piece of equipment are $300, $320, and $350 respectively. What is its mean cost using weighting factors that approximate β-distribution?

15.27   Everywhere-But-Here Trucking Company is evaluating the following projects for investment. Each project has a 5-year useful life and no salvage value. For MARR of 12% and a capital outlay of $20,000, which projects should be funded under ROR?

Project	1	2	3	4
First Cost	$7,500	$3,000	$8,000	$4,500
Annual Benefit	$ 800	$ 300	$2,500	$1,000

15.28   A seismic meter is acquired by an oil exploration company for $23,000. Its useful life is 8 years and it has a salvage value of $3,000. The annual savings resulting from this meter is $1,200. Prepare the cashflow table if the annual inflation during the useful life of the meter is 3.5%.

15.29   For 12% MARR, determine the present worth of the following cashflow considering (a) no inflation and (b) 3% annual inflation.

Year	Deluxe
0	-$70,000
1	$30,000
2	$28,500
3	$27,000
4	$25,500
5	$24,000

15.30   A new machine for the manufacture of plastic cutlery costs $158,000. The machine will increase the monthly production by 20,000 packs. Each pack costs 50 cents to manufacture and is sold for 75 cents. What is the payback period for the machine if the annual inflation rate in 4%?

15.31   Determine the payback period of a project whose cashflows are:

Year	0	1	2	3	4
Cashflow	-$3,500	$1,750	$1,200	$1,050	$950

Assume an annual maintenance fee of $200 and an inflation rate of 3.5%.

15.32   A new hydraulic drilling machine is being considered for purchase. The lead engineer says that the drill will cost $12,000. After talking to several insurance experts, the engineer determined the most likely, pessimistic, and optimistic values for the machine.

	Optimistic	Most Likely	Pessimistic
Net Annual Benefit	$4,500	$3,000	$1,200
Useful Life (yrs)	9	7	4
Salvage Value	$1,000	$750	$0

Should the machine be funded if the investment payback period is not to exceed 4 years and the MARR is 10%? Assume that the data follow β-distribution.

15.33 Automating the production line of a bread baking company will positively benefit the bottom line of the bread company. Suppose that the automation will save the bread company 30% of labor cost. Compute the benefits of automating the production line.

Cost Item	Current Process	After Automation
Value of Part	$0.25	$0.25
Annual Production	$650,000	$800,000
Labor Cost	$175,000	
Labor Per Part		

15.34 Chuck D earns $20,000 as an annual bonus and invests it for 5 years at an annual interest rate of 10%, compounded semiannually. If the annual inflation is projected to be 4% per year during the next 5 years and 5% thereafter, how much will he get as (a) actual dollars, (b) current dollars?

FE EXAM PREP QUESTIONS

15.35 Annual labor costs divided by annual production yields:
   a. laboring units
   b. labor per unit
   c. labor costs
   d. none of the above

15.36 Twenty shares of company common stock were purchased for $15 each. Ten years later those same shares of common stock are sold for $450. What is the rate of return on the investment?
   a. 15%
   b. –5%
   c. 4%
   d. 3%

15.37 Now suppose that the stock in Problem 15.36 was split in year 5 of the investment and still sold for $450. The rate of return on the shares is closest to which of the following?
   a. 10%
   b. 7%
   c. 25%
   d. 9%

15.38 A company sells rights to an innovative technology for $400,000. If the technology is capable of earning $70,000 in revenue over the next 4 years, you would sell the technology if the interest rate was
   a. greater than 5%

b. greater than 10%

c. between 5% and 9%

d. none of the above

15.39 An investment is worth $200,000 per year at time zero. If the inflation rate is 4% per year and the market interest rate is 8% per year, the worth of the contract at year zero is closest to which? Assume 8-year useful life.

a. $1.2 million

b. $750,000

c. $0

d. $1.35 million

15.40 Two 5-year contracts are offered by a business partner. Contract one includes payments increasing from $0 in year 1 by $75,000 each subsequent year. Contract two offers $100,000 per year. Assuming 8% annual interest rate, which contract would be more beneficial?

a. fixed-rate contract

b. increasing rate

c. both contracts are equivalent

d. none of the above

15.41 An air-conditioned trailer costs $16,600 today. If the inflation rate is 4% per year, its cost 2 years from now will be

a. $23,546

b. $25,000

c. $17,955

d. $18,366

15.42 A diamond ring costs $45,000 today. If the inflation rate is 3.5% per year, its cost 3 years from now will be

a. $51,638

b. $49,567

c. $60,000

d. $53,412

15.43 Larry borrows $65,000 to buy a jaws-of-life hydraulic cutter. Beginning next month he will pay the lender $700 per month for the next 20 years. What ROR will the lender be enjoying if the annual inflation rate is 5%?

a. 12%

b. 15%

c. none of the above

d. can't be calculated

15.44 The payback period method

a. is exact

b. is simple

c. accounts for the time value of money

d. considers the salvage value in the analysis

15.45 What is the payback period of a $14,500 piece of equipment if it generates $3,200 of revenue each year?

a. 3.13

b. 4.53

c. 2

d. 5.1

Bibliography

As an introductory text, the focus in this book has all along been on the basic principles of engineering economics. For readers interested in furthering their knowledge, textbooks dealing with advanced topics and/or in-depth analysis of engineering economics are listed as follows.

American Telephone and Telegraph (AT&T) Co. *Engineering Economy*. 3rd ed., McGraw-Hill, 1977.

Blank, Leland T., & Anthony J. Tarquin. *Engineering Economy*. 8th ed., McGraw-Hill, 2017.

Eschenbach, Ted G. *Engineering Economy*. 3rd ed., Irwin, 2010.

Fleischer, Gerald A. *Introduction to Engineering Economy*. PWS Publishing Company, 1994.

Grant, Eugene L., W.G., Ireson, & R.S. Leavenworth. *Principles of Engineering Economy*. 8th ed., John Wiley & Sons, 1990.

McCright, Grady E. *Engineering Economy*. 2nd ed., 1999.

Newnan, Donald G., & Jerome P. Lavelle. *Engineering Economic Analysis*. 13th ed., Engineering Press, 2017.

Park, Chan S. *Contemporary Engineering Economics*. 6th ed., Addison-Wesley, 2016.

Park, Chan S., & Gunter P. Sharp-Bette. *Advanced Engineering Economics*. Wiley, 1990.

Riggs, James L., David D. Bedworth, & Sabah U. Randhawa. *Engineering Economics*. 4th ed., McGraw-Hill, 1996.

Sepulveda, Jose A., William E. Souder, & Byron S. Gottfried. *Engineering Economics*. 1st ed., McGraw-Hill, 1984.

Smith, Gerald W. *Engineering Economy*, 1988.

Steiner, H.M. *Engineering Economic Principles*. McGraw-Hill, 1992.

Sullivan, William G., James A. Bontadelli, & Elin M. Wicks. *Engineering Economy*. 17th ed., Prentice-Hall, 2018.

Thuesen, Gerald J., & Wolter J. Fabrycky. *Engineering Economy*. 9th ed., Prentice-Hall, 2000.

White, John A., Kenneth E. Case, & David B. Pratt. *Principles of Engineering Economic Analysis*. 6th ed., Wiley, 2012.

Young, Donovan. *Modern Engineering Economy*. Wiley, 1993.

Index